CW00762077

Novel Catalyst Materials for Bioelectrochemical Systems: Fundamentals and Applications

ACS SYMPOSIUM SERIES **1342**

Novel Catalyst Materials for Bioelectrochemical Systems: Fundamentals and Applications

Lakhveer Singh, Editor
Faculty of Civil and Environmental Engineering, Universiti Malaysia Pahang (UMP)
Kuantan, Pahang, Malaysia

Durga Madhab Mahapatra, Editor
Biological and Ecological Engineering, Oregon State University
Corvallis, Oregon, United States

Hong Liu, Editor
Biological and Ecological Engineering, Oregon State University
Corvallis, Oregon, United States

American Chemical Society, Washington, DC

Library of Congress Cataloging-in-Publication Data

Names: Singh, Lakhveer, editor. | Mahapatra, Durga Madhab, 1983- editor. | Liu, Hong, 1973- editor.
Title: Novel catalyst materials for bioelectrochemical systems : fundamentals and applications / Lakhveer Singh, editor ; Durga Madhab Mahapatra, editor ; Hong Liu, editor.
Description: Washington, DC : American Chemical Society, [2020] | Series: ACS symposium series ; 1342 | Includes bibliographical references and index.
Identifiers: LCCN 2019048139 (print) | LCCN 2019048140 (ebook) | ISBN 9780841236684 (hardcover) | ISBN 9780841236677 (ebook other)
Subjects: LCSH: Bioelectrochemistry. | Catalysts.
Classification: LCC QP517.B53 N68 2019 (print) | LCC QP517.B53 (ebook) | DDC 572/.437--dc23
LC record available at https://lccn.loc.gov/2019048139
LC ebook record available at https://lccn.loc.gov/2019048140

The paper used in this publication meets the minimum requirements of American National Standard for Information Sciences—Permanence of Paper for Printed Library Materials, ANSI Z39.48n1984.

Copyright © 2020 American Chemical Society

All Rights Reserved. Reprographic copying beyond that permitted by Sections 107 or 108 of the U.S. Copyright Act is allowed for internal use only, provided that a per-chapter fee of $40.25 plus $0.75 per page is paid to the Copyright Clearance Center, Inc., 222 Rosewood Drive, Danvers, MA 01923, USA. Republication or reproduction for sale of pages in this book is permitted only under license from ACS. Direct these and other permission requests to ACS Copyright Office, Publications Division, 1155 16th Street, N.W., Washington, DC 20036.

The citation of trade names and/or names of manufacturers in this publication is not to be construed as an endorsement or as approval by ACS of the commercial products or services referenced herein; nor should the mere reference herein to any drawing, specification, chemical process, or other data be regarded as a license or as a conveyance of any right or permission to the holder, reader, or any other person or corporation, to manufacture, reproduce, use, or sell any patented invention or copyrighted work that may in any way be related thereto. Registered names, trademarks, etc., used in this publication, even without specific indication thereof, are not to be considered unprotected by law.

PRINTED IN THE UNITED STATES OF AMERICA

Foreword

The ACS Symposium Series is an established program that publishes high-quality volumes of thematic manuscripts. For over 40 years, the ACS Symposium Series has been delivering essential research from world leading scientists, including 36 Chemistry Nobel Laureates, to audiences spanning disciplines and applications.

Books are developed from successful symposia sponsored by the ACS or other organizations. Topics span the entirety of chemistry, including applications, basic research, and interdisciplinary reviews.

Before agreeing to publish a book, prospective editors submit a proposal, including a table of contents. The proposal is reviewed for originality, coverage, and interest to the audience. Some manuscripts may be excluded to better focus the book; others may be added to aid comprehensiveness. All chapters are peer reviewed prior to final acceptance or rejection.

As a rule, only original research papers and original review papers are included in the volumes. Verbatim reproductions of previous published papers are not accepted.

ACS Books

Contents

Indexes

Preface

Bioelectrochemical systems evolved from microbial fuel cells that generate electricity from catalytic electrode (both anodic and cathodic) processes. The paramount potential of such systems is the range of materials (nature and type) of the catalytic materials that are increasingly used with varied substrates for energy recovery coupled with waste treatment. Such elasticities have resulted in a massive expansion of the scope of bioelectrochemical systems from energy generation to a large number of future applications. These developments are possible due to the nano intercessions and advances in fast-evolving catalysis. The present book provides the fundamental advances in state-of-the-art catalytic nano interventions in improvising the efficacy of bioelectrochemical systems that provide benefits for a number of applications in water energy nexus. This book specifically emphasizes the key developments and the evolution of constructing novel hybrid catalytic nanomaterials by amalgamating noble and non-noble metals, photocatalysis, and catalytic interfaces.

In light of recent trends and developments, novel hybrid electrocatalytic materials and their application in improving electrode configurations have been explored involving design and fabrication of advanced C–based nanocomposite, 3D porous configuration, and their applications to combat the present-day challenges in energy production, storage, and environmental problems. Thermodynamic properties, composition, morphology, catalytic structure (3D), surface modifications, improved surfaces for microbial attachments, and effective electron transfer efficiency have been discussed to understand the various routes and mechanisms for improving the overall performance efficiency. This book thoroughly covers topics on the frontiers of advanced catalysis that involve developing electrode material (cathode and anode) with novel nanoengineered material–based catalysts (e.g., modified stainless steel as anode materials, catalysis in photosynthetic bioelectrochemical systems, oxygen reduction reaction electrocatalysts, metal-organic framework–derived catalysts, bacterially generated nanocatalysts, redox enzyme systems, and state of the art catalysts targeted for hydrogen evolution). We also emphasize nafion–modified electrode surfaces, selective bioelectrocatalytic H_2O_2 reduction, and sensing functions in addition to novel bioelectrocatalytic strategies based on immobilized redox metalloenzymes on tailored electrodes.

This book will aid in the understanding of critical constraints of the conventional electrode materials in bioelectrochemical system such as lower current densities; limitations in substrate delivery, dead zones, and clogging; electrode reaction kinetics; and high investments in electrode materials for scale-up and large-scale bioelectrochemical system. Additionally, the various strategies in the present book also address one of the most important and grossly ignored environmental issues caused by the disposal of heavy metal–containing catalysts, which minimize environmental externalities from an environmental perspective.

Cover information: Reproduced with permission from *ACS Appl. Energy Mater.* **2018**, *1*, 319. DOI 10.1021/acsaem.7b00249. Copyright 2018 American Chemical Society.

Dr. Lakhveer Singh, Associate Professor
Faculty of Technology
Universiti Malaysia Pahang (UMP)
Lebuhraya Tun Razak
26300 Gambang, Kuantan, Pahang, Malaysia

Dr. Durga Madhab Mahapatra, Research Associate
College of Agricultural Sciences
Biological and Ecological Engineering
Oregon State University
116 Gilmore Hall
Corvallis, Oregon 97331, United States

Dr. Hong Liu, Professor
Biological and Ecological Engineering
Oregon State University
116 Gilmore Hall
Corvallis, Oregon 97331, United States

Chapter 1

Recent Development in Cathodic Catalyst towards Performance of Bioelectrochemical Systems

Arya Das, Mamata Mohapatra, and Suddhasatwa Basu*

CSIR-Institute of Minerals and Materials Technology, Bhubaneswar, India 751013

***E-mail: sbasu@iitd.ac.in.**

In recent years, bioelectrochemical systems (BESs) have grabbed ample attention as renewable energy sources employed for various applications ranging from energy production to wastewater treatment to desalination. A BES as an energy-efficient technology still has a lot of constraints for practical implementation, particularly with balancing the need to increase the efficiency with the cost effectiveness of the system. Thus, arises the need for effective catalysts to have high catalytic activity, stability, and economic feasibility. Different types of cathodic catalysts including carbon-based catalysts, metal-based catalysts, metal-nitrogen-carbon (M-N-C) complexes, and metal organic frameworks (MOFs) have been developed to enhance the overall efficiency of BESs. This chapter will give a comprehensive insight into the fundamentals of a BES, the different types of cathodic catalysts used, and their contribution to enhancing the performance of the system that would provide a future direction for their practical implementation.

Introduction

Rapid depletion of water and energy resources is a major concern worldwide. In this context, intensive resources from industrial, municipal, and agricultural wastes are being utilized not only to meet a part of the annual energy demand of around 23 TW by the year 2050 (*1*) but also to create a beneficial strategy to simultaneously tackle the growing water scarcity and take care of the environmental impact. Therefore, there is a demand for an innovative solution for production and recovery of energy and industrial materials with adequate treatment methods for waste treatment. In this regard, bioelectrochemical systems (BESs) with hybrid phenomena of water-energy nexuses were born. These systems have created a new avenue for clean, energy-efficient, and resource-recovering technologies to make the world a low carbon economy. Renewable biomass and wastewater are the two most important inputs utilized in these systems to cater to the exponentially increasing demand of new energy sources. However, such systems have major challenging prospects due to the highly complex mixture at a very low level concentration of feed. This makes them very

© 2020 American Chemical Society

difficult to concentrate and purify in an economically sustainable manner. These dynamic systems offer promising opportunities for ECF (electricity, chemical, and fuels) generation due to redox reactions of biodegradable organic compounds (substrates) catalyzed by pure cultures, bacteria communities, or isolated enzymes. Thus, the vital part is the interaction between microbes and solid electron acceptors or donors (e.g., an electrode) to accomplish the removal of contaminants in wastewater with simultaneous electricity generation (*2*). As compared to conventional fuel cells, BESs are cost effective, operate under relatively mild conditions (room temperature and neutral pH), and do not use expensive catalyst metals without any adverse environmental impact (*3–6*). Based on the various operational modes and desired end applications, notably remote electricity generation and wastewater treatment, resource recovery, sensor applications, desalination, and electrochemical reduction of CO_2 into high-value chemical compounds (*7*), a number of BESs have been trialed with promising results and become a current research hotspot. Among these, electricity generator through microbial fuel cells (MFCs), hydrogen production through microbial electrolysis cells (MECs), synthesis of value-added chemicals through microbial electrosynthesis (MES), desalination through microbial desalination cells (MDCs), and removing contaminants through microbial remediation cells (MRCs) are the basic systems (*8*). Currently many coupled systems have been explored as efficient technologies as shown in Figure 1. The organic matter source used in anolytes may be a simple organic compound (such as acetate or glucose), or more complex substrates such as domestic and industrial wastewater from oil and petroleum refinery wastewater, pharmaceutical, dairy, food-processing, leachate, brewery, and winery industries (*8–22*) may also be used for energy generation. In addition, there is great interest in employing BESs for recovery of valuable resources from wastewater including nutrient, energy, and water (NEW) (*23*).

Microbial Fuel Cells (MFCs)

An MFC is a low-cost and sustainable technology that generates an electrical current via microbial oxidation of organic matter to harvest energy and treat wastewater (*24, 25*). An MFC consists of an anode and cathode in aqueous solutions in discrete chambers, separated by a membrane to avoid the migration of electrolytes from one chamber to the other (*26*) with an external circuit connecting anode and cathode electrodes. Bacteria are used as catalysts in the anode compartment to break down organic matter, oxidize the fuel, and generate electrons and protons. Further electrons are then transferred to the cathode via an external circuit where metals can be electrochemically reduced and eventually recovered from the cathode surface (*27*). While the protons diffuse through the proton exchange membrane (PEM) (*25*). Electrons and protons may have been consumed in the cathode chamber, reducing oxygen to water thus generating electricity (*28, 29*). Mostly carbon-based materials such as graphite and modified carbon materials with high surface area such as carbon foam and sponges are used as anode and cathode comprises (i.e., coated graphite, coated carbon cloth, and catalyst-based electrodes) (*30*). They can be categorized on the basis of catholyte (either abiotic catholytes like potassium ferricyanide and nutrient medium or biotic catholytes like algae and cyanobacteria) depending on the type of electrolyte, the current density passing through an electrode, electrode properties, the presence of mediators (shuttle), the conductivity of an electrolyte, the pH of an electrolyte, operational temperature, and type of exoelectrogens (*31*). Therefore, fluctuation in any of these parameters may cause loss of current generation (*32*). In order to minimize these losses, various approaches, like maximizing electrode surface area and improving exoelectrogen activity by using exoelectrogenic consortia, can be adapted (*33*). In a single MFC reactor, the maximum theoretical voltage is ≈ 1.2 V when the redox potential of bacteria respiration enzymes and O_2 have been taken into account. The typical values realized in the lab range from 0.3 V to 0.5 V (*34*). Installing multiple MFC reactors in a series and in parallel can increase the produced voltage and current or a combination of both could be utilized to optimize the final power output. The power density of MFCs can be increased by several orders of magnitude. The maximum power density reported for an air cathode MFC is 1.55 kW/m^3, and a further power density of 2.87 KW/m^3 was achieved with a novel configuration of double cloth electrode assembly (*35*). Usually the concentrations of organic matter treated by MFCs are in the range of domestic wastewater (COD $\leq$ 125 mg/L).

Microbial Electrolysis Cell (MECs)

MECs rely on bacteria to convert chemical energy to electrical energy and electrolysis of water. An external power is applied to a BES for flow of electrons from the anode to the cathode, which subsequently leads to hydrogen generation at the cathode resources (*36, 37*). Unlike MFCs, anaerobic conditions at the cathode chamber enable hydrogen production (*38, 39*). The additional voltage applied helps generate higher electrical currents. MFC can be deployed as a power source to provide energy required for MEC operation (*40*). The hydrogen synthesized in MECs can further initiate production of other chemicals biochemically. In an MEC, electrochemically active bacteria oxidize organic matter and generate CO_2, electrons, and protons. The bacteria transfer the electrons to the anode and the protons are released in the solution. The electrons then travel through a wire to a cathode and combine with the free protons in solution. However, this does not occur spontaneously. In order to produce hydrogen at the cathode from the combination of these protons and electrons, MEC reactors require an externally supplied voltage ($\geq$0.2 V) under biologically assisted conditions of pH = 7, T = 30 °C, and P = 1 atm (1.01 $\times$ 10^5 Pa) (*37, 41*). This is done by an input voltage provided via a power supply. However, MECs require a relatively low energy input (0.2–0.8 V) compared to typical water electrolysis (1.23–1.8 V). A reverse electro dialysis stack may be used in between the electrodes in order to operate the MEC. The electrochemical performance, microbial community structure, and theoretical maximum energy gain by exoelectrogens for their growth can be controlled by setting different anode potentials (*42*). The real energy gain by exoelectrogens depends on the redox potential of the terminal electron-transferring component (e.g., outer membrane protein) that serves as the electron donor to the anode (*43*). At low anode potentials, lower biomass accumulation occurs due to less energy gain by exoelectrogens during growth, which results in slower development of the anode biofilm community and a more delayed start-up of current production (*44*). At higher anode potentials, exoelectrogens could gain more energy for growth if they have the capability to capture this additional energy (*44*). An MEC can be used directly in an anaerobic digester to improve performance and increase the methane concentration in the product gas (*45–48*). The performance of MECs is known to be impacted in terms of current production for simple substrates such as acetate, although the community structure is relatively unchanged for different SAPs. The SAP influences the biocatalytic activity, electrochemical performance, substrate degradation, and microbial community structure. It can also alter the syntrophic interactions between the organisms (*49*). In the case of microbial electrolysis to produce hydrogen, homoacetogens could be responsible for the hydrogen sinking to impair hydrogen yield and energy efficiency (*50*). However, Wang et al. gave insight into the role of hydrogen partial pressure along with hydrogen scavengers in the cathodic biofilm on the performance of such an MEC. The other physical parameters, such as electrode spacing (*51, 52*), anode surface area, and the physical interaction of the solution with the electrodes, was reported to improve overall MEC performance. Therefore, MECs are currently focused on designing the cell by manipulating internal resistance to get the desired electrical current and hydrogen production for a given input voltage. However, the internal resistance distribution depends on the type and nature of cathode materials and their reaction. Thus, recent investigations into the electrochemical performance of such MECs were reported by considering the effect of anode surface area, cathode surface area, buffer concentration, electrode spacing, and mixing condition. The specific internal resistances can be tuned by choosing a proper electrode material, microbial community, and operational condition to model the practical designs of a wide range of bioelectrochemical devices (*53*).

Microbial Electrosynthesis (MES)

MES is a type of bioelectrochemical device that involves electricity-driven microbial reduction of carbon dioxide into a variety of useful multicarbon compounds (*54–58*). In this system, the combination of a chemical anode with a biocatalyzed cathode (biocathode), or with both a biocatalyzed anode and cathode, is used for many investigations (*54–61*). Wastewater or sludge hydrolysates have a multitude of organic substances and contaminants that can be used as substrates at the anode when a biocatalyzed anode (bioanode) is used in an MES (*56, 59*). They function as the electron source and undergo oxidation to produce electrons and protons, which can then be subsequently used for microbial reduction at the cathode. The external power required for microbial oxidation of wastewater organics at the bioanode is lower than that required for water splitting at the abiotic anode (*61*). This allows for recovery of useful energy in the form of chemicals from waste and also allows for the use of low-cost carbon materials as electrodes compared to chemical anodes. Several products, including methane, acetic acid, propanol, butanol, and ethanol, can be microbially synthesized from CO_2 reduction at different potentials (*55, 59, 62*).

This system is also capable of converting waste gas streams into valuable organic molecules and is a sustainable route for the microbial fermentation of synthesis gas (syngas), whereby organisms such as *Clostridium* spp. (e.g., *Clostridium ljungdahlii* and *Clostridium autoethanogenum*) metabolize CO_2 or CO into acetate and ethanol using H_2 or CO as the electron donor for CO_2 reduction (*63, 64*). When gas streams that are rich in CO_2 but poor in H_2 and CO are used, reducing equivalents can be provided by the cathode electrode of an electrochemical system, either directly in the form of electrons or indirectly as H_2 produced through water electrolysis (*64, 65*). The culturing conditions with the pH of the medium decide the acetate-to-ethanol ratio (*66, 67*). With a pH pH of >6, acetate is the main product; whereas with a pH of <5, ethanol is the main product (*66, 67*). Furthermore, in the presence of both acetate and ethanol and under a neutral pH, organisms such as *Clostridium kluyveri* can produce short- and medium-chain carboxylates (SCCs and MCCs) (*68*) through the reverse β-oxidation carbon chain elongation pathway. Under a mildly acidic pH, higher alcohols can be obtained through solventogenesis of the SCCs and MCCs (*69*). The production of higher alcohols can be approached by culturing different *Clostridium* spp. capable of all production steps in a single reactor system operated at a relatively low pH (*69, 70*). The current horizon of applications for this system for desired requirements can be obtained using particular biocatalytic processes.

Microbial Desalination Cells (MDCs)

This system is like a conventional MFC where the electric voltage produced by exoelectrogen bacteria is used as the driving force to remove ions from water instead of any other external power source, making the technology energetically self-sustainable. The organic matter in wastewater serves as the energy source to drive the desalination by means of bacterial metabolism. The anode chamber is fed by a wastewater stream and the bacteria attached to the anode metabolizes it, thus producing electrons and protons (oxidation of the organic matter). The anode is connected via wire to the air cathode, which is located in the cathode chamber where the electrons travel to react with oxygen and protons to produce water. This process produces the electric power responsible for driving the desalination taking place in the middle chamber. The desalination chamber is fed with a different stream containing just salt, and the electric potential gradient drives salt ions to the opposite-charged electrode through the ion exchange membranes in order to maintain the charge balance in the electrode chambers. The salinities used in this technique are lower than sea water (35 g/L). Desalination of 63% can be achieved with 5 g/L of NaCl saline stream. For the system to have sodium acetate as a carbon source, the desalination efficiency drops when the salinity is closer to sea water (*71*). This is due to the growth of bacteria being affected by the increase in Cl^- ion concentration in the anode chamber. At the same time, when using wastewaters with high organic matter content, the pH in the anode chamber will decrease and deplete the bacterial growth. The composition of saline water can also influence the performance (*72*) of the MDC device as the chemical constituent precipitate in the ion exchange membrane's surface decreases the power output. The use of two different wastewater streams reduces its applicability to coastal regions where domestic wastewater and sea water coexist, but not for industrial or other saline water sources. However, the performance can be enhanced by the use of multiple stacked desalination cells (*73*). A comprehensive review of different MDC modifications including their desalination performance and generated power have been reported recently (*74*). Various operational parameters affecting the MDC performance in terms of desalination and generated coulombic efficiencies (CEs) have been analyzed.

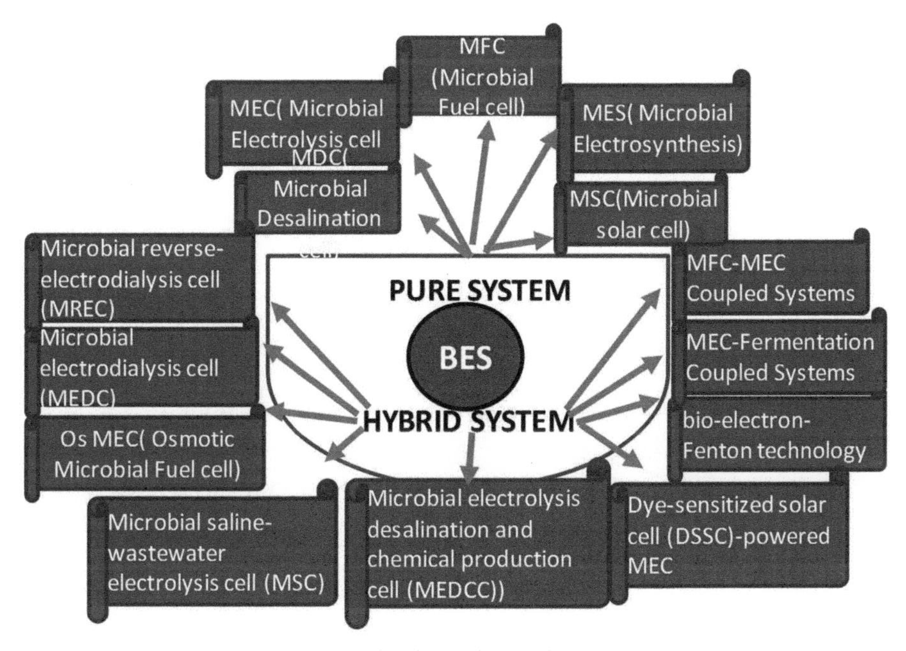

Figure 1. Various bioelectrochemical systems (BESs).

Working Principles of a BES

The end application of the BES depends on the electrode interaction with these microorganisms. Microorganisms used in BES can act both as electron donors to an electrode (*75*) generating an electrical current (electro genic) and as electron acceptors requiring an electric current (electrotrophic) (*76*). The proposed mechanism of microorganisms interacting with the electrode and mode of electron transfer between microorganismis is shown in Figure 2 (*77*). A number of terminologies (Figure 2) have been used to describe bacteria while interacting with electrodes in various BESs that guides the overall efficiency of the system. A lot of diversified mechanisms are used by microorganisms to transfer electrons to the anode (Figure 2). Studies on electron transfer to anodes for pure culture *Geobacter sulfurreducens* were carried out to explore the electron transfer mechanism across the electrode. Studies revealed transfer of electrons across the electrode takes place via c-type cytochromes (*77–79*). *G. sulfurreducens* can also alternately form thick anode biofilms (>50 mm), and electron transfer takes place though electrically conductive microbial nanowires termed as pili (*80*) to harness the current (*81–83*). The anode biofilms are conductive contrary to previous insulating biofilms and release of cytochrome into the biofilm matrix, which may also contribute energy generation. As reported earlier, through modeling studies current production observed in *G. sulfurreducens* fuel cells can only be feasible with a conductive biofilm (*84*, *85*) and its high capacitance is endorsed to the abundant c-type cytochromes, which provide extensive electron storage capacity (*86*). In contrast, in studies with a *Shewanella oneidensis* electrode reducer, soluble electron shuttles are the mediators of electron transfer across the electrode (*87*). *S. oneidensis* also produces microbial nanowires (*88*), but direct transfer through nanowires are not the sole mediators and c-type cytochromes on the outer surface are beneficial for enhanced transfer and current production (*89*). The difference in interactions of *S. oneidensis* and *G. sulfurreducens* with smooth gold electrodes further confirms their different electron transfer mechanisms (*90*), but this may reflect a requirement for the cytochromes for electron shuttle reduction. Many of the cells are planktonic, and electron transfer over such long distances is only conceivable with electron shuttles (*91*). Similar observations were seen in *Escherichia coli* which releases metabolites as electron shuttles (*92*). Alternatively, a different mechanism from electrodes to *G. sulfurreducens* is also reported (*93*).

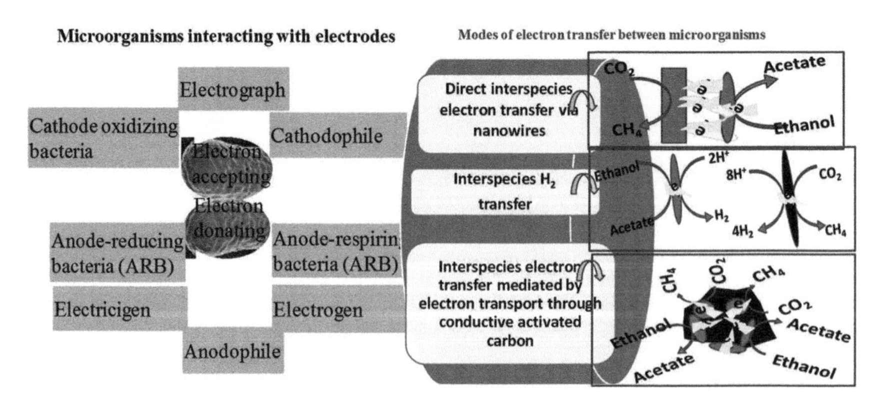

Figure 2. Microorganisms interacting with the electrode and mode of electron transfer between microorganisms.

Factor Affecting the Performance of BES

Though there are enormous R & D efforts going on to develop suitable BESs, there are still no practical applications due tolow power density and high cost. The rate at which energy is recovered is measured from the anodic current density, which reflects the specific conversion rate and power density (*94*). Other than these performance parameters, two other types of efficiencies are used to evaluate MFC performance: the CE and the voltage efficiency (VE) (*95*). Both a high voltage and a CE are responsible for the enhancement of available energy from the substrate, which need to be transferred to an eventual product (*96*). The CE shows a fraction of the electrons obtained from oxidizable substrates present in the wastewater, which are recovered at the anode indicating the efficiency of the conversion of a substrate into electrical current. When a continuous system has reached equilibrium and no change over time of both influent and effluent concentrations and measured current occurs, the CE can be calculated. The VE is related to the internal losses of the system, which can be enhanced by reducing the internal resistance of the system (*97, 98*). For example, low anodic overpotential of anodes close to the thermodynamic potential of the substrate is one of the criteria for increasing the performance of the BES. This can be achieved by minimizing the activation and concentration overpotential (*99*). The low power generation may also be attributed to ohmic polarization, reaction kinetics, and mass transport. The slow kinetics of electron transfer between the bacteria and the anode is due to poor biocompatibility, small specific surface area, and high overpotential between the bacteria and the anode (*95*). Parameters like operating conditions, electrode materials and their structure, cell design, and microbial inoculation as shown in Figure 3 (*100–113*) are the major focus for achieving the optimal efficiency. For example, the overall efficiency of the system highly relies on the electron transfer across the surface, which in turn is highly dependent on composition, morphology, and surface properties of anode materials. In real scenarios, the practical implementable potential achieved is lower than theoretical potentials due to several forms of ohmic losses and energy losses of the electrode. The efficiency of the practical system is both dependent on both anode and cathode energy losses termed as overpotentials. The electrode overpotentials include activation losses, bacterial metabolism losses, and concentration or mass transport losses. Ohmic losses mainly come from the interconnection of all the components which may lead to voltage drop and ionic losses owing to exchange membrane. Minimizing the losses to enhance the performance of BESs in the anode compartment can be achieved by minimizing interelectrode spacing, enhancing conductivity of anode materials, and decreasing the contact resistance. Activation losses that correspond to electron transfer from microorganisms to the surface for initiation of oxidation reduction reaction (ORR) at the electrode can be minimized by increasing the available surface area of the anode material used and augmenting the electron transfer between the microorganisms and the anode. Bacterial metabolism depends on the anode potential and the surface (*114*). A low anode potential may result in a high metabolic energy gain from bacteria, but a much lower potential can inhibit the electron transfer. In summary, the anode material properties highly influence the operating efficiency of the system, and thus proper modification can be useful to gain higher outputs. The semi-cell reaction of the anode relies on exoelectrogens (*115*). Optimizing the ability of anodic exoelectrogens is thus a key factor in boosting the overall efficiency of BESs (*116*). Recently, the magnetic MFC concept was implemented to obtain a 71.0–105% increase in voltage production and a 42.9–104% increase in power density compared with non-magnetic MFCs (*116*). Other performance-limiting factors include biofouling (leading to electrode surface blockage and ultimately a reduction in surface area), catalyst inactivation (if present), and excessive biofilm growth, which could possibly lead to the production of non-conductive debris arising while trying to enhance the performance of an MFC for industrial and social applications (*117*). The production of non-conductive debris, such as polymeric substances or dead cells, can isolate the electrochemically active biofilms from the electrode surface or, with more porous electrodes, become entrapped in the three-dimensional (3-D) architecture; this leads to a potential reduction in available surface area and ultimately a reduction in current generation (*117–119*). A study conducted in 2017 used cell viability counts and field emission scanning electron microscopy analysis to show that an increase in high polarization resistance correlated with the formation of a dead layer of cells (*119*). Furthermore, this study also revealed that the use of ultrasonic treatment was a verified method for controlling biofilm thickness and enhanced cell viability, thus maintaining stable power generation (*119*). There have been other biofilm-related factors that are thought to contribute to the performance of an MFC. In a study conducted by Sun et al. (*119*), it was revealed that when the predominant bacteria in an MFC set-up was *Geobacter anodireducens*, a two-layered biofilm developed over time with an inner dead core and an outer layer of live cells. Results suggest that the outer layer was responsible for current generation and the dead inner layer continued as an electrically conductive matrix (*120*). It could be speculated that this continued electrochemical activity could be dependent upon the mechanism of electron transfer. For example *Geobacter* spp. are well known for their electrochemical activity due to nanowires, which may still have a viable connection to the electrode surface, even through the non-conductive debris. Other attributing factors that can have a detrimental effect on both the power outputs and the efficacy of an MFC include the inactivation of electrocatalysts (if present) and the crossover of organic compounds or electron acceptors from the anode to the cathode (and vice versa). The crossover of electron acceptors from the cathodic compartment into the anode has been shown in a previous study to disrupt biofilm formation and lead to biofilm inactivation, which can considerably decrease MFC performance due to the flow of internal currents and the formation of mixed potentials (i.e., a system that is short-circuited) (*121, 122*). Catalysts, such as inorganic materials, used in microbial cells for both oxidative and reductive reactions can couple these reactions to make products of interest at high thermodynamic efficiency. In MECs, the microbial catalysts perform oxidative reactions and the inorganic material catalyzes the reductive reaction (*120–124*). The reverse phenomena occur in MES, biohybrid, or artificial photosynthesis systems (*125–136*). In both the systems, distinct chemical environments are required for catalysts to optimally function (*137–139*). To achieve and sustain these chemically distinct environments, the biotic and abiotic catalysts are separated either locally or through the membrane, which leads to crippling ohmic losses on the order of 25% of the cell voltage that impairs system scale (*135, 140*). Recently, a new concept has been adopted for completing the redox couple between a microbial catalyst and inorganic catalyst at the nanoscale while separating the incompatible anodic and cathodic reaction environments by tailoring silica membranes coupled with inorganic catalysts and microbial catalysts on the shortest possible scale length scale: nanometers (*141*). This electron transport occurs only when the energetics of the microbial catalyst and molecular wires are matched, and it occurs rapidly enough to allow the microbial catalyst to maintain biomass. Many research currently focuses on modeling the parameters to get the highest performance out of such systems. Li et al. (*142*) reported that domestic wastewater, in comparison with swine wastewater and landfill leachate, was the most suitable for MFC and MEC systems because of the higher power generation in MFCs, greater biohydrogen generation in MECs, and better treatment efficiency in terms of COD removal. The optimal temperature was found to be 25–35 °C, while the optimal pH was 6–7 and optimal external resistance was 100–1000 W. However, carbon cloth used as electrodes is not economical, which demands that other existing carbonaceous materials must be explored for an economic point of view. In general, performance and stability is needed to scale up and commercialize BESs, which could possibly be achieved with improved system designs.

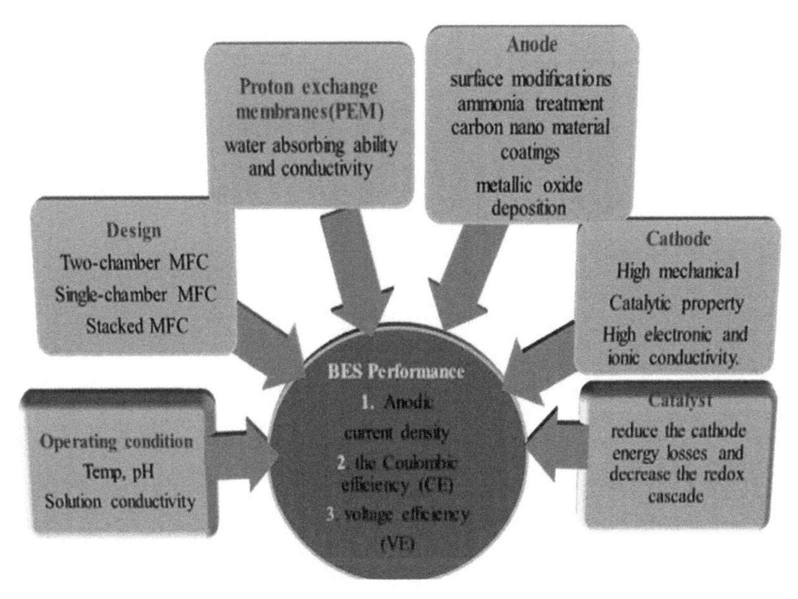

Figure 3. Parameters responsible for the performance of BESs.

Catalysts in BESs

As discussed earlier, a BES is an electrochemical system employing catalysts and biocatalysts to catalyze electrochemical reactions at either one or both of the electrodes accompanied by substrate oxidation at the anode and reduce reactions (e.g., ORR and HER) at the cathode. Electrons flow from a lower potential anode to a higher potential cathode through an external circuit which harnesses energy in the form of electricity (MFC) or chemical fuels (MEC). The electrode potential across the anode and cathode influences electric flow without an external power source in cases of MFCs and driven by an external supply in cases of MECs. The ORR and hydrogen evolution reaction (HER) cathodic reactions contribute to the net energy production. Due to the optimal operating conditions of a BES operation, an apt catalyst is needed to perform under such conditions with good stability. Presently, the low energy efficiency of BESs limits the practical applications of such systems; thus, the need for improved catalysts with better understanding of their reactions with the electrode is essential.

The reaction kinetics of ORR and HER are very slow, which in turn affects the overall performance of BESs and is highly dependent on the catalyst activity. Catalysts play a very vital role in augmenting the overall performance of the system. There is a need for novel catalysts with superior activity, stability, and cost effectiveness for real implementation. The need for alternate cost effective systems depends on stable cost effective and highly active catalysts. The synthesis and development of such catalysts to enhance the overall performance of BESs (Figure 4) have aroused great interest in the scientific community.

Cathodic Catalysts Employed in BESs

The cathodic catalysts to be employed in BESs should be cost effective, highly active, and stable. Simple and large-scale synthesis of cathodic catalysts can further help in practical implementation with an eye on the economic viability. The stability and long-term performance of the cathodic catalyst is a major challenge. The constant exposure of the cathode to intermediate products and blockage of O_2 transport by biofilms can largely hinder the energy output (*143, 144*). With respect to these requirements, the cathodic catalysts must possess tolerance to poisoning and recyclability for better performance of BESs. This section will discuss the different types of abiotic cathode catalysts and compare the performance of these different types of catalysts giving insight into the present scenario and the challenges needed for future researchers to increase the performance for practical implementation of BESs. The different types of cathodic catalysts are discussed below.

Carbon-Based Catalysts

The cost effectiveness supported by high surface area, stability, and higher conductivity makes carbon material a potential candidate to be used as a cathode catalyst (*145*). Hence, carbon materials are used as cathode catalysts such as activated carbon (AC) (*146*), carbon nanofibers (*147*), modified carbon blacks (CB) (*148*), graphene (*149*), and carbon nanotubes (*150*). Carbon blacks are mostly used as a support material for composite catalysts when pyrolized and doped with nitrogen and fluorine (N and F) atoms and can reach a power density of 672 mW m^{-2} (*151*). Activated charcoal, also known as porous carbon, can be utilized as a catalyst due to the avaibility of high surface area. AC produced from different precursors will possess different active sites and surface area. Nitridation of AC is one of the most adopted ways to enhance the catalytic activity of AC. AC nitrided using ammonia gas at 700 °C yielded a maximum power density of 2450 mW m^{-2} (*152*); yet, while being treated with cyanimide as a nitrogen precursor, the AC yielded a power density of 650 mW m^{-2} (*153*) and also reflected an excellent life cycle degrading only 30% of its current density compared to the 73% drop in the commercial Pt/C cathodic electrode (*153*). Carbon nanofibers and nanotubes are graphene sheets stacked in different manners to provide excellent surface area and activity to be used as cathode catalysts. Zhou et al. reported that carbon nanofibers synthesized from natural spider silk pyrolysis could achieve a power density of 1800 mW m^{-2} with a minimal decrease in power density even after continuous use for 90 days (*154*). Graphite and graphene have been found to be effective catalysts owing to high surface area and excellent electrical conductivity. N-doped graphene implanted in graphitic carbonitride yielded a power density of 1618 mW m^{-2} (*155*). Santoro et al. reported a maximum power density of 2.059 Wm^{-2} (*156*) when graphene sheets are loaded with activated charcoal. Recently, Chen et al. reported a novel rotating graphite fiber brush collector prepared by coating nitrogen- and phosphorous- (N and P)doped carbon to a graphite fiber brush current collector achieving a power density as high as 879 mW m^{-2} (*157*).

Metal-Based Catalyst

Platinum (Pt) is the most active catalyst (*158*), but, keeping an eye on the cost effectiveness, several efforts are being made to reduce the Pt loading by introducing alternate inexpensive metals (Fe, Co, Ni) (*159, 160*) without affecting the performance or enhancing the overall efficiency of the system. Often referred to as Pt alloys, this metal loading can generate higher power densities compared to the commercial counterpart (*161*). Although higher power densities are achieved, the metal alloys are unstable and are found to leach away in adverse acidic environments (*162*). A green, cost effective approach of using scrap metals achieved a power density of 422mW m^{-2} (*163*). Metal oxides with good electrochemical properties, natural abundance, and low cost supported with good catalytic activity also make for potential cathode catalysts and have been studied extensively. MnO_2

with a different stoichiometry (MnOX), which contributes differently as a catalyst, has been also studied (*164*). Other oxides have also been explored as potential catalysts (i.e., vanadium oxides (*165*), cobalt oxides (*166*), zirconium oxides (*167*), and various perovskite oxides (*168*)). Among these, perovskite oxides can be found to achieve power densities of 405mW m^{-2} (*169*). Recently, Bhowmick et al. reported a Bi-doped TiO_2 photocathode catalyst with a power density of 222 mW m^{-2} (*170*) suggesting that modifications in the oxides can successfully enhance the performance of the cell. Metal and metal oxide composites with carbon materials can also act as effective cathodic catalysts for BESs. An aluminum-nickel system doped with nitrogen and dispersed with carbon nanofibers achieved a power density of 1850 mW m^{-2} (*171*). Recently, Noori et al. reported a graphene-supported V_2O_5 nanorod as its based cathodic catalyst reflecting a peak power density of 533 mW m^{-2}, which was found to be a considerably superior metal oxide–free counterpart (*165*). Very recently, Farahani et al. reported multivalent MnOx supported on nitrogen-doped carbon (C(N)/MnOx-SP) and reduced graphene oxide (rGO(N)/ MnOx-SP) synthesized through a solid state method exhibiting a peak power density of 467mWm^{-2} (*172*), establishing them as viable catalysts for BES cathodes.

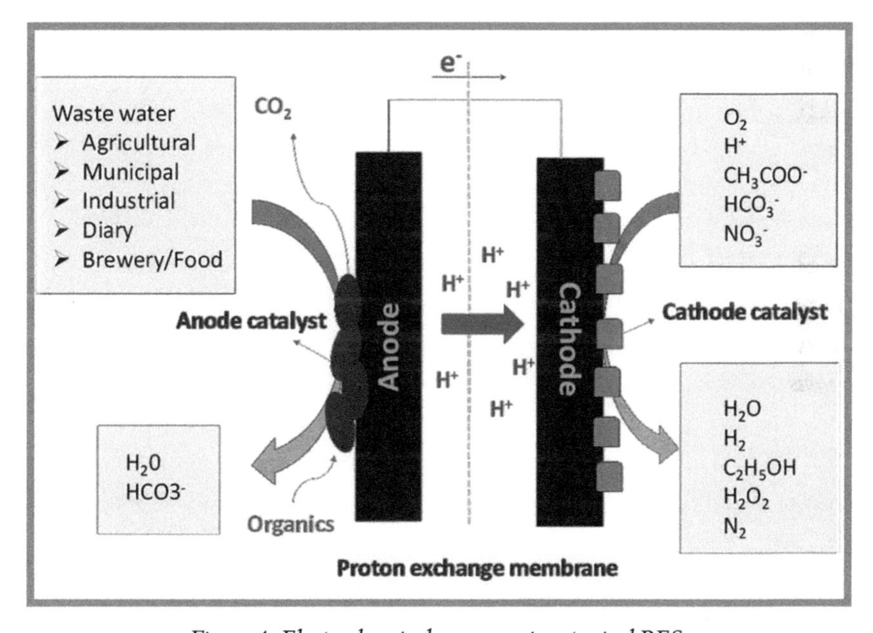

Figure 4. Electrochemical processes in a typical BES.

Metal–Nitrogen–Carbon (M-N-C) Complexes and Metal Organic Frameworks (MOFs)

M-N-C complexes comprised of metals coordinated with nitrogen groups in carbon matrices with carbon support provide the desired surface area and nitrogen group for the desired activity (*172*). Metal macrocycles that fall under the radar of M-N-C complexes exhibit excellent catalytic properties but are found to be unstable in acidic environments. Recent studies have shown pyrolized metal macrocycles can enhance the stability and the catalytic activity of the metal macrocycles (*173*). An FePc composite with polyindole and carbon nanotubes without any heat treatment provides a power density of 799 mW m^{-2} (*174*). The number of active M-N-C sites available can be enhanced by forming an MOF making them potential cathodic catalysts (*175*). Santoro et al. achieved a power density of 214 ± 5 μWcm^{-2} by using a six-step synthesized Fe-N-C catalyst with a sacrificial support method (SSM) technique (*176*). Recently Kodali et al. studied iron aminoantipyrine (Fe-AAPyr) and graphene nanosheet (GNS)–derived catalysts; their physical mixture (Fe-AAPyr-GNS) was simultaneously synthesized and investigated as a cathode catalyst for an ORR. The addition of GNS and Fe-AAPyr concurrently led to the higher output of the investigation that was 235 ± 1 μW cm^{-2} (*177*). Zhang et al. observed a Cu (II)-based MOF (Cu-bipy-BTC, bipy = 2,2′-bipyridine, BTC = 1,3,5-tricarboxylate) pyrolized at 800 °C that achieves a power density of (326 ± 11.2 mW m^{-2}), which is 2.6 times that of its unpyrolized counterpart (126 ± 7 mW m^{-2}) (*178*). Recently, Gracia et al. reported a power density of 104.5 ± 0.0 μW cm^{-2} for MFC fed with human urine with a novel Fe-based catalyst of Fe–streptomycin (STR), a low cost nitrogen source using an SSM (*179*), which was found to be highly stable for up to three months and is highly cost effective. The recent trials for enhancing the power density using various cathodic catalysts is summarized in Table 1.

Table 1. Power Densities of Various Catalysts Used in BESs

Cathode type	*Power density*	*Ref*
Carbon-based		
CB (N & F)	672 mW m^{-2}	(*151*)
AC-nitrided at -700 °C	2450 mW m^{-2}	(*152*)
CNF's spider silk	1800 mW m^{-2}	(*154*)
Graphene sheets/AC	2.059 Mw m^{-2}	(*156*)
N & P–doped graphite brush fiber	879 mW m^{-2}	(*157*)
Metal-based		
	422 mW m^{-2}	(*163*)
Scrap metal	405 mW m^{-2}	(*169*)
Prevoskite oxides	222 mW m^{-2}	(*170*)
Bi-TiO_2	1850 mW m^{-2}	(*171*)
Al/Ni/N/CNFs	533 mW m^{-2}	(*165*)
Graphene/V2O5	467 mW m^{-2}	(*172*)
rGO/(N)/MnOx-Sp		
M-N-C complexes – MOFs		
FePc-polyindole/CNT	799 mW m^{-2}	(*174*)
Fe-N-C (SSM)	214 μW Cm^{-2}	(*176*)
Fe-AAPyr-GNS	235 μW Cm^{-2}	(*177*)
MOF-800 °C	326 mW m^{-2}	(*178*)
Fe-STR	104 μW Cm^{-2}	(*179*)

Conclusion

BESs are effective energy storage and conversion systems that highly rely on the electrochemical processes taking place. The electrochemical processes are in turn dependent on the electrode materials, electrolyte, anode and cathode potentials, mediators, mass, and charge transfers. The use of a catalyst can effectively enhance the electrochemical kinetics leading to a better performance. In this context, a lot of catalysts and their role in enhancing energy production have been studied. This chapter describes the different types of bioelectrical systems that have been developed which also includes hybrid devices that have been implemented. The different factors affecting the performance of the system have been elucidated. The role of cathodic catalysts in enhancing the performance has been well presented. Carbon-based materials with high surface areas show high activity that establishes them as effective catalysts. Metal-based catalysts with modifications show excellent catalytic activity compared to their metal counterparts, but they also lack stability compared to carbon-based materials. Metal-carbon composites can utilize the properties of both to render high activity and enhanced stability. M–N–C complexes and MOFs possess high stability and activity owing to active sites and can be effectively employed as catalysts in BESs. BESs remain a viable alternate efficient energy source considering the exponentially decreasing energy landscape around the world. It is well expected that BESs will serve as sustainable versatile technologies in the future, and there is a need to identify potential cost effective catalysts to enhance their performance to be practically implemented on a large scale.

References

1. Chae, K. J.; Choi, M. J.; Kim, K. Y.; Ajayi, F. F.; Chang, I. S.; Kim, I. S. A Solar-Powered Microbial Electrolysis Cell with a Platinum Catalyst-Free Cathode to Produce Hydrogen. *Environ. Sci. Technol.* **2009**, *43*, 9525–9530.
2. Pant, D.; Singh, A.; Van Bogaert, G.; Olsen, S. I.; Singh Nigam, P.; Diels, L.; Vanbroekhoven, K. Bioelectrochemical Systems (BES) for Sustainable Energy Production and Product Recovery from Organic Wastes and Industrial Wastewaters. *RSC Adv.* **2012**, *2*, 1248–1263.
3. Tee, P. F.; Abdullah, M. O.; Tan, I. A. W.; Amin, M. A. M.; Nolasco-Hipolito, C.; Bujang, K. Effects of Temperature on Wastewater Treatment in an Affordable Microbial Fuel Cell-Adsorption Hybrid System. *Environ. Chem. Eng.* **2017**, 5, 178–188.
4. Ye, Y. L.; Zhu, X. P.; Logan, B. E. Effect of Buffer Charge on Performance of Air-Cathodes Used in Microbial Fuel Cells. *Electrochim. Acta* **2016**, *194*, 441–447.
5. Foley, J. M.; Rozendal, R. A.; Hertle, C. K.; Lant, P. A.; Rabaey, K. Life Cycle Assessment of High-Rate Anaerobic Treatment, Microbial Fuel Cells, and Microbial Electrolysis Cells. *Environ. Sci. Technol.* **2010**, *44*, 3629–3637.
6. Pant, D.; Singh, A.; Bogaert, G. V.; Gallego, Y. A.; Diels, L.; Vanbroekhoven, K. An Introduction to the Life Cycle Assessment (LCA) of Bioelectrochemical Systems (BES) for Sustainable Energy and Product Generation: Relevance and Key Aspects. *Renewable Sustainable Energy Rev.* **2011**, *15*, 1305–1313.
7. Bajracharya, S.; Sharma, M.; Mohanakrishna, G.; Benneton, X. D.; Strik, D. P. B. T. B.; Sarma, P. M.; Pant, D. An Overview on Emerging Bioelectrochemical Systems (BESs): Technology for Sustainable Electricity, Waste Remediation, Resource Recovery, Chemical Production and Beyond. *Renew. Energy* **2016**, *98*, 153–170.

8. Gadkari, S.; Gu, S.; Sadhukhan, J. Towards Automated Design of Bioelectrochemical Systems: A Comprehensive Review of Mathematical Models. *Chem. Eng. J.* **2018**, *343*, 303–316.
9. Faria, A.; Gonçalves, L.; Peixoto, J. M.; Peixoto, L.; Brito, A. G.; Martins, G. Resources Recovery in the Dairy Industry: Bioelectricity Production Using a Continuous Microbial Fuel Cell. *J. Cleaner Prod.* **2017**, *140*, 971–976.
10. Cecconet, D.; Molognoni, D.; Callegari, A.; Capodaglio, A. G. Agro-Food Industry Wastewater Treatment with Microbial Fuel Cells: Energetic Recovery Issues. *Int. J. Hydrogen Energy* **2018**, *43*, 500–511.
11. Callegari, A.; Cecconet, D.; Molognoni, D.; Capodaglio, A. G. Sustainable Processing of Dairy Wastewater: Long-Term Pilot Application of a Bio-Electrochemical System. *J. Cleaner Prod.* **2018**, *189*, 563–569.
12. Molognoni, D.; Chiarolla, S.; Cecconet, D.; Callegari, A.; Capodaglio, A. G. Industrial Wastewater Treatment with a Bioelectrochemical Process: Assessment of Depuration Efficiency and Energy Production. *Water Sci. Technol.* **2018**, *77*, 134–144.
13. Colombo, A.; Schievano, A.; Trasatti, S. P.; Morrone, R.; D'Antona, N.; Cristiani, P. Signal Trends of Microbial Fuel Cells Fed with Different Food-Industry Residues. *Int. J. Hydrogen Energy* **2017**, *42*, 1841–1852.
14. Iskander, S. M.; Novak, J. T.; Brazil, B.; He, Z. Simultaneous Energy Generation and UV Quencher Removal from Landfill Leachate Using a Microbial Fuel Cell. *Environ. Sci. Pollut. Res. Int.* **2017**, *24*, 26040–26048.
15. Sonawane, J. M.; Adeloju, S. B.; Ghosh, P. C. Landfill Leachate: A Promising Substrate for Microbial Fuel Cells. *Int. J. Hydrogen Energy* **2017**, *42*, 23794–23798.
16. Cecconet, D.; Molognoni, D.; Callegari, A.; Capodaglio, A. G. Biological Combination Processes for Efficient Removal of Pharmaceutically Active Compounds from Wastewater: A Review and Future Perspectives. *J. Environ. Chem. Eng.* **2017**, *5*, 3590–3603.
17. Feng, Y.; Wang, X.; Logan, B. E.; Lee, H. Brewery Wastewater Treatment Using Air-Cathode Microbial Fuel Cells. *Appl. Microbiol. Biotechnol.* **2008**, *78*, 873–880.
18. Lu, M.; Chen, S.; Babanova, S.; Phadke, S.; Salvacion, M.; Mirhosseini, A.; Chan, S.; Carpenter, K.; Cortese, R.; Bretschger, O. Long-Term Performance of a 20-L Continuous Flow Microbial Fuel Cell for Treatment of Brewery Wastewater. *J. Power Sources* **2017**, *356*, 274–287.
19. Cusick, R. D.; Kiely, P. D.; Logan, B. E. A Monetary Comparison of Energy Recovered from Microbial Fuel Cells and Microbial Electrolysis Cells Fed Winery or Domestic Wastewaters. *Int. J. Hydrogen Energy* **2010**, *35*, 8855–8861.
20. Yu, N.; Xing, D.; Li, W.; Yang, Y.; Li, Z.; Li, Y.; Ren, N. Electricity and Methane Production from Soybean Edible Oil Refinery Wastewater Using Microbial Electrochemical Systems. *Int. J. Hydrogen Energy.* **2017**, *42*, 96–102.
21. Srikanth, S.; Kumar, M.; Singh, D.; Singh, M. P.; Das, B. P. Electro-Biocatalytic Treatment of Petroleum Refinery Wastewater Using Microbial Fuel Cell (MFC) in Continuous Mode Operation. *Bioresource Technol.* **2016**, *221*, 70–77.
22. Cecconet, D.; Callegari, A.; Capodaglio, A. G. Bioelectrochemical Systems for Removal of Selected Metals and Perchlorate from Groundwater: A Review. *Energies* **2018**, *11*, 2643.
23. Jain, A.; He, Z. "NEW" Resource Recovery from Wastewater Using Bioelectrochemical Systems: Moving Forward with Functions. *Front. Environ. Sci. Eng.* **2018**, *12*, 1–13.

24. Chouler, J.; Padgett, G. A.; Cameron, P. J.; Preuss, K.; Titirici, M. M.; Ieropoulos, I.; Di Lorenzo, M. Towards Effective Small Scale Microbial Fuel Cells for Energy Generation from Urine. *Electrochim. Acta.* **2016**, *192*, 89–98.
25. Ezziat, L.; Elabed, A.; Ibnsouda, S.; Abed, S. Challenges of Microbial Fuel Cell Architecture on Heavy Metal Recovery and Removal From Wastewater. *Front. Energy Res.* **2019**, 7, 1–13.
26. Ho, N. A. D.; Babel, S.; Sombatmankhong, K. Bio-Electrochemical System for Recovery of Silver Coupled with Power Generation and Wastewater Treatment from Silver(I) Diammine Complex. *J. Water Process Eng.* **2018**, *23*, 186–194.
27. Ucar, D.; Zhang, Y.; Angelidaki, I. An Overview of Electron Acceptors in Microbial Fuel Cells. *Front. Microbiol.* **2017**, *8*, 643–657.
28. Xiao, L.; He, Z. Applications and Perspectives of Phototrophic Microorganisms for Electricity Generation from Organic Compounds in Microbial Fuel Cells. *Ren. Sust. Energ. Rev.* **2014**, 37, 550–559.
29. Rabaey, K.; Girguis, P.; Nielsen, L. K. Metabolic and Practical Considerations on Microbial Electrosynthesis. *Curr. Opin. Biotechnol.* **2011**, *22*, 371–377.
30. Tharali, A.; Sain, N.; Osborne, W. Microbial Fuel Cells in Bioelectricity Production. *Front. Life Sci.* **2016**, *9*, 252–266.
31. Call, T. P.; Carey, T.; Bombelli, P.; Smith, D. J. L.; Hooper, P.; Howe, C. J.; Torrisi, F. Platinum-Free, Graphene Based Anodes and Air Cathodes for Single Chamber Microbial Fuel Cells. *J. Mater Chem. A* **2017**, 5, 23872–23886.
32. Saba, B.; Christy, A. D.; Yu, Z.; Co, A. C. Sustainable Power Generation from Bacterio-Algal Microbial Fuel Cells (MFCs): An Overview. *Renewable and Sustainable Energy Rev.* **2017**, 73, 75–84.
33. Olmo, O. M. *Microbial Fuel Cells Under Extreme Salinity*. Ph.D. Thesis, Rice University, Houston, Texas, USA, 2016.
34. Kim, Y.; Hatzell, M. C.; Hutchinson, A. J.; Logan, B. E. Capturing Power at Higher Voltages from Arrays of Microbial Fuel Cells Without Voltage Reversal. *Energy Environ. Sci.* **2011**, *4*, 4662–4667.
35. Fan, Y.; Han, S. K.; Liu, H. Improved Performance of CEA Microbial Fuel Cells with Increased Reactor Size. *Energy Environ. Sci.* **2012**, 5, 8273–8280.
36. Meda, U. S.; Rakesh, S. S. N.; Raj, M. A. L. A. Bio-Hydrogen Production in Microbial Electrolysis Cell Using Waste Water From Sugar Industry. *Int. J. Eng. Sci. Res. Technol.* **2015**, *4*, 452–458.
37. Kadier, A.; Simay, Y.; Abdeshahian, P.; Azman, N. F.; Chandrasekhar, K.; Kalil, M. A Comprehensive Review of Microbial Electrolysis Cells (MEC) Reactor Designs and Configurations for Sustainable Hydrogen Gas Production. *Alexandria Eng. J.* **2016**, 55, 427–443.
38. Logan, B. E.; Hamelers, B.; Rozendal, R.; Schröder, U.; Keller, J.; Freguia, S.; Aelterman, P.; Verstraete, W.; Rabaey, K. Microbial Fuel Cells: Methodology and Technology. *Environ Sci Technol.* **2006**, *40*, 5181–5192.
39. Jeremiasse, A. W.; Hamelers, H. V. M.; Buisman, C. J. N. Microbial Electrolysis Cell with a Microbial Biocathode. *Bioelectrochemistry* **2010**, 78, 39–43.

40. Wang, A.; Sun, D.; Cao, G.; Wang, H.; Ren, N.; Wu, W. M.; Logan, B. E. Integrated Hydrogen Production Process from Cellulose by Combining Dark Fermentation, Microbial Fuel Cells, and a Microbial Electrolysis Cell. *Bioresour. Technol.* **2011**, *102*, 4137–4143.
41. Liu, H.; Grot, S.; Logan, B. E. Electrochemically Assisted Microbial Production of Hydrogen from Acetate. *Environ. Sci. Technol.* **2005**, *39*, 4317–4320.
42. Hari, A.; Katuri, K.; Logan, B. E.; Saikaly, P. Set Anode Potentials Affect the Electron Fluxes and Microbial Community Structure in Propionate-Fed Microbial Electrolysis Cells. *Sci. Rep.*. **2016**, *6*, 38690.
43. Kim, Y.; Logan, B. E. Hydrogen Production from Inexhaustible Supplies of Fresh and Salt Water Using Microbial Reverse-Electrodialysis Electrolysis Cells. *Proc. Natl. Acad. Sci.* **2011**, *108*, 16176–16181.
44. Wagner, R. C.; Call, D. F.; Logan, B. E. Optimal Set Anode Potentials Vary in Bioelectrochemical Systems. *Environ. Sci. Technol.* **2010**, *44*, 6036–6041.
45. Yan, H. J.; Yates, M. D.; Regan, J. M. Effects of Constant or Dynamic Low Anode Potentials on Microbial Community Development in Bioelectrochemical Systems. *Microbiol. Biotechnol.* **2015**, *99*, 9319–9329.
46. Ishii, S.; Suzuki, S.; Norden-Krichmar, T. M.; Phan, T.; Wanger, G.; Nealson, K. H.; Sekiguchi, Y.; Gorby, Y. A.; Bretschger, O. Microbial Population and Functional Dynamics Associated with Surface Potential and Carbon Metabolism. *ISME J.* **2014**, *8*, 963–978.
47. Cai, W.; Han, T.; Guo, Z.; Varrone, C.; Wang, A.; Liu, W. Methane Production Enhancement by an Independent Cathode in Integrated Anaerobic Reactor with Microbial Electrolysis. *Bioresource Technol.* **2016**, *208*, 13–18.
48. Feng, Y.; Liu, Y.; Zhang, Y. Enhancement of Sludge Decomposition and Hydrogen Production from Waste Activated Sludge in a Microbial Electrolysis Cell with Cheap Electrodes. *Environ. Sci.: Water Res. Technol.* **2015**, *1*, 761–768.
49. Sun, D.; Call, D. F.; Kiely, P. D.; Wang, A.; Logan, B. E. Syntrophic Interactions Improve Power Production in Formic Acid Fed MFCs Operated with Set Anode Potentials or Fixed Resistances. *Biotechnol. Bioeng.* **2012**, *109*, 405–414.
50. Wang, L.; Singh, L.; Liu, H. Revealing the Impact of Hydrogen Production-Consumption Loop Against Efficient Hydrogen Recovery in Single Chamber Microbial Electrolysis Cells (MECs). *Int. J. Hydrogen Energy* **2018**, *43*, 13064–13071.
51. Kadier, A.; Kalil, M. S.; Abdeshahian, P.; Chandrasekhar, K.; Mohamed, A.; Azman, N. F.; Logroño, W.; Simayi, Y.; Hamid, A. A. Recent Advances and Emerging Challenges in Microbial Electrolysis Cells (MECs) for Microbial Production of Hydrogen and Value-Added Chemicals. *Renew. Sust. Energ. Rev.* **2016**, *61*, 501–525.
52. Ki, D.; Popat, S. C.; Torres, C. I. Reduced Overpotentials in Microbial Electrolysis Cells Through Improved Design, Operation, and Electrochemical Characterization. *Chem. Eng. J.* **2016**, *287*, 181–188.
53. Miller, A.; Singh, L.; Wang, L.; Liu, H. Linking Internal Resistance with Design and Operation Decisions in Microbial Electrolysis Cells. *Environment International* **2019**, *126*, 611–618.
54. Gadkari, S.; Shemfe, M.; Modestra, J.; Mohan, S. V.; Sadhukhan, J. Understanding the Interdependence of Operating Parameters in Microbial Electrosynthesis: A Numerical Investigation. *Phys. Chem. Chem. Phys.* **2019**, *21*, 10761–10772.

55. Rabaey, K.; Rozendal, R. A. Microbial Electrosynthesis - Revisiting the Electrical Route for Microbial Production. *Nat. Rev. Microbiol.* **2010**, *8*, 8706–8716.
56. Sadhukhan, J.; Lloyd, J. R.; Scott, K.; Premier, G. C.; Eileen, H. Y.; Curtis, T.; Head, I. M. A Critical Review of Integration Analysis of Microbial Electrosynthesis (MES) Systems with Waste Biorefineries for the Production of Biofuel and Chemical from Reuse of CO_2. *Renewable Sustainable Energy Rev.* **2016**, *56*, 116–132.
57. Jiang, Y.; May, H. D.; Lu, L.; Liang, P.; Huang, X.; Ren, Z. J. Carbon Dioxide and Organic Waste Valorization by Microbial Electrosynthesis and Electro-Fermentation. *Water Res.* **2019**, *149*, 42–55.
58. Modestra, J. A.; Mohan, S. V. Microbial Electrosynthesis of Carboxylic Acids Through CO2 Reduction with Selectively Enriched Biocatalyst: Microbial Dynamics. *J. CO2 Util.* **2017**, *20*, 190–199.
59. Cheng, S.; Xing, D.; Call, D. F.; Logan, B. E. Direct Biological Conversion of Electrical Current into Methane by Electromethanogenesis. *Environ. Sci. Technol.* **2009**, *43*, 3953–3958.
60. Gong, Y.; Ebrahim, A.; Feist, A. M.; Embree, M.; Zhang, T.; Lovely, D.; Zengler, K. Sulfide-Driven Microbial Electrosynthesis. *Environ. Sci. Technol.* **2012**, *47*, 568–573.
61. Xiang, Y.; Liu, G.; Zhang, R.; Lu, Y.; Luo, H. High-Efficient Acetate Production from Carbon Dioxide Using a Bioanode Microbial Electrosynthesis System with Bipolar Membrane. *Bioresour. Technol.* **2017**, *233*, 227–235.
62. Desloover, J.; Arends, J. B.; Hennebel, T.; Rabaey, K. Operational and Technical Considerations for Microbial Electrosynthesis. *Biochem. Soc. Trans.* **2012**, *40*, 1233–1238.
63. Vassilev, I.; Kracke, F.; Freguia, S.; Keller, J.; Krömer, J. O.; Ledezma, P.; Virdis, B. Microbial Electrosynthesis System with Dual Biocathode Arrangement for Simultaneous Acetogenesis, Solventogenesis and Carbon Chain Elongation. *Chem. Commun.* **2019**, *55*, 4351–4354.
64. Rabaey, K.; Rozendal, R. A. Microbial Electrosynthesis - Revisiting the Electrical Route for Microbial Production. *Nat. Rev. Microbiol.* **2010**, *8*, 706–716.
65. Vassilev, I.; Hernandez, P. A.; Batlle-Vilanova, P.; Freguia, S.; Krömer, J. O.; Keller, J.; Ledezma, P.; Virdis, B. Microbial Electrosynthesis of Isobutyric, Butyric, Caproic Acids, and Corresponding Alcohols from Carbon Dioxide. *ACS Sustainable Chem. Eng.* **2018**, *6*, 8485–8493.
66. Liu, C.; Luo, G.; Wang, W.; He, Y.; Zhang, R.; Liu, G. The Effects of pH and Temperature on the Acetate Production and Microbial Community Compositions by Syngas Fermentation. *Fuel* **2018**, *224*, 537–544.
67. Ganigué, R.; Sánchez-Paredes, P.; Bañeras, L.; Colprim, J. Low Fermentation pH Is a Trigger to Alcohol Production, but a Killer to Chain Elongation. *Front Microbiol.* **2016**, *7*, 702.
68. Kucek, L. A.; Spirito, C. M.; Angenent, L. T. High N-Caprylate Productivities and Specificities from Dilute Ethanol and Acetate: Chain Elongation with Microbiomes to Upgrade Products from Syngas Fermentation. *Energy Environ. Sci.* **2016**, *9*, 3482–3494.
69. Richter, H.; Molitor, B.; Diender, M.; Sousa, D. Z.; Angenent, L. T. A Narrow pH Range Supports Butanol, Hexanol, and Octanol Production from Syngas in a Continuous Co-Culture of *Clostridium ljungdahlii* and *Clostridium kluyveri* with In-Line Product Extraction. *Front Microbiol.* **2016**, *7*, 1773.

70. Diender, M.; Stams, A. J.; Sousa, D. Z. Production of Medium-Chain Fatty Acids and Higher Alcohols by a Synthetic Co-Culture Grown on Carbon Monoxide or Syngas. *Biotechnol. Biofuels.* **2016**, *9*, 82.
71. Mehanna, M.; Saito, T.; Yan, J.; Hickner, M.; Cao, X.; Huang, X.; Logan, B. E. Using Microbial Desalination Cells to Reduce Water Salinity Prior to Reverse Osmosis. *Energy Environ. Sci.* **2010**, *3*, 1114–1120.
72. Luo, H.; Xub, P.; Jenkins, P. E.; Rena, Z. Ionic Composition and Transport Mechanisms in Microbial Desalination Cells. *J. Membr. Sci.* **2012**, *409*, 16–23.
73. Chen, X.; Xia, X.; Liang, P.; Cao, X.; Sun, H.; Huang, X. Stacked Microbial Desalination Cells to Enhance Water Desalination Efficiency. *Environ. Sci. Technol.* **2011**, *45*, 2465–2470.
74. Mamun, A. A.; Ahmad, W.; Baawain, M. S.; Khadem, M.; Dhār, B. R. A review of Microbial Desalination Cell Technology: Configurations, Optimization and Applications. *J. Cleaner Prod.* **2018**, *183*, 458–480.
75. Bond, D. R.; Lovley, D. R. Electricity Production by *Geobacter sulfurreducens* Attached to Electrodes. *Appl. Environ. Microbiol.* **2003**, *69*, 1548–1555.
76. Carbajosa, S.; Malki, M.; Caillard, R.; Lopez, M. F.; Palomares, F. J.; Martín-Gago, J. A.; Rodríguez, N.; Amils, R.; Fernández, V. M.; De Lacey, A. L. Electrochemical Growth of *Acidithiobacillus ferrooxidans* on a Graphite Electrode for Obtaining a Biocathode for Direct Electrocatalytic Reduction of Oxygen. *Biosens. Bioelectron.* **2010**, *26*, 877–880.
77. Philips, J.; Verbeeck, K.; Rabaey, R.; Arends, J. B. A. Electron Transfer Mechanisms in Biofilms. In *Microbial Electrochemical and Fuel Cells*; Woodhead Publishing: Gent, Belgium, 2015; Vol. 11, pp 67–113.
78. Logan, B. E.; Regan, J. M. Electricity-Producing Bacterial Communities in Microbial Fuel Cells. *Trends Microbiol.* **2006**, *14*, 512–518.
79. Rozendal, R. A.; Hamelers, H. V. M.; Euverink, G. J. W.; Metz, S. J.; Buisman, C. J. N. Principle and Perspectives of Hydrogen Production Through Biocatalyzed Electrolysis. *Int. J. Hydrogen Energy.* **2006**, *31*, 1632–1640.
80. Holmes, D. E.; Chaudhuri, S. K.; Nevin, K. P.; Mehta, T.; Methé, B. A.; Liu, A.; Ward, J. E.; Woodard, T. L.; Webster, J.; Lovley, D. R. Microarray and Genetic Analysis of Electron Transfer to Electrodes in *Geobacter sulfurreducens*. *Environ. Microbiol.* **2006**, *8*, 1805–1815.
81. Kim, B. C.; Postier, B. L.; DiDonato, R. J.; Chaudhuri, S. K.; Nevin, K. P.; Lovley, D. R. Insights into Genes Involved in Electricity Generation in *Geobacter sulfurreducens* via Whole Genome Microarray Analysis of the OmcF-Deficient Mutant. *Bioelectrochem.* **2008**, *73*, 70–75.
82. Reguera, G.; Nevin, K.; Nicoll, J.; Covalla, S.; Woodard, T.; Lovley, D. R. Biofilm and Nanowire Production Leads to Increased Current in *Geobacter sulfurreducens* Fuel Cells. *Appl. Environ. Microbiol.* **2006**, *72*, 7345–7348.
83. Nevin, K.; Richter, H.; Covalla, S.; Johnson, J.; Woodard, T.; Jia, H.; Zhang, M.; Lovley, D. R. Power Output and Coulombic Efficiencies from Biofilms of *Geobacter sulfurreducens* Comparable to Mixed Community Microbial Fuel Cells. *Environ. Microbiol.* **2008**, *10*, 2505–2514.
84. Picioreanu, C.; Head, I.; Katuri, K.; Loosdrecht, M.; Scott, K. Mathematical Model for Microbial Fuel Cells with Anodic Biofilms and Anaerobic Digestion. *Water. Res.* **2007**, *41*, 2921–2940.

85. Marcus, A.; Torres, C.; Rittmann, B. Conduction-Based Modeling of the Biofilm Anode of a Microbial Fuel Cell. *Biotechnol. Bioeng.* **2007**, *98*, 1171–82.
86. Esteve-Núñez, A.; Sosnik, J.; Visconti, P.; Lovley, D. R. Fluorescent Properties of c-Type Cytochromes Reveal Their Potential Role as an Extracytoplasmic Electron Sink in *Geobacter sulfurreducens*. *Appl. Environ. Microbiol.* **2010**, *76*, 4080–4084.
87. Von Canstein, H.; Ogawa, J.; Shimizu, S.; Lloyd, J. R. Secretion of Flavins by *Shewanella* species and Their Role in Extracellular Electron Transfer. *Appl. Environ. Microbiol.* **2008**, *74*, 615–623.
88. Gorby, Y. A.; Yanina, S.; McLean, J. S.; Rosso, K. M.; Moyles, D.; Dohnalkova, A.; Beveridge, T. J.; Chang, I. S.; Kim, B. H.; Kim, K. S.; Culley, D. E.; Reed, S. B.; Romine, M. F.; Saffarini, D. A.; Hill, E. A.; Shi, L.; Elias, D. A.; Kennedy, D. W.; Pinchuk, G.; Watanabe, K.; Ishii, S.; Logan, B.; Nealson, K. H.; Fredrickson, J. K. Electrically Conductive Bacterial Nanowires Produced by *Shewanella oneidensis* strain MR-1 and Other Microorganisms. *Proc. Natl. Acad. Sci. U. S. A.* **2006**, *103*, 11358–11363.
89. Richter, H.; McCarthy, K.; Nevin, K. P.; Johnson, J. P.; Rotello, V. M.; Lovley, D. R. Electricity Generation by *Geobacter sulfurreducens* Attached to Gold Electrodes. *Langmuir.* **2008**, *24*, 4376–4379.
90. Bretschger, O.; Obraztsova, A.; Stumm, C. A.; Chang, I. S.; Gorby, Y. A.; Reed, S. B.; Culley, D. E.; Reardon, C. L.; Barua, S.; Romine, M. F.; Zhou, J.; Beliaev, A. S.; Bouhenni, R.; Saffarini, D.; Mansfeld, F.; Kim, B. H.; Fredrickson, J. K.; Nealson, K. H. Current Production and Metal Oxide Reduction by *Shewanella oneidensis* MR-1 Wild Type and Mutants. *Appl. Environ. Microbiol.* **2007**, *73*, 7003–7012.
91. Lanthier, M.; Gregory, K. B.; Lovley, D. R. Growth with High Planktonic Biomass in *Shewanella oneidensis* Fuel Cells. *FEMS Microbiol. Lett.* **2008**, *278*, 29–35.
92. Mevers, E.; Su, L.; Pishchany, G.; Baruch, M.; Cornejo, J.; Hobert, E.; Dimise, E.; Franklin, C. M. A.; Clardy, J. An Elusive Electron Shuttle from a Facultative Anaerobe. *eLife* **2019**, *8*, 48–54.
93. Eggleston, C. M.; Vörös, J.; Shi, L.; Lower, B. H.; Droubay, T. C.; Colberg, P. J. S. Binding and Direct Electrochemistry of OmcA, an Outer-Membrane Cytochrome from an Iron Reducing Bacterium, with Oxide Electrodes: A Candidate Biofuel Cell System. *Inorg. Chim. Acta* **2008**, *361*, 769–777.
94. Zhu, D.; Wang, D.; Song, T.; Guo, T.; Ouyang, P.; Wei, P.; Xie, J. Effect of Carbon Nanotube Modified Cathode by Electrophoretic Deposition Method on the Performance of Sediment Microbial Fuel Cells. *Biotechnol. Lett.* **2015**, *5*, 101–107.
95. Yang, J.; Cheng, S.; Sun, Y.; Li, C. Improving the Power Generation of Microbial Fuel Cells by Modifying the Anode with Single-Wall Carbon Nanohorns. *Biotechnol. Lett.* **2017**, *39*, 1515–152.
96. Hamelers, H. V. M.; Sleutels, T. H. J. A.; Jeremiasse, A. W.; Post, J. W.; Strik, D. P. B.; Rozendal, T. B.; Rabaey, R. A.; Angenent, K.; Schroder, L. T.; U., Keller, J. *Bio-Electrochemical Systems: From Extracellular Electron Transfer to Biotechnological Application*; IWA Publishing: London, 2009; Vol. 8, p 524.
97. Sleutels, T. H. J. A.; Darus, L.; Hamelers, H. V. M.; Buisman, C. J. N. Effect of Operational Parameters on Coulombic Efficiency in Bioelectrochemical Systems. *Bioresour. Technol.* **2011**, *102*, 11172–11176.

98. Clauwaert, P.; Aelterman, P.; Pham, T. H.; DeSchamphelaire, L.; Carballa, M.; Rabaey, K.; Verstraete, W. Minimizing Losses in Bio-Electrochemical Systems: The Road to Applications. *Appl. Microbiol. Biotechnol.* **2008**, *79*, 901–913.
99. Larrosa-Guerrero, A.; Scott, K.; Head, I. M.; Mateo, F.; Ginesta, A.; Godinez, C. Effect of Temperature on the Performance of Microbial Fuel Cells. *Fuel* **2010**, *89*, 3985.
100. Sleutels, T. H. J. A.; Molenaar, S. D.; Heijne, A. T.; Buisman, C. J. N. Low Substrate Loading Limits Methanogenesis and Leads to High Coulombic efficiency in bioelectrochemical systems. *Microorganisms* **2016**, *4*, 7.
101. Torres, C. I.; Marcus, A. K.; Rittman, B. E. Proton Transport Inside the Biofilm Limits Electrical Current Generation by Anode-Respiring Bacteria. *Biotechnol. Bioeng.* **2008**, *100*, 872.
102. Kim, B. H.; Chang, I. S.; Gadd, G. M. Challenges in Microbial Fuel Cell Development and Operation. *Appl. Microbiol. Biotechnol.* **2007**, *76*, 485.
103. Yang, W.; Logan, B. E. Immobilization of a Metal–Nitrogen–Carbon Catalyst on Activated Carbon with Enhanced Cathode Performance in Microbial Fuel Cells. *ChemSusChem.* **2016**, *9*, 2226.
104. Rossi, R.; Yang, W.; Setti, L.; Logan, B. E. Assessment of a Metal-Organic Framework Catalyst in Air Cathode Microbial Fuel Cells Over Time with Different Buffers and Solutions. *Bioresour. Technol.* **2017**, *233*, 399.
105. Mustakeem, M. Electrode Materials for Microbial Fuel Cells: Nanomaterial Approach. *Mater. Renew. Sustain. Energy.* **2015**, *4*, 22.
106. Guo, K.; Prevoteau, A.; Patil, S. A.; Rabaey, K. Engineering Electrodes for Microbial Electrocatalysis. *Curr. Opin. Biotechnol.* **2015**, *33*, 149.
107. Wei, J.; Liang, P.; Huang, X. Recent Progress in Electrodes for Microbial Fuel Cells. *Bioresour. Technol.* **2011**, *102*, 9335.
108. Wang, Z.; Mahadevan, G. D.; Wu, Y.; Zhao, F. Progress of Air-Breathing Cathode in Microbial Fuel Cells. *J. Power Sources* **2017**, *356*, 245.
109. Liew, K. B.; Daud, W. R. W.; Ghasemi, M.; Leong, J. X.; Lim, S. S.; Ismail, M. Non-Pt Catalyst as Oxygen Reduction Reaction in Microbial Fuel Cells: A Review. *Int. J. Hydrogen Energy* **2014**, *39*, 4870.
110. Janicek, A.; Fan, Y.; Liu, H. Design of Microbial Fuel Cells for Practical Application: A Review and Analysis of Scale-Up Studies. *Biofuels* **2014**, *5*, 79.
111. Rahimnejad, M.; Adhami, A.; Darvari, S.; Zirepour, A.; Oh, S. E. Microbial Fuel Cell as New Technology for Bioelectricity Generation: A Review. *Alexandria. Eng. J.* **2015**, *54*, 745.
112. Pocaznoi, D.; Erable, B.; Etcheverry, L.; Delia, M. L.; Bergel, A. Towards an Engineering-Oriented Strategy for Building Microbial Anodes for Microbial Fuel Cells. *Phys. Chem. Chem. Phys.* **2012**, *14*, 13332.
113. Ewing, T.; Thi Ha, P.; Babauta, J. T.; Trong Tang, N.; Heo, D.; Beyenal, H. Scale-up of Sediment Microbial Fuel Cells. *J. Power Sources* **2014**, *272*, 311.
114. Rabaey, K.; Boon, N.; Höfte, M.; Verstraete, W. Microbial Phenazine Production Enhances Electron Transfer in Biofuel Cells. *Environ. Sci. Technol.* **2005**, *39*, 3401.
115. Harnisch, F.; Schröder, U. From MFC to MXC: Chemical and Biological Cathodes and Their Potential for Microbial Bioelectrochemical Systems. *Chem. Soc. Rev.* **2010**, *39*, 4433.

116. Zhou, H.; Mei, X.; Liu, B.; Xie, G.; Xing, D. Magnet Anode Enhances Extracellular Electron Transfer and Enrichment of Exoelectrogenic Bacteria in Bioelectrochemical Systems. *Biotechnol. Biofuels.* **2019**, *12*, 133.
117. Sun, M.; Zhai, L. F.; Li, W. W.; Yu, Q. H. Harvest and Utilization of Chemical Energy in Wastes by Microbial Fuel Cells. *Chem. Soc. Rev.* **2016**, *45*, 2847.
118. Blanchet, E.; Erable, B.; De Solan, M. L.; Bergel, A. Two-Dimensional Carbon Cloth and Three-Dimensional Carbon Felt Perform Similarly to Form Bioanode Fed with Food Waste. *Electrochem. Commun.* **2016**, *66*, 38.
119. Islam, M. A.; Woon, C. W.; Ethiraj, B.; Cheng, C. K.; Yousuf, A.; Khan, M. M. R. Ultrasound Driven Biofilm Removal for Stable Power Generation in Microbial Fuel Cell. *Energy Fuel.* **2016**, *31*, 968.
120. Sun, D.; Cheng, S.; Wang, A.; Li, F.; Logan, B. E.; Cen, K. Temporal-Spatial Changes in Viabilities and Electrochemical Properties of Anode Biofilms. *Environ. Sci. Technol.* **2015**, *49*, 5227.
121. Harnisch, F.; Wirth, S.; Schröder, U. Effects of Substrate and Metabolite Crossover on the Cathodic Oxygen Reduction Reaction in Microbial Fuel Cells: Platinum vs. Iron(II) Phthalocyanine Based Electrodes. *Electrochem. Commun.* **2009**, *11*, 2253.
122. Winfield, J.; Ieropoulos, I.; Rossiter, J.; Greenman, J.; Patton, D. Biodegradation and Proton Exchange Using Natural Rubber in Microbial Fuel Cells. *Biodegradation* **2013**, *24*, 733.
123. Call, D.; Logan, B. E. Hydrogen Production in a Single Chamber Microbial Electrolysis Cell Lacking a Membrane. *Environ. Sci. Technol.* **2008**, *42*, 3401.
124. Rozendal, R. A.; Leone, E.; Keller, J.; Rabaey, K. Efficient Hydrogen Peroxide Generation from Organic Matter in a Bioelectrochemical System. *Electrochemistry Commun.* **2009**, *11*, 1752.
125. Rabaey, K.; Butzer, S.; Brown, S.; Keller, J.; Rozendal, R. A. High Current Generation Coupled to Caustic Production Using a Lamellar Bioelectrochemical System. *Environ. Sci. Technol.* **2010**, *44*, 4315.
126. Li, C.; Lesnik, K. L.; Liu, H. Stay Connected: Electrical Conductivity of Microbial Aggregates. *Biotech. Adv.* **2017**, *35*, 669–680.
127. Lovley, D. R. Happy Together: Microbial Communities that Hook Up to Swap Electrons. *ISME J.* **2017**, *11*, 327–336.
128. Nevin, K. P.; Woodard, T. L.; Franks, A. E.; Summers, Z. M.; Lovley, D. R. Microbial Electrosynthesis: Feeding Microbes Electricity to Convert Carbon Dioxide and Water to Multicarbon Extracellular Organic Compounds. *mBio.* **2010**, *1*, 103–110.
129. Nevin, K. P.; Hensley, S. A.; Franks, A. E.; Summers, Z. M.; Ou, J.; Woodard, T. L.; Snoeyenbos-West, O. L.; Lovley, D. R. Electrosynthesis of Organic Compounds from Carbon Dioxide is Catalyzed by a Diversity of Acetogenic Microorganisms. *Appl. Environ.Microbiol.* **2011**, *77*, 2882.
130. Marshall, C. W.; Ross, D. E.; Fichot, E. B.; Norman, R. S.; May, H. D. Electrosynthesis of Commodity Chemicals by an Autotrophic Microbial Community. *Appl. Environ. Microbiol.* **2012**, *788*, 412.
131. Li, H.; Opgenorth, P. H.; Wernick, D. G.; Rogers, S.; Wu, T. Y.; Higashide, W.; Malati, P.; Huo, Y. X.; Cho, K. M.; Liao, J. C. Integrated Electromicrobial Conversion of CO2 to Higher Alcohols. *Science* **2012**, *335*, 1596.

132. Liu, C.; Gallagher, J. J.; Sakimoto, K. K.; Nichols, E. M.; Chang, C. J.; Chang, M. C. Y.; Yang, P. Nanowire–Bacteria Hybrids for Unassisted Solar Carbon Dioxide Fixation to Value-Added Chemicals. *Nano. Lett.* **2015**, *15*, 3634.
133. Nichols, E. M.; Gallagher, J. J.; Liu, C.; Su, Y.; Resasco, J.; Yu, Y.; Sun, Y.; Yang, P.; Chang, M. C. Y.; Chang, C. J. Hybrid Bioinorganic Approach to Solar-to-Chemical Conversion. *Proc. Natl. Acad. Sci. U.S.A.* **2015**, *112*, 11461.
134. Sakimoto, K. K.; Wong, A. B.; Yang, P. Self-Photosensitization of Nonphotosynthetic Bacteria for Solar-to-Chemical Production. *Science* **2016**, *351*, 71–74.
135. Torella, J. P.; Gagliardi, C. J.; Chen, J. S.; Bediako, D. K.; Colón, B.; Way, J. C.; Silver, P. A.; Nocera, D. G. Efficient Solar-to-Fuels Production from a Hybrid Microbial-Water-Splitting Catalyst System. *Proc. Natl. Acad. Sci. U.S.A.* **2015**, *112*, 2337.
136. Liu, C.; Colón, B. C.; Ziesack, M.; Silver, P. A.; Nocera, D. G. Water Splitting–Biosynthetic System with CO2 Reduction Efficiencies Exceeding Photosynthesis. *Science.* **2016**, *352*, 1210–1213.
137. Sakimoto, K. K.; Kornienko, N.; Yang, P. Cyborgian Material Design for Solar Fuel Production: The Emerging Photosynthetic Biohybrid Systems. *Acc. Chem. Res.* **2017**, *50*, 476–481.
138. Rosenbaum, M. A.; Franks, A. E. Microbial Catalysis in Bioelectrochemical Technologies: Status Quo, Challenges and Perspectives. *Appl. Microbiol. Biotechnol.* **2014**, *98*, 509–518.
139. Lu, L.; Ren, Z. Microbial Electrolysis Cells for Waste Biorefinery: A State of the Art Review. *Bioresour. Technol.* **2016**, *215*, 254–264.
140. Logan, B. E. *Microbial Fuel Cells*, 1st ed.; John Wiley & Sons: Hoboken, 2008; pp 85–110
141. Cornejo, J. A.; Sheng, H.; Edri, E.; Franklin, C. M. A.; Frei, H. Nanoscale Membranes that Chemically Isolate and Electronically Wire Up the Abiotic/Biotic Interface. *Nat. Commun.* **2018**, *9*, 2263.
142. Simeng, L.; Chen, G. Factors Affecting the Effectiveness of Bioelectrochemical System Applications: Data Synthesis and Meta-Analysis. *Batteries.* **2018**, *4*, 34.
143. Liu, B.; Bruckner, C.; Lei, Y.; Cheng, Y.; Santoro, C.; Li, B. Cobalt Porphyrin-Based Material as Methanol Tolerant Cathode in Single Chamber Microbial Fuel Cells (SCMFCs). *J. Power Sources.* **2014**, *257*, 246–253.
144. Yuan, Y.; Zhou, S.; Tang, J. In Situ Investigation of Cathode and Local Biofilm Microenvironments Reveals Important Roles of OH- and Oxygen Transport in Microbial Fuel Cells. *Environ. Sci. Technol.* **2013**, *47*, 4911–4917.
145. Liu, M.; Zhang, R.; Chen, W. Graphene-Supported Nanoelectrocatalysts for Fuel Cells: Synthesis, Properties, and Applications. *Chem. Rev.* **2014**, *114*, 5117–5160.
146. Srikanth, S.; Pant, D.; Dominguez-Benetton, X. I.; Genn, K.; Vanbroekhoven, P.; Vermeiren; Alvarez-Gallego, Y. Gas Diffusion Electrodes Manufactured by Casting Evaluation as Air Cathodes for Microbial Fuel Cells (MFC). *Materials* **2016**, *9*, 601.
147. Santoro, C.; Stadlhofer, A.; Hacker, V.; Squadrito, G.; Schroder, U.; Li, B. Activated Carbon Nanofibers (ACNF) as Cathode for Single Chamber Microbial Fuel Cells (SCMFCs). *J. Power Sources.* **2013**, *243*, 499–507.

148. Duteanu, N.; Erable, B.; Senthil Kumar, S. M.; Ghangrekar, M. M.; Scott, K. Effect of Chemically Modified Vulcan XC-72R on the Performance of Air-Breathing Cathode in a Single-Chamber Microbial Fuel Cell. *Bioresour. Technol.* **2010**, *101*, 5250–5255.
149. Yuan, H.; He, Z. Graphene-Modified Electrodes for Enhancing the Performance of Microbial Fuel Cells. *Nanoscale* **2015**, 7, 7022–7029.
150. Feng, L.; Yan, Y.; Chen, Y.; Wang, L. Nitrogen-Doped Carbon Nanotubes as Efficient and Durable Metal-Free Cathodic Catalysts for Oxygen Reduction in Microbial Fuel Cells. *Energy Environ. Sci.* **2011**, *4*, 1892–1899.
151. Meng, K.; Liu, Q.; Huang, Y.; Wang, Y. Facile Synthesis of Nitrogen and Fluorine Co-Doped Carbon Materials as Efficient Electrocatalysts for Oxygen Reduction Reactions in Air-Cathode Microbial Fuel Cells. *J. Mater. Chem. A.* **2015**, *3*, 6873–6877.
152. Watson, V. J.; Nieto Delgado, C.; Logan, B. E. Improvement of Activated Carbons as Oxygen Reduction Catalysts in Neutral Solutions by Ammonia Gas Treatment and Their Performance in Microbial Fuel Cells. *J. Power Sources.* **2013**, *242*, 756–761.
153. Zhang, B.; Wen, Z.; Ci, S.; Mao, S.; Chen, J.; He, Z. Synthesizing Nitrogen-Doped Activated Carbon and Probing its Active Sites for Oxygen Reduction Reaction in Microbial Fuel Cells. *ACS Appl. Mater. Interfaces.* **2014**, *6*, 7464–7470.
154. Zhou, L.; Fu, P.; Cai, X.; Zhou, S.; Yuan, Y. Naturally Derived Carbon Nanofibers as Sustainable Electrocatalysts for Microbial Energy Harvesting: A New Application of Spider Silk. *Appl. Catal., B.* **2016**, *188*, 31–38.
155. Feng, L.; Chen, Y.; Chen, L. Easy-to-Operate and Low-Temperature Synthesis of Gram-Scale Nitrogen-Doped Graphene and Its Application as Cathode Catalyst in Microbial Fuel Cells. *ACS Nano.* **2011**, *5*, 9611–9618.
156. Santoro, C.; Kodali, M.; Kabir, S.; Soavi, F.; Serov, A.; Atanassov, P. Three-Dimensional Graphene Nanosheets as Cathode Catalysts in Standard and Supercapacitive Microbial Fuel Cell. *J. Power Sources.* **2017**, *356*, 371–380.
157. Chen, S.; Patil, S.; Schröder, U. A High-Performance Rotating Graphite Fiber Brush Air-Cathode for Microbial Fuel Cells. *Appl. Energy.* **2018**, *211*, 1089–1094.
158. Lima, F. H. B.; Zhang, J.; Shao, M. H.; Sasaki, K.; Vukmirovic, M. B.; Ticianelli, E. A.; Adzic, R. R. Catalytic Activity−d-Band Center Correlation for the O2 Reduction Reaction on Platinum in Alkaline Solutions. *J. Phys. Chem. C.* **2007**, *111*, 404–410.
159. Zhang, J. N.; You, S. J.; Yuan, Y. X.; Zhao, Q. L.; Zhang, G. D. Efficient Electrocatalysis of Cathodic Oxygen Reduction with Pt–Fe Alloy Catalyst in Microbial Fuel Cell. *Electrochem. Commun.* **2011**, *13*, 903–905.
160. Chang, Y. Y.; Zhao, H. Z.; Zhong, C.; Xue, A. Effects of Different Pt-M (M = Fe, Co, Ni) Alloy as Cathodic Catalyst on the Performance of Two-Chambered Microbial Fuel Cells. *Russ. J. Electrochem.* **2014**, *50*, 885–890.
161. Yan, Z.; Wang, M.; Lu, Y.; Liu, R.; Zhao, J. Ethylene Glycol Stabilized NaBH4 Reduction for Preparation Carbon-Supported Pt–Co Alloy Nanoparticles Used as Oxygen Reduction Electrocatalysts for Microbial Fuel. *Cells J. Solid State Electrochem.* **2014**, *18*, 1087–1097.
162. Lefebvre, O.; Tan, Z.; Shen, Y.; Ng, H. Y. Optimization of a Microbial Fuel Cell for Wastewater Treatment Using Recycled Scrap Metals as a Cost-Effective Cathode Material. *Bioresour. Technol.* **2013**, *127*, 158–164.

163. Thackeray, M. M.; Rossouw, M. H.; de Kock, A. A.; de la Harpe, P.; Gummow, R. J.; Pearce, K.; Liles, D. C. The Versatility of MnO2 for Lithium Battery Applications. *J. Power Sources.* **1993**, *43*, 289–300.
164. Stoerzinger, K. A.; Risch, M.; Han, B.; Shao-Horn, Y. Recent Insights into Manganese Oxides in Catalyzing Oxygen Reduction Kinetics. *ACS Catal.* **2015**, *5*, 6021–6031.
165. Noori, M. T.; Ghangrekar, M. M.; Mukherjee, C. K. V2O5 Microflower Decorated Cathode for Enhancing Power Generation in Air-Cathode Microbial Fuel Cell Treating Fish Market Wastewater. *Int. J. Hydrogen Energy.* **2016**, *41*, 3638–3645.
166. Gong, X. B.; You, S. J.; Wang, X. H.; Zhang, J. N.; Gan, Y.; Ren, N. Q. A Novel Stainless Steel Mesh/Cobalt Oxide Hybrid Electrode for Efficient Catalysis of Oxygen Reduction in a Microbial Fuel Cell. *Biosens. Bioelectron.* **2014**, *55*, 237–241.
167. Mecheri, B.; Iannaci, A.; D'Epifanio, A.; Mauri, A.; Licoccia, S. Carbon-Supported Zirconium Oxide as a Cathode for Microbial Fuel Cell Applications. *ChemPlusChem.* **2016**, *81*, 80–85.
168. Dong, H.; Yu, H.; Wang, X.; Zhou, Q.; Sun, J. Carbon-Supported Perovskite Oxides as Oxygen Reduction Reaction Catalyst in Single Chambered Microbial Fuel Cells. *J. Chem. Technol. Biotechnol.* **2013**, *88*, 774–778.
169. Bhowmick, G. D.; Noori, M. T.; Das, I.; Neethu, B.; Ghangrekar, M. M.; Mitra, A. Bismuth Doped TiO2 as an Excellent Photocathode Catalyst to Enhance the Performance of Microbial Fuel Cell. *Int. J. Hydrogen Energy.* **2018**, *43*, 7501–7510.
170. Singh, S.; Verma, N. Graphitic Carbon Micronanofibers Asymmetrically Dispersed with Alumina-Nickel Nanoparticles: A Novel Electrode for Mediatorless Microbial Fuel Cells. *Int. J. Hydrogen Energy.* **2015**, *40*, 5928–5938.
171. Modi, A.; Singh, S.; Verma, N. In Situ Nitrogen-Doping of Nickel Nanoparticle-Dispersed Carbon Nanofiber-Based Electrodes: Its Positive Effects on the Performance of a Microbial Fuel Cell. *Electrochim. Acta.* **2016**, *190*, 620–627.
172. Farahania, F.; Mecherib, B.; Majidia, M.; Oliveira, M.; D'Epifanio, A.; Zurlo, F.; Placidic, E.; Arciprete, F.; Licoccia, S. MnOx-Based Electrocatalysts for Enhanced Oxygen Reduction in Microbial Fuel Cell Air Cathodes. *J. Power Sources* **2018**, *390*, 45–53.
173. Tang, C.; Zhang, Q. Can metal–Nitrogen–Carbon Catalysts Satisfy Oxygen Electrochemistry? *J. Mater. Chem. A* **2016**, *4*, 4998–5001.
174. Zhao, F.; Harnisch, F.; Schroder, U.; Scholz, F.; Bogdanoff, P.; Herrmann, I. Application of Pyrolysed Iron(II) Phthalocyanine and CoTMPP Based Oxygen Reduction Catalysts as Cathode Materials in Microbial Fuel Cells. *Electrochem. Commun.* **2005**, 7, 1405–1410.
175. Deng, L.; Zhou, M.; Liu, C.; Liu, L.; Liu, C.; Dong, S. Development of High Performance of Co/Fe/N/CNT Nanocatalyst for Oxygen Reduction in Microbial Fuel Cells. *Talanta* **2010**, *81*, 444–448.
176. Santoro, C.; Rojas-Carbone, S.; Awais, R.; Gokhale, R.; Kodali, M.; Serov, A.; Artyushkova, K.; Atanassov, P. Influence of Platinum Group Metal-Free Catalyst Synthesis on Microbial Fuel Cell Performance. *J. Power Sources* **2018**, *375*, 11–20.
177. Kodali, M.; Herrera, S.; Kabir, S.; Serov, A.; Santoro, C.; Ieropoulos, I.; Atanassov, P. Enhancement of Microbial Fuel Cell Performance by Introducing a Nano-Composite Cathode Catalyst. *Electrochim. Acta.* **2018**, *265*, 56–64.

178. Zhang, L.; Hua, Y.; Chena, J.; Huang, W.; Chen, J.; Chen, Y. A Novel Metal Organic Framework-Derived Carbon-Based Catalyst for Oxygen Reduction Reaction in a Microbial Fuel Cell. *J. Power Sources.* **2018**, *384*, 98–106.
179. Garcia, M.; Santoro, C.; Kodali, M.; Serov, A.; Artyushkova, K.; Atanassov, P.; Ieropoulos, I. Iron-Streptomycin Derived Catalyst for Efficient Oxygen Reduction Reaction in Ceramic Microbial Fuel Cells Operating with Urine. *J. Power Sources.* **2019**, *425*, 50–59.

Chapter 2

H_2 Evolution Catalysts for Microbial Electrolysis Cells

Sidan Lu, Guangcai Tan, and Xiuping Zhu*

Department of Civil and Environmental Engineering, Louisiana State University, Baton Rouge, Louisiana 70803, United States

***E-mail: xzhu@lsu.edu.**

Microbial electrolysis cells (MECs) can efficiently produce H_2 gas using organics-containing wastewater. However, traditional cathode catalysts such as platinum (Pt) are too expensive for large-scale uses. Currently, many studies have reported efficient and cheap alternatives to Pt; however, few reviews have focused on the cathode catalyst materials and their possible future development. In this chapter, we review the currently known cathode catalysts based on their performance to catalyze hydrogen evolution reaction (HER) and their material costs. The review starts by introducing mechanisms of MECs and the functions of H_2 evolution catalysts, followed by presenting current types of HER catalysts and the important properties of cathode materials that affect the HER process. Greater interest is placed on the H_2 production abilities and energy efficiencies of the catalysts in MECs. Moreover, the challenges and future perspectives of developing novel cathode materials have been thoroughly discussed.

Introduction

Microbial Electrolysis Cells

H_2 is an attractive energy carrier with high potential for large-scale economical and clean applications. However, most of the reported H_2 generation methods are more expensive compared to fossil fuels. Presently, there is a high demand to explore economically efficient ways to produce H_2 for sustainable development.

Microbial electrolysis cells (MECs) are promising bioelectrochemical systems that use electroactive microbes to oxidize organic matters and generate electricity and H_2 (*1*). Generally, MECs have microorganisms in the anode chamber that oxidize organics in the wastewater and release H^+, electrons (e^-), and CO_2 to the electrolyte and anode (*2*). The H^+ then diffuses to the cathode, joins with e^- that are also transferred from the anode through the external circuit, and becomes H_2.

© 2020 American Chemical Society

Compared to water electrolysis that needs a minimum applied voltage of 1.23 V (Anode: $H_2O \rightarrow 2H^+ + ½O_2 + 2e^-$, $E_{Anode} = 0.82$ V vs NHE; Cathode: $2H^+ + 2e^- \rightarrow H_2$, $E_{cathode} = -0.41$ V vs. NHE), MECs theoretically only require a 0.12 V applied voltage for a hydrogen evolution reaction (HER) (Figure 1; Anode: $C_2H_4O_2 + 2H_2O \rightarrow 2CO_2 + 8e^- + 8H^+$, $E_{anode} = -0.29$ V vs NHE; Cathode: $8H^+ + 8e^- \rightarrow 4H_2$, $E_{Cathode} = -0.41$ V vs NHE) (2, 3). However, larger voltages (0.7~1.2 V) are usually used in practice to compensate for the cathode overpotentials of MECs. The overpotentials can result from electrolyte resistance, concentration overpotential between the electrolyte and the cathode, and activation overpotential (HER activation barrier) (3). Decreasing the cathode overpotential is important for improving the energy efficiency of MECs.

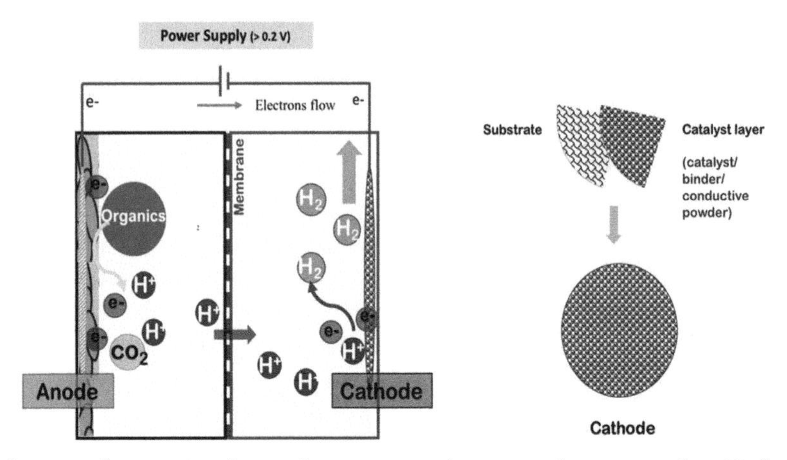

Figure 1. Schematic of an MEC with or without an ion exchange membrane to produce H_2 from organics, and the layered structure of the cathode.

Cathode Materials

Cathodes of MECs usually have layered structures consisting of a catalyst layer on a conductive substrate (4). The catalyst layer is a mixture of catalyst, binder, and conductive powder (e.g., carbon black [CB], activated carbon), and the substrate is usually made of metal or carbon (C) that supports the shape of the cathode and collects the current. The cathodes are connected to the power supply via an external wire (Figure 1).

Effective catalysts can reduce the activation barrier of the HER process and determine the efficiency of H_2 generation (5). Important factors, such as the H_2 production rate, coulombic efficiency, overall hydrogen recovery, and long-term durability, should be considered when evaluating a cathode (6). Coulombic efficiency is the ratio of the moles of H_2 that could be recovered based on the measured currents to the theoretical H_2 production based on the organics. Overall hydrogen recovery refers to the moles of H_2 actually recovered at the cathode compared to the moles that theoretically could be produced from the organics. In addition, cost of catalysts is also important when developing cathodes. Pt is known as an efficient HER catalyst. For example, in a single-chamber MEC with a Pt cathode, the H_2 production rate reached 0.33–0.68 m^3 $H_2/m^3/d$ (7) and the coulombic efficiency can be 78–81% (2, 8). However, Pt is too expensive for large-scale applications (\$500–2000 per m^2) (8). Approximately 47% of the costs come from the Pt cathodes of MECs, which need to be reduced by 90% to achieve a real application of the systems (9). Recently, researchers have been exploring more cost-effective catalysts, and in Table 1, current HER catalysts for MECs are listed and the costs of some typical catalysts are also shown in Table 2.

It should be noted that except for efficient and low-cost HER catalysts, an effective MEC system also relies on many other factors, such as the operation modes, temperatures, electrolyte solution, and byproducts (*10*), but are not considered in this review.

Catalytic Mechanisms

To judge the performance of HER catalysts, the available active sites of the catalyst and the binding free energy of hydrogen atoms (ΔG_{H^*}) are important (*11*). HER occurs either via the Volmer-Tafel mechanism or the Volmer–Heyrovsky mechanism. The Volmer reaction is the H adsorption step by the catalyst (equations 1 and 2). The Tafel (equation 3) and Heyrovsky (equations 4 and 5) reactions are the H_2 desorption steps from the catalyst (*11*). In the equations, the * is an empty active site of catalysts and the H* represents the absorbed H atom onto the site.

$$H_3O^+ + e^- + {}^* \rightarrow H^* + H_2O\text{, Volmer reaction in acidic solution} \quad (1)$$

$$H_3O^+ + 2e^- + {}^* \rightarrow 2H^* + OH^-\text{, Volmer reaction in alkali solution} \quad (2)$$

$$H^* + H^* \rightarrow H_2 + 2^*\text{, Tafel reaction} \quad (3)$$

$$H^* + H_3O + e^- \rightarrow H_2 + H_2O\text{, Heyrovsky reaction in acidic solution} \quad (4)$$

$$2H^* + H_3O^+ + 2e^- \rightarrow 2H_2 + OH^-\text{, Heyrovsky reaction in alkali solution} \quad (5)$$

Volcano plots of the exchange current density as a function of Gibbs free energy of adsorbed hydrogen atoms (ΔG_{H^*}) make it easier to compare the activities of different catalysts for H_2 evolution (*11*). According to Figure 2, Pt, Pd, Rh, Re, Ir, MoS_2, and Ni are strongly able to absorb hydrogen atoms, and consistently their metals and complexes have been found to have excellent qualities for the HER process.

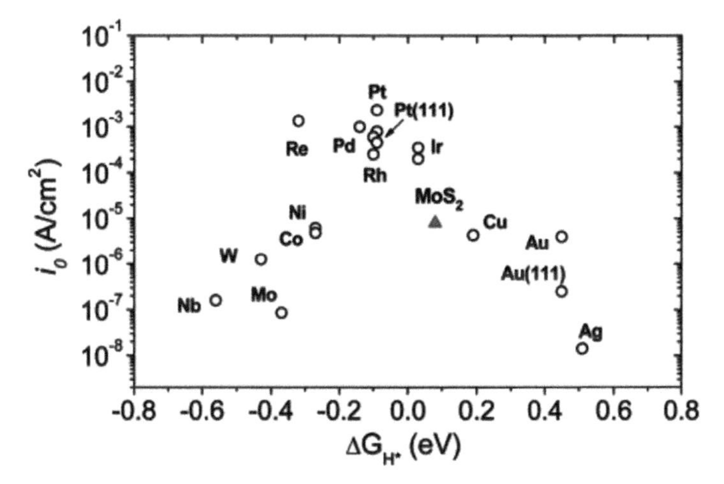

Figure 2. The volcano plot of the exchange current density as a function of Gibbs free energy of adsorbed hydrogen atoms. Reproduced with permission from reference (12). Copyright 2007 AAAS.

Table 1. HER Catalysts Reported in Recent Years with the Substrate, the Anode, the Electrolyte, the Applied Voltage, the System Configuration, the H_2 Production Rate, the Current Density, the Coulombic Efficiency, and the Published Year

HER Catalysts	*Substrate*	*Anode*	*Electrolyte*	*Applied Voltage (V)*	*Configuration*	*H_2 Production Rate ($m^3 H_2/m^3/d$)*	*Current Density (A/m^2)*	*Coulombic Efficiency (%)*	*Published Year*
Pt (0.5 mg/cm^2)	C paper	C paper	50 mM PBS	0.85	Single-chamber	-	0.9 (unknown)	78	2005 (*2*)
Pt	Titanium (Ti) mesh	Graphite felt	PBS	1.0	Single-chamber	$0.3^{Avg.}$	$2.2\text{-}2.4^{Avg.}$	23	2007 (*12*)
Pt (0.5 mg/cm^2)	C cloth	C fiber brush	50 mM PBS	0.6	Single-chamber no membrane	$1.4^{Max.}$	-	70	2009 (*6*)
Biofilm	Graphite felt	Graphite felt	100 mM PBS	0.7	Two-chamber	$0.03^{Avg.}$	$1.2^{Max.}$	49	2007 (*13*)
Carbon felt	C felt	Graphite spheres	100 mM PBS	1.0	Up-flow single-chamber	$0.6^{Avg.}$	$3.3^{Max.}$	60	2009 (*14*)
Fe (stainless steel [SS] brush)	-	Graphite fiber brush	50 mM PBS	0.6	Single-chamber	-	$2.5\text{-}3.9^{Max.}$	2009 (*8*)	
Fe/Fe3C nanorods (N-doped)	Graphite carbon	Carbon brush	100 mM PBS	0.8	Two-chamber	$0.02^{Avg.}$	$2.6^{Max.}$	43.6	2012 (*15*)
Pd nanoparticles	C paper	Activated carbon fibers	50 mM PBS	0.6	Two-chamber	-	-	56	2011 (*16*)
MoS2	C nanotubes (CNTs)	C brush	PBS	0.8	Two-chamber	$0.01^{Avg.}$	$2.3^{Max.}$	25.7	2014 (*17*)
NiMo	C cloth	C cloth	100 mM PBS	0.6	Single-chamber	-	$12.0^{Max.}$	-	2009 (*18*)
Ni-hollow fiber membranes (Ni-HFMs)	Hollow fiber membrane	Graphite fiber brush	PBS	0.9	Single-chamber	$0.2^{Avg.}$	$3.7^{Max.}$	57	2014 (*19*)

Table 1. (Continued). HER Catalysts Reported in Recent Years with the Substrate, the Anode, the Electrolyte, the Applied Voltage, the System Configuration, the H_2 Production Rate, the Current Density, the Coulombic Efficiency, and the Published Year

HER Catalysts	*Substrate*	*Anode*	*Electrolyte*	*Applied Voltage (V)*	*Configuration*	*H_2 Production Rate (m^3 $H_2/m^3/d$)*	*Current Density (A/m^2)*	*Coulombic Efficiency (%)*	*Published Year*
Ni mesh (electroformed)	-	Graphite plates	PBS	1.1	Continuous flow	4.2 $^{Max.}$	-	75	2015 (20)
NiFe layered double hydroxide (NiFe LDH)	Nickel foam	C brush	Brewery wastewater	0.8	Single-chamber	2.0 $^{Max.}$	35.6 $^{Max.}$	77	2016 (21)
$NiCl_2$	SS mesh	Graphite brushes	50 mM PBS	0.9	Two-chamber	0.3 $^{Avg.}$	2.7 $^{Max.}$	-	2018 (22)
Ni_5P_4-NiP_2 nanosheet matrix (3D biphasic)	-	C fiber	50 mM PBS	0.8	Single-chamber	-	-	-	2018 (23)
Ni-Graphene oxide (Ni-GO)	C film		10 mM PBS	0.8	Single-chamber	4.2 $^{Max.}$	1.2 $^{Max.}$	94	2019 (24)

$^{Max.}$ the maximum values of the H_2 production rates and the current densities based on the peak performance of the operation cycle; $^{Avg.}$ the average values of the H_2 production rates and the current densities based on the whole operation cycle.

Table 2. Costs of the Materials of Typical HER Catalysts

Catalysts	*Substrates*	*Cathode Cost*
Pt/C (10 wt%)	SS mesh with Nafion binder	$ 690 /m² (*25*)
Pt/C (10 wt%)	C cloth with Nafion binder	$ 2300/m² (*8*)
AC-Ni	Nickel	$ 10/m² (*22*)
Ni sheets (Ni 210)	-	$ 2.82/m² (*26*)
Ni sheets (Ni 625)	-	$ 370/m² (www.mcmaster.com)
SS sheets (SS 304)	-	$ 73/m² (www.mcmaster.com)
SS mesh	-	$ 46/m² (*25*)
SS wool	-	$ 300/m² (*25*)
N–Fe/Fe_3C	Graphite carbon	$ 115/m² (*15*)
MoS_2 (bare)	-	$ 57/m² (*27*)
MoS_2	Carbon black (CB)	$ 15/m² (*27*)

Types of HER Catalysts

Until now, a wide range of HER catalysts have been explored in MECs, including pure metals, metal complexes with metal and non-metal substrates, non-metal materials, and biocathodes with biofilms growing on C surfaces. The details of these catalysts are described in the following ways.

Transit Metals

Many transit metals are active in catalyzing H_2 evolution. Some of them are expensive (e.g., Pt, rhodium [Rh], iridium [Ir], and palladium [Pd]), while some can be relatively cheap (e.g., iron [Fe], copper [Cu], nickel [Ni], tungsten [W], and molybdenum [Mo]). To improve their performance and decrease their costs, the metals can be combined with non-metal species, such as carbon (C), nitrogen (N), oxides (O), phosphides (P), and sulfides (S) to make cost-effective catalysts. The performance of commonly used transit metal-based catalysts are discussed as follows.

Platinum (Pt)

Pt is a well-known noble metal that can efficiently catalyze the H_2 evolution. It provides active sites for cathodic evolution of H_2 with a very low overpotential (-0.05 V at 15 A/m²) (*28*). The typical method to fabricate a Pt cathode for MECs is to apply 0.5 mg/cm² of Pt powder mixed with activated carbon (Pt/C) or CB onto a piece of C cloth (*29*). Pt-based electrodes, such as Pt-coated C felt, brushes, cloths, rods, plates, graphite, and Ti mesh are commercially available (*12*, *30*). However, there are some disadvantages of Pt cathodes, including the high cost, their inactivation with chemicals such as sulfides, and the negative impacts on the environment (*31*).

While the applications of Pt catalysts in MECs have been widely explored, their performance varied in a large range even with the same kinds of cathodes. For example, in one study, a Pt/C cathode with a loading of 0.5 mg/cm^2 of Pt achieved an H_2 production rate of 0.01 m^3 $H_2/m^3/d$ (*17*), while another study reported an H_2 production rate of 3.1 m^3 $H_2/m^3/d$ with the same loading (*32*). Therefore, when comparing the performance of different cathodes, other factors, such as the electrolytes, anode materials, surface ratios of the anode and cathode electrodes, applied voltages, operation modes, and microbial interactions, should also be considered (*33*). Development of mathematical models based on these multiple factors could be an effective method for obtaining a comprehensive evalution (*33*).

Nickel (Ni)

Ni is a metal with high HER catalytic activities. Ni in many forms (e.g., powder, foam, alloy, or nano-assembly) has been used as catalysts in MECs. For example, pure Ni alloy (99.99 m/m%) showed inexpensive and highly efficient advantages in MECs compared to typical Pt/C catalysts with an overall hydrogen recovery rate of 62.9% (*20*). Moreover, a fed-batch MEC with Ni mesh cathodes obtained a maximum H_2 production rate of 4.18 m^3 $H_2/m^3/d$ and a coulombic efficiency of 75 % (*34*). Ni foam-graphene showed a higher HER ability with an H_2 production rate of 1.31 m^3 $H_2/m^3/d$ higher than some flat materials (metal cloth or paper) due to the three-dimensional (3D) structure (*35*).

Besides pure Ni metals, the combinations of Ni with other active elements also make efficient HER catalysts. It was observed that Ni alloys, such as NiMo, NiW, NiFeP, NiFeMo, NiFeCoP, NiCr, and Ni particles, can be chemically deposited on C substrates as MEC catalysts, and the H_2 production rate was 2.8-3.7 m^3 $H_2/m^3/d$ (*36*). In another study that directly grew NiFe-layered double hydroxide electrocatalysts on Ni foam, an H_2 production rate of 2.01–2.12 m^3 $H_2/m^3/d$ was obtained, which was comparable to Pt catalysts (*21*). More recently, a study researched the H_2 gas produced using different Ni alloys as the cathode compared to Pt catalysts, and the results suggested that the Ni alloy (Ni 210)– CB catalyst achieved the most similar results to the catalytic abilities of Pt (1.3 vs 1.6 m^3 $H_2/m^3/d$).[4a]

Molybdenum (Mo)

MoS_2 is a typical catalyst for HER because the position on the volcano plot is closed to Pt (Figure 2) (*11, 37*). It has been shown that simple mixtures of MoS_2 and CB were effective as HER catalysts (*27*). Furthermore, the nanocomposite of MoS_2 was coated on highly conductive CNTs, and the H_2 production rate of the MECs with MoS_2@CNT cathodes was comparable to that of the Pt-based catalysts (0.01 m^3 $H_2/m^2/d$) (*17*).

Besides MoS_2, other Mo-based catalysts were also developed. At an applied voltage of 0.6 V, the MECs with NiMo cathodes accomplished an H_2 production rate of 2.0 m^3 $H_2/m^3/d$ at a current density of 270 A/m^3 (12 A/m^2), which was 33% higher than that of the MECs with NiW cathodes and slightly lower than the rate of the Pt catalyst (2.3 m^3 $H_2/m^3/d$) (*18*). Another study demonstrated that Mo_2C/N-doped graphene (Mo_2C/N-rGO) was a good HER catalyst in a series stacked MFC-ammonia electrolytic cell system (*38*). Moreover, Co-Mo alloys were reported to have a high theoretical H_2 production rate of 50 m^3 $H_2/m^3/d$ in MECs. The extremely high H_2 production was calculated based on the measured current densities (*39*). Most recently, an MoP catalyst was mixed with Nafion and applied on C cloth as an MEC cathode. After improving the anode-to-cathode surface area ratio from 1:1 to 4:1, a high H_2 production rate of 3.7 m^3 $H_2/m^2/d$ was obtained in the reactors (*40*).

Iron (Fe)

SS brushes with high surface areas have been discovered as effective cathode catalysts in MECs (*41*). The H_2 gas was produced at a rate of 1.7 m^3 $H_2/m^3/d$ at a current density of 188 A/m^3 with an applied voltage of 0.6 V. SS wool also showed a superior H_2 production rate (1.3 m^3 $H_2/m^3/d$) compared to the plate SS cathodes due to the larger surface area (*25*). Besides pure SS catalysts, SS mesh and SS wool coated in other species (such as Pt or MoS_2) have been widely used in MECs (*42*).

More recently, researchers reported a novel nitrogen-containing Fe-based catalyst on C cloth (N-Fe/Fe_3C@C) in MECs, which was not as efficient as the Pt catalysts (0.018 vs 0.023 m^3 $H_2/m^2/d$). The overall hydrogen recovery was 34.8% vs 32.4% of that of Pt/C, however the cost was only 5% of Pt/C catalysts, thus suggesting promises for large-scale applications (*15*).

According to a study that compared SS and Ni alloys as low-cost cathodes in MECs (*8*), Ni 625 (0.79 m^3 $H_2/m^2/d$) and SS A286 (1.5 m^3 $H_2/m^2/d$) achieved higher H_2 production rates than Pt (0.68 m^3 $H_2/m^2/d$) (*8*).

Copper (Cu), Titanium (Ti), Palladium (Pd), and Aluminum (Al)

Currently, few studies have reported the use of Cu as the HER catalyst in MECs. A Cu/C cathode showed an H_2 production rate of 0.053 m^3 $H_2/m^3/d$ at an applied voltage of 0.4 V (*43*). Although inexpensive, the catalytic activity and the H_2 production ability are not as desirable as Pt/C (0.065 m^3 $H_2/m^3/d$) or Ni (0.068 m^3 $H_2/m^3/d$) at the same conditions.

Ti mesh cathodes were reported in an MEC study with a slightly lower H_2 production rate of 0.23 m^3 $H_2/m^3/d$, compared to 0.29 m^3 $H_2/m^3/d$ for Pt/C and 0.28 m^3 $H_2/m^3/d$ for Ni (*44*). However, the coulombic efficiencies of the three cathodes were similar (74.8%–77.6%).

Pd is a rare and lustrous metal used for catalyzing H_2 generation in hydrogen fuel cells (*45*). C paper coated with Pd nanoparticles showed a lower cathode overpotential in MECs than C paper coated with Pt black, and thus the overall hydrogen recovery of the Pd catalyst (46.4%) was slightly higher than that of the Pt (41.6%) (*16*).

Similarly, defect-rich porous Al alloy electrodes were also reported in electrochemical systems for efficient H_2 evolution (*46*).

Nonmetal Materials

Nonmetal species not only serve as supportive structures that improve charge transfer, they also influence the adsorption and reduction of H^+ and thus enhance the activity of catalysts in the presence of special sites.[4b] Currently, C, N, and P species are the most widely explored nonmetal HER catalysts.

Carbon (C)

In 2009, an upflow-single chamber MEC was reported using a C felt cathode designed for H_2 production (*14*). The obtained H_2 production rate of 0.57 m^3 $H_2/m^3/d$ was higher than that of some Pt catalysts. Another recent study reported a sequence of coulombic efficiencies: Pt/C (80%) > Pt/Ni/C (77%) > PtCu/C (73%) > Ni/C (72.6%) > Cu/C (66%) > C nanoparticles (CNPs) (47%) > CNTs (38.9%) > plain C felt (38.7%) (*43*). Although the cost of C catalysts is low, they are usually less effective HER catalysts compared to metals.

Nitrogen (N)

As an abundant element in natural environments and wastewater, N has great potential to be utilized in MECs due to its ability to facilitate the Volmer step in HER (the adsorption of H^+ to catalysts). For example, N-doped activated C obtained an H_2 production rate of 0.003 m^3 $H_2/m^3/d$ (*47*). Another three-dimensional (3D) hybrid of layered MoS_2 and nitrogen-doped graphene nanosheet aerogels achieved an H_2 production rate of 0.19 m^3 $H_2/m^3/d$ at an applied voltage of 0.8 V (*48*).

Phosphate (P)

P is also a widely distributed element that can facilitate HER catalyzation. A rotating disk SS electrode with the presence of P species in MECs showed a multiplied current density of H_2 evolution comparable to the control cathode that had the same catalyst composition as P, and the H_2 evolution rate reached 4.9 $L/h/m^2$ under 0.8 V (*49*). Moreover, Ni-P coated Ni foam was also tested as an alternative catalyst to Pt in a MEC, and its H_2 production rate of 2.29 m^3 $H_2/m^3/d$ was superior to Ni foam and SS mesh at 0.9 V with an overall hydrogen recovery of 86.8% (*50*).

Biocathodes

Besides abiotic HER catalysts, the biocathode was proposed as a sustainable method to catalyze H_2 evolution. The electrons are transported from the anode to the biofilms of the cathode and used by certain microbes (called electrophs) to generate H_2 (*51*). Biocathodes are an attractive option as no metals are needed and the biofilm can be reused multiple times; thus, the cost of biocathodes is low for applications (*52*). Additionally, the microbial species can be naturally selected from the cultures of electrochemically active microbes without an extra cost. Phyla such as *Proteobacteria, Firmicutes, Bacteroidetes,* and *Desulfovibrio* spp. have been reported to be responsible for H_2 production (*53*).

However, the biofilm of a biocathode has difficulty acclimating in MECs because the low oxygen level is not ideal for most microbes to live, and only some anoxic electrophs can grow. To incubate the biocathodes, acetate-hydrogen oxidizing bioanodes (e.g., graphite brush, C felt, or C cloth) were first produced in MFCs and then reversely switched to produce H_2 in the cathodes of MECs (*54*). In some studies, chemicals, such as ferrocyanide, were fed to the biocathodes to help with the acclimation, and other steps including adding hydrogen or organic substrates were also completed (*51, 53*). Although these processes were successful in biofilm formation, they unfortunately resulted in higher operational costs. Currently, sediment microbial fuel cells are used to prepare the cathodes for MECs (*55*); however, the loss of H_2 gas and the growth of hydrogenotrophic methanogens that convert H_2 into CH_4 need better solutions. As of today, there is still space for biocathode MECs to exceed MECs with metallic catalysts.

Physical Properties of Cathode Materials

Except for the compositions of the HER catalysts, the physical properties of the cathode materials can also affect H_2 generation. These properties include the surface area, structure, porosity, and synergistic effects between catalysts and supports.

Surface Area

The surface area of cathodes is significant to the H_2 generation. Previously, it was determined that an SS brush cathode with a high surface area generated comparable H_2 gas to Pt/C C cloth cathodes with an H_2 production rate of 1.7 m^3 $H_2/m^3/d$ at an applied voltage of 0.6 V (*41*). Also, SS wool with a large surface area performed better than other SS catalysts (mesh, fiber felt, and brushes) in H_2 production (*25*). Furthermore, using conductive hollow fiber membranes in cathodes to treat organic strength solutions in tubular and rectangle reactor configurations was proposed. The design of hollow fiber membrane cathodes increased the surface area of the catalysts, which consequently benefited the H_2 generation (*19*, *56*).

Structure

Recently, a Ni-GO catalyst with 3D micropillars fabricated on C film have been tested in MECs. The catalyst's performance was comparable to Pt catalysts with an H_2 production rate of 4.22 m^3 $H_2/m^3/d$ measured at 0.8 V, and the overall hydrogen recovery reached 72.2% vs 68.8% of Pt/C (*24*). The reason for the high H_2 production was that the 3D structure enhanced the penetration of ions through an increased electrochemical active surface area. In another study where Pt catalysts were coated on the supportive structure (3D Ni foam), the H_2 production rate reached 0.71 m^3 $H_2/m^3/d$ at an applied voltage of 0.8 V (*57*).

Porosity

Porosity of cathodes determines the active sites to some extent. It was discovered that a mipor Ti tube (mipor diameter ~10 μm, inner diameter of 20 mm, outer diameter of 30 mm, and lenght of 50 mm) coated with Pt significantly improved the H_2 production in a single-chamber MEC (*10*). Another study determined that the HER activity of MoN was enhanced by creating porous MoN on N-doped C (MoN-NC) in electrochemical conditions. The catalyst showed high catalytic activity and a small overpotential of 62 mV at a current density of 10 mA cm^{-2} with a low Tafel slope of a 54-mV decay in 0.5 M H_2SO_4 of aqueous solution (*58*). However, the performance of this catalyst is still unknown in MECs.

Synergistic Effects between Catalysts and Supports

H_2 generation can be enhanced by improving the electronic communication between the active catalytic sites and the conductive substrates (*59*). Currently, nanocarbon materials and graphene are demonstrated as effective catalysts with an increased charge transfer ability between the catalytic component and the substrate (*60*). but no such tests have been reported in MECs. On the other hand, H_2 generation in MECs is a two-phase reaction that happens between the solid catalyst and the liquid electrolyte. With more H^+ reaching the active sites of the catalysts, more H_2 can be produced. Thus, the binder of cathodes is also important for determining performance. Previously, Nafion has been commonly used as a cathode binder; however, the costs are very high. In recent years, novel materials such as sulfonated poly(arylene ether sulfone) copolymers (BPSH40) were reported as good alternatives to Nafion (*5*).

Challenges and Future Perspectives

Although current applications of MECs for H_2 generation and wastewater treatment are still limited, and their cost is difficult to overcome compared to traditional technologies (such as catalytic steam reforming, partial oxidation, and coal gasification) (*61*), MECs are anticipated to be promising for the future if more economic, efficient, and stable cathode materials for MECs are developed. However, there are some challenges for future development of these cathode materials.

Development of Cost-Effective Cathode Materials

Presently, Pt is still the most widely used catalyst in MECs; however, the high price hampers its large-scale uses. Non-noble materials, such as Ni mesh and alloys, SS mesh and wool, and C, are much more cost-effective compared to Pt, but the H_2 production rates are usually lower. Fortunately, with the help of nanotechnologies, comparable performances to Pt can be achieved in MECs. For example, novel NiFe LDH catalysts had similar H_2 production rates as Pt (2.01 m^3 $H_2/m^3/d$) and higher overall hydrogen recovery (76%) than that of Pt/C (55%) (*21*). The catalysts were fabricated through a simple hydrothermal growth method where the aligned nanoplate film of the 3D structure grew vertically on a Ni foam surface (*21*). However, further development of effective and low-cost HER catalysts are still needed.

Furthermore, it is worth noting that cathodes may still be costly even if the catalysts are inexpensive. For example, high temperatures (200–900 °C) are required for fabricating Ni-based catalysts to firmly combine Ni particles with the catalyst supports, which consume a lot of heating energy. Additionally, N-doping is a process that requires a large amount of ammonia (NH_3) as the carrier gas, which reduces the economic benefits of N replacing expensive metals in the catalyst. Therefore, effective and less expensive preparation methods for cathodes also need to be explored. A recently reported regenerable Ni-functionalized activated C cathode might be a good example. By using the economic Ni adsorption method, the amount of Ni loading and the cost to make the cathode are greatly reduced (22). In the future, a comprehensive consideration of the costs of the catalyst and cathode preparation as well as the H_2 production rate, energy efficiency, and long-term durability need to be taken into account.

Prevention of CH_4 Formation

The loss of H_2 in MECs is a problem for H_2 recovery, especially in membrane-less reactors where H_2 is easily converted to CH_4 in the presence of CO_2 ($CO_2 + H_2 \rightarrow CH_4$) by methanogens (*52*). Previously, a scaled-up continuous-flow MEC obtained the maximum H_2 production rate of 0.07 $m^3\ H_2/m^3/d$ for 22–43 days (*62*). However, the CH_4 production became dominant (51%) after 43 days, and near the end (during days 52–97), the system produced very little H_2 due to the growth of methanogens (86% CH_4 in the biogas) (*62*). This suggests that methods have to be developed to suppress methanogen growth or quickly isolate H_2 from the reactor.

To prevent CH_4 formation in the system, a research group discovered a method to operate MECs at 15 °C that effectively prevented the generation of CH_4 (*63*). However, this method is not suitable for wastewater at higher temperatures as its cool down process is energy-intensive. Another study reported an active gas collection method that collected H_2 using a vacuum and a gas-permeable hydrophobic membrane, which successfully stopped CH_4 production in the system (*64*). However, this process is still energy intensive. In the opinion of the authors, it would be promising to use materials that could affect the microbial community compositions of the electrodes and inhibit the growth of methanogens (*63, 65*). Cathodes to prevent H_2 from transferring to the electrolytes (e.g., gas diffusion and ion selective electrodes) are also favorable. Biofilms formed through bioparticles depositing in bioelectrochemical processes on the cathode can not only consume H_2 to generate CH_4, but also cause biofouling to increase the overpotentials of the cathodes. In the future, abiotic cathode materials to prevent the formation of biofilm would be helpful.

Gaps between Electrochemical and Bioelectrochemical Systems

Currently, many advanced HER catalysts have been tested in electrochemical systems with alkali or acidic electrolytes. However, their performance in MECs is still not known, and the differences of pH between electrochemical and bioelectrochemical systems are the main challenges. For HER catalysts, the catalytic activities are pH dependent, which means they work better under alkali (e.g., 0.1M KOH) or acidic (e.g., 0.1M $HClO_4$) conditions compared to neutral solutions (*61*). This phenomenon is caused by complex effects between Volmer-Tafel and Heyrovsky reactions and the catalysts at different pH levels, (*61*), such as the activation barrier for the discharge of H_3O^+ on sites of catalysts and the exposed edge sites of catalysts that determine the formation of H* (*61*). However, most microorganisms only survive when the pH fluctuates ~7 in MECs. Variations of microbial community compositions, toxicities such as sulfides, the activities of catalysts, and biofouling among other activities, make the conditions more complex. Therefore, the suitability of the currently known electrochemical catalysts for MECs should be a focus of future research.

References

1. Liu, H.; Logan, B. E. Electricity generation using an air-cathode single chamber microbial fuel cell in the presence and absence of a proton exchange membrane. *Environ. Sci. Technol.* **2004**, *38* (14), 4040–4046.
2. Liu, H.; Grot, S.; Logan, B. E. Electrochemically assisted microbial production of hydrogen from acetate. *Environ. Sci. Technol.* **2005**, *39* (11), 4317–4320.

3. Rozendal, R. A.; Hamelers, H. V.; Euverink, G. J.; Metz, S. J.; Buisman, C. J. Principle and perspectives of hydrogen production through biocatalyzed electrolysis. *Int. J. Hydrogen Energ.* **2006,** *31* (12), 1632–1640.
4. Selembo, P. A.; Merrill, M. D.; Logan, B. E. Hydrogen production with nickel powder cathode catalysts in microbial electrolysis cells. *Int. J. Hydrogen Energ.* **2010,** *35* (2), 428–437(a). Zheng, Y.; Jiao, Y.; Zhu, Y.; Li, L. H.; Han, Y.; Chen, Y.; Du, A.; Jaroniec, M.; Qiao, S. Z. Hydrogen evolution by a metal-free electrocatalyst. *Nat. Commun.* **2014,** 5, 3783(b).
5. Ivanov, I.; Ahn, Y.; Poirson, T.; Hickner, M. A.; Logan, B. E. Comparison of cathode catalyst binders for the hydrogen evolution reaction in microbial electrolysis cells. *Int. J. Hydrogen Energ.* **2017,** *42* (24), 15739–15744.
6. Lu, L.; Ren, N.; Xing, D.; Logan, B. E. Hydrogen production with effluent from an ethanol–H2-coproducing fermentation reactor using a single-chamber microbial electrolysis cell. *Biosens. Bioelectron.* **2009,** *24* (10), 3055–3060.
7. Logan, B. E.; Rossi, R.; Ragab, A. a.; Saikaly, P. E. Electroactive microorganisms in bioelectrochemical systems. *Nat. Rev. Microbiol.* **2019,** 1.
8. Selembo, P. A.; Merrill, M. D.; Logan, B. E. The use of stainless steel and nickel alloys as low-cost cathodes in microbial electrolysis cells. *J. Power Sources* **2009,** *190* (2), 271–278.
9. Kundu, A.; Sahu, J. N.; Redzwan, G.; Hashim, M. An overview of cathode material and catalysts suitable for generating hydrogen in microbial electrolysis cell. *Int. J. Hydrogen Energ.* **2013,** *38* (4), 1745–1757.
10. Kadier, A.; Simayi, Y.; Abdeshahian, P.; Azman, N. F.; Chandrasekhar, K.; Kalil, M. S. A comprehensive review of microbial electrolysis cells (MEC) reactor designs and configurations for sustainable hydrogen gas production. *Alexandria Eng. J.* **2016,** *55* (1), 427–443.
11. Jaramillo, T. F.; Jørgensen, K. P.; Bonde, J.; Nielsen, J. H.; Horch, S.; Chorkendorff, I. Identification of active edge sites for electrochemical H2 evolution from MoS2 nanocatalysts. *Science* **2007,** *317* (5834), 100–102.
12. Rozendal, R. A.; Hamelers, H. V.; Molenkamp, R. J.; Buisman, C. J. Performance of single chamber biocatalyzed electrolysis with different types of ion exchange membranes. *Water Res.* **2007,** *41* (9), 1984–1994.
13. Jeremiasse, A. W.; Hamelers, H. V.; Buisman, C. J. Microbial electrolysis cell with a microbial biocathode. *Bioelectrochemistry* **2010,** *78* (1), 39–43.
14. Lee, H.-S.; Torres, C. I.; Parameswaran, P.; Rittmann, B. E. Fate of H2 in an upflow single-chamber microbial electrolysis cell using a metal-catalyst-free cathode. *Environ. Sci. Technol.* **2009,** *43* (20), 7971–7976.
15. Xiao, L.; Wen, Z.; Ci, S.; Chen, J.; He, Z. Carbon/iron-based nanorod catalysts for hydrogen production in microbial electrolysis cells. *Nano Energy* **2012,** *1* (5), 751–756.
16. Huang, Y.-X.; Liu, X.-W.; Sun, X.-F.; Sheng, G.-P.; Zhang, Y.-Y.; Yan, G.-M.; Wang, S.-G.; Xu, A.-W.; Yu, H.-Q. A new cathodic electrode deposit with palladium nanoparticles for cost-effective hydrogen production in a microbial electrolysis cell. *Int. J. Hydrogen Energ.* **2011,** *36* (4), 2773–2776.
17. Yuan, H.; Li, J.; Yuan, C.; He, Z. Facile synthesis of MoS2@ CNT as an effective catalyst for hydrogen production in microbial electrolysis cells. *ChemElectroChem* **2014,** *1* (11), 1828–1833.

18. Hu, H.; Fan, Y.; Liu, H. Hydrogen production in single-chamber tubular microbial electrolysis cells using non-precious-metal catalysts. *Int. J. Hydrogen Energ.* **2009**, *34* (20), 8535–8542.
19. Katuri, K. P.; Werner, C. M.; Jimenez-Sandoval, R. J.; Chen, W.; Jeon, S.; Logan, B. E.; Lai, Z.; Amy, G. L.; Saikaly, P. E. A novel anaerobic electrochemical membrane bioreactor (AnEMBR) with conductive hollow-fiber membrane for treatment of low-organic strength solutions. *Environ. Sci. Technol.* **2014**, *48* (21), 12833–12841.
20. Kadier, A.; Simayi, Y.; Chandrasekhar, K.; Ismail, M.; Kalil, M. S. Hydrogen gas production with an electroformed Ni mesh cathode catalysts in a single-chamber microbial electrolysis cell (MEC). *Int. J. Hydrogen Energ.* **2015**, *40* (41), 14095–14103.
21. Lu, L.; Hou, D.; Fang, Y.; Huang, Y.; Ren, Z. J. Nickel based catalysts for highly efficient H2 evolution from wastewater in microbial electrolysis cells. *Electrochim. Acta* **2016**, *206*, 381–387.
22. Kim, K.-Y.; Yang, W.; Logan, B. E. Regenerable Nickel-Functionalized Activated Carbon Cathodes Enhanced by Metal Adsorption to Improve Hydrogen Production in Microbial Electrolysis Cells. *Environ. Sci. Technol.* **2018**, *52* (12), 7131–7137.
23. Cai, W.; Liu, W.; Sun, H.; Li, J.; Yang, L.; Liu, M.; Zhao, S.; Wang, A. Ni5P4-NiP2 nanosheet matrix enhances electron-transfer kinetics for hydrogen recovery in microbial electrolysis cells. *Applied Energy* **2018**, *209*, 56–64.
24. Yadav, A.; Verma, N. Efficient hydrogen production using Ni-graphene oxide-dispersed laser-engraved 3D carbon micropillars as electrodes for microbial electrolytic cell. *Renew. Energy* **2019**, *138*, 628–638.
25. Kim, K.-Y.; Zikmund, E.; Logan, B. E. Impact of catholyte recirculation on different 3-dimensional stainless steel cathodes in microbial electrolysis cells. *Int. J. Hydrogen Energ.* **2017**, *42* (50), 29708–29715.
26. Yuan, H.; He, Z. Platinum Group Metal–free Catalysts for Hydrogen Evolution Reaction in Microbial Electrolysis Cells. *Chem. Rec.* **2017**, *17* (7), 641–652.
27. Tokash, J. C.; Logan, B. E. Electrochemical evaluation of molybdenum disulfide as a catalyst for hydrogen evolution in microbial electrolysis cells. *Int. J. Hydrogen Energ.* **2011**, *36* (16), 9439–9445.
28. Grzeszczuk, M.; Poks, P. The HER performance of colloidal Pt nanoparticles incorporated in polyaniline. *Electrochimica Acta* **2000**, *45* (25-26), 4171–4177.
29. Kim, K.-Y.; Logan, B. E. Nickel powder blended activated carbon cathodes for hydrogen production in microbial electrolysis cells. *International Journal of Hydrogen Energy* **2019**, *44* (26), 13169–13174.
30. Sangeetha, T.; Guo, Z.; Liu, W.; Cui, M.; Yang, C.; Wang, L.; Wang, A. Cathode material as an influencing factor on beer wastewater treatment and methane production in a novel integrated upflow microbial electrolysis cell (Upflow-MEC). *Int. J. Hydrogen Energ.* **2016**, *41* (4), 2189–2196.
31. Logan, B. E.; Call, D.; Cheng, S.; Hamelers, H. V.; Sleutels, T. H.; Jeremiasse, A. W.; Rozendal, R. A. Microbial electrolysis cells for high yield hydrogen gas production from organic matter. *Environmental Science & Technology* **2008**, *42* (23), 8630–8640.
32. Call, D.; Logan, B. E. Hydrogen production in a single chamber microbial electrolysis cell lacking a membrane. *Environ. Sci. Technol.* **2008**, *42* (9), 3401–3406.

33. Yahya, A. M.; Hussain, M. A.; Abdul Wahab, A. K. Modeling, optimization, and control of microbial electrolysis cells in a fed-batch reactor for production of renewable biohydrogen gas. *Int. J. Hydrogen Energ.* **2015**, *39* (4), 557–572.
34. Kadier, A.; Logroño, W.; Rai, P. K.; Kalil, M. S.; Mohamed, A.; Hasan, H. A.; Hamid, A. A. None-platinum electrode catalysts and membranes for highly efficient and inexpensive H2 production in microbial electrolysis cells (MECs): A review. *Iran. J. Catal.* **2017**, 7 (2), 89–102.
35. Cai, W.; Liu, W.; Han, J.; Wang, A. Enhanced hydrogen production in microbial electrolysis cell with 3D self-assembly nickel foam-graphene cathode. *Biosens. Bioelectron.* **2016**, *80*, 118–122.
36. Manuel, M.-F.; Neburchilov, V.; Wang, H.; Guiot, S.; Tartakovsky, B. Hydrogen production in a microbial electrolysis cell with nickel-based gas diffusion cathodes. *J. Power Sources* **2010**, *195* (17), 5514–5519.
37. Li, Y.; Wang, H.; Xie, L.; Liang, Y.; Hong, G.; Dai, H. MoS2 nanoparticles grown on graphene: an advanced catalyst for the hydrogen evolution reaction. *J. Am. Chem. Soc.* **2011**, *133* (19), 7296–7299.
38. Zhang, G.; Zhou, Y.; Yang, F. Hydrogen production from microbial fuel cells-ammonia electrolysis cell coupled system fed with landfill leachate using Mo2C/N-doped graphene nanocomposite as HER catalyst. *Electrochim. Acta* **2019**, *299*, 672–681.
39. Jeremiasse, A. W.; Bergsma, J.; Kleijn, J. M.; Saakes, M.; Buisman, C. J.; Stuart, M. C.; Hamelers, H. V. Performance of metal alloys as hydrogen evolution reaction catalysts in a microbial electrolysis cell. *Int. J. Hydrogen Energ.* **2011**, *36* (17), 10482–10489.
40. Miller, A.; Singh, L.; Wang, L.; Liu, H. Linking internal resistance with design and operation decisions in microbial electrolysis cells. *Environment International* **2019**, *126*, 611–618.
41. Call, D. F.; Merrill, M. D.; Logan, B. E. High surface area stainless steel brushes as cathodes in microbial electrolysis cells. *Environ. Sci. Technol.* **2009**, *43* (6), 2179–2183.
42. Ribot-Llobet, E.; Nam, J.-Y.; Tokash, J. C.; Guisasola, A.; Logan, B. E. Assessment of four different cathode materials at different initial pHs using unbuffered catholytes in microbial electrolysis cells. *Int. J. Hydrog. Energy* **2013**, *38* (7), 2951–2956.
43. Choi, M.-J.; Yang, E.; Yu, H.-W.; Kim, I. S.; Oh, S.-E.; Chae, K.-J. Transition metal/carbon nanoparticle composite catalysts as platinum substitutes for bioelectrochemical hydrogen production using microbial electrolysis cells. *Int. J. Hydrogen Energ.* **2019**, *44* (4), 2258–2265.
44. Farhangi, S.; Ebrahimi, S.; Niasar, M. S. Commercial materials as cathode for hydrogen production in microbial electrolysis cell. *Biotechnol. Lett.* **2014**, *36* (10), 1987–1992.
45. Zhang, F.; Cheng, S.; Pant, D.; Van Bogaert, G.; Logan, B. E. Power generation using an activated carbon and metal mesh cathode in a microbial fuel cell. *Electrochem. Commun.* **2009**, *11* (11), 2177–2179.
46. Periasamy, A. P.; Sriram, P.; Chen, Y.-W.; Wu, C.-W.; Yen, T.-J.; Chang, H.-T. Porous aluminum electrodes with 3D channels and zig-zag edges for efficient hydrogen evolution. *Chem. Commun.* **2019**, *55* (38), 5447–5450.
47. Zhang, B.; Wen, Z.; Ci, S.; Chen, J.; He, Z. Nitrogen-doped activated carbon as a metal free catalyst for hydrogen production in microbial electrolysis cells. *RSC Adv.* **2014**, *4* (90), 49161–49164.

48. Hou, Y.; Zhang, B.; Wen, Z.; Cui, S.; Guo, X.; He, Z.; Chen, J. A 3D hybrid of layered MoS2/ nitrogen-doped graphene nanosheet aerogels: an effective catalyst for hydrogen evolution in microbial electrolysis cells. *J. Mat. Chem.* **2014**, *2* (34), 13795–13800.
49. Munoz, L. D.; Erable, B.; Etcheverry, L.; Riess, J.; Basséguy, R.; Bergel, A. Combining phosphate species and stainless steel cathode to enhance hydrogen evolution in microbial electrolysis cell (MEC). *Electrochem. Commun.* **2010**, *12* (2), 183–186.
50. Li, F.; Liu, W.; Sun, Y.; Ding, W.; Cheng, S. Enhancing hydrogen production with Ni–P coated nickel foam as cathode catalyst in single chamber microbial electrolysis cells. *Int. J. Hydrogen Energ.* **2017**, *42* (6), 3641–3646.
51. Rozendal, R. A.; Jeremiasse, A. W.; Hamelers, H. V.; Buisman, C. J. Hydrogen production with a microbial biocathode. *Environ. Sci. Technol.* **2007**, *42* (2), 629–634.
52. Lu, L.; Xing, D.; Ren, N. Pyrosequencing reveals highly diverse microbial communities in microbial electrolysis cells involved in enhanced H2 production from waste activated sludge. *Water Res.* **2012**, *46* (7), 2425–2434.
53. Croese, E.; Pereira, M. A.; Euverink, G.-J. W.; Stams, A. J.; Geelhoed, J. S. Analysis of the microbial community of the biocathode of a hydrogen-producing microbial electrolysis cell. *Appl. Environ. Microbiol.* **2011**, *92* (5), 1083–1093.
54. Jafary, T.; Daud, W. R. W.; Ghasemi, M.; Kim, B. H.; Jahim, J. M.; Ismail, M.; Lim, S. S. Biocathode in microbial electrolysis cell; present status and future prospects. *Renew. Sust. Energ. Rev.* **2015**, *47*, 23–33.
55. Pisciotta, J. M.; Zaybak, Z.; Call, D. F.; Nam, J.-Y.; Logan, B. E. Enrichment of microbial electrolysis cell biocathodes from sediment microbial fuel cell bioanodes. *Appl. Environ. Microbiol.* **2012**, *78* (15), 5212–5219.
56. Werner, C. M.; Katuri, K. P.; Hari, A. R.; Chen, W.; Lai, Z.; Logan, B. E.; Amy, G. L.; Saikaly, P. E. Graphene-coated hollow fiber membrane as the cathode in anaerobic electrochemical membrane bioreactors–Effect of configuration and applied voltage on performance and membrane fouling. *Environ. Sci. Technol.* **2016**, *50* (8), 4439–4447.
57. Wang, L.; Liu, W.; He, Z.; Guo, Z.; Zhou, A.; Wang, A. Cathodic hydrogen recovery and methane conversion using Pt coating 3D nickel foam instead of Pt-carbon cloth in microbial electrolysis cells. *Int. J. Hydrogen Energ.* **2017**, *42* (31), 19604–19610.
58. Zhu, Y.; Chen, G.; Xu, X.; Yang, G.; Liu, M.; Shao, Z. Enhancing electrocatalytic activity for hydrogen evolution by strongly coupled molybdenum nitride@ nitrogen-doped carbon porous nano-octahedrons. *ACS Catalysis* **2017**, *7* (5), 3540–3547.
59. Voiry, D.; Salehi, M.; Silva, R.; Fujita, T.; Chen, M.; Asefa, T.; Shenoy, V. B.; Eda, G.; Chhowalla, M. Conducting MoS2 nanosheets as catalysts for hydrogen evolution reaction. *Nano Letters* **2013**, *13* (12), 6222–6227.
60. Zhao, M.; Zhang, J.; Xiao, H.; Hu, T.; Jia, J.; Wu, H. Facile in situ synthesis of a carbon quantum dot/graphene heterostructure as an efficient metal-free electrocatalyst for overall water splitting. *Chem. Commun.* **2019**, *55* (11), 1635–1638.
61. Staszak-Jirkovský, J.; Malliakas, C. D.; Lopes, P. P.; Danilovic, N.; Kota, S. S.; Chang, K.-C.; Genorio, B.; Strmcnik, D.; Stamenkovic, V. R.; Kanatzidis, M. G. Design of active and stable Co–Mo–S x chalcogels as pH-universal catalysts for the hydrogen evolution reaction. *Nat. Mater.* **2016**, *15* (2), 197.

62. Cusick, R. D.; Bryan, B.; Parker, D. S.; Merrill, M. D.; Mehanna, M.; Kiely, P. D.; Liu, G.; Logan, B. E. Performance of a pilot-scale continuous flow microbial electrolysis cell fed winery wastewater. *Appl. Microbiol. Biotechnol.* **2011**, *89* (6), 2053–2063.
63. Lu, L.; Xing, D.; Ren, N. Bioreactor performance and quantitative analysis of methanogenic and bacterial community dynamics in microbial electrolysis cells during large temperature fluctuations. *Environ. Sci. Technol.* **2012**, *46* (12), 6874–6881.
64. Lu, L.; Hou, D.; Wang, X.; Jassby, D.; Ren, Z. J. Active H2 harvesting prevents methanogenesis in microbial electrolysis cells. *Environ. Sci. Technol. Lett.* **2016**, *3* (8), 286–290.
65. Sun, R.; Zhou, A.; Jia, J.; Liang, Q.; Liu, Q.; Xing, D.; Ren, N. Characterization of methane production and microbial community shifts during waste activated sludge degradation in microbial electrolysis cells. *Bioresour. Tech.* **2015**, *175*, 68–74.

Chapter 3

Novel Nanoengineered Materials-Based Catalysts for Various Bioelectrochemical Systems

Udaratta Bhattacharjee[*,1] and Lalit M. Pandey[*,1,2]

[1]Center for the Environment, Indian Institute of Technology Guwahati, Guwahati 781039, India

[2]Department of Biosciences and Bioengineering, Indian Institute of Technology Guwahati, Guwahati 781039, India

[*]E-mail: udaratta@iitg.ac.in; lalitpandey@iitg.ac.in.

Bioelectrochemical systems (BES) such as microbial fuel cells (MFCs) and enzymatic fuel cells (EFCs) are an evolving technology that needs further optimization of definite parameters to be used at a large scale for a viable application. Although many improvements have been made for these systems in terms of different configurations, designs, and materials, they lack commercialization and adequate power output because of several factors involved in the process. Exploration of novel catalyst materials is an important factor for expanding the application of these systems with good performance and reasonable prices to bring them to the commercial platform. Platinum-based catalysts have been widely used as cathode catalysts for decades. Inexpensive metal-doped catalysts, metal and carbon-based catalysts, and biocatalysts have been investigated for increasing the proficiency of these systems because of their ability to enhance the oxygen reduction rate kinetics with greater power capacity and stability. MFC technology has been the most promising innovation, which involves the metabolic pathway of microbes for electricity generation. In a similar way, EFCs employ enzymes for power output with or without the use of mediators. Therefore, a framework of other technologies employing similar configurations to MFCs were developed for specific applications such as production of hydrogen, hydrogen peroxide, methane, and ethanol as well as desalination. This chapter focuses on the recent trends of using novel catalyst materials to enhance the activity of various BES, their applications, limitations, and future scope.

Introduction

Environmental pollution and extensive use of fossil fuels are the major problems associated with the various factors related to the depletion of sustainable life for the entire ecosystem. Many

© 2020 American Chemical Society

substantial factors associated with the existing systems are the discharge and disposal of sewage, dumping of toxic particles, mining, fertilizers, and different industrial wastewaters. Some of the conventional methods used to treat waste constituents include physiochemical adsorption and biological treatment (*1–6*). However, bioelectrochemical systems (BES) are gaining significance because of their sustainability, higher power efficiency, and economic benefits. These systems utilize microorganisms and are capable of breaking down toxic organic matter such as nitrates and heavy metal ions present in wastewater into less harmful components. They are also used in the simultaneous removal of organics and pollutants, development of toxic compounds and biochemical oxygen demand sensors (*7, 8*), and hydrogen or electricity production (*9, 10*). The concept of BES has gained a lot of interest among researchers since it has certain advantages over conventional fuel cells in which expensive metal catalysts are used. BES can be used as an innovative method, and many applications and developments are being made with respect to electrodes, catalysts, and membranes or separators (*11*). A pictorial representation of a BES for metal recovery and removal is shown in Figure 1. In this figure, the organic compounds present in the wastewater have been oxidized by the microbes in the anodic chamber and converted to carbon dioxide, protons, and electrons. On the basis of biocatalyst, BES can be classified as microbial fuel cells (MFCs) and enzymatic fuel cells (EFCs), whereas on the basis of application, they can be divided as microbial electrolysis cells (MECs), microbial desalination cells (MDCs) and microbial solar cells (MSCs) (*12*). Usually, in MFCs under certain conditions, the biofilm developed may also serve as a biocatalyst, although the exact mechanism between the catalyst and the microbes is not yet known (*13, 14*). Typically, BES consist of cathode and anode chambers in the presence and absence of a membrane (*15, 16*). An oxidation reaction takes place at the anode followed by a reduction reaction on the cathode to complete the redox process.

For the improvement of the oxygen reduction reaction (ORR) kinetics and decrease in the activation energy in the cathodic chamber, metal catalysts such as platinum (Pt) are often used. These catalysts suffer many drawbacks such as high cost, difference in catalytic activity, and potential poisoning due to presence of sulphur and carbon monoxide. Development of stable and efficient cathode catalysts has always been a challenge. Hence, many catalysts have been explored for use in various BES through the integration of various other compounds in order to increase both activity and stability. Biocathodes utilizing electroactive species were also reported (*17, 18*). A wide range of inexpensive catalysts such as metal-based catalysts, carbon-based catalysts, metal–nitrogen–carbon complexes, metal–carbon hybrids, nanomaterials, and biocatalysts were established for various bioelectrochemical applications (*13, 14, 19, 20*). In addition, as the cathodic chamber undergoes a reduction reaction by the electron acceptors and hence it is well-studied for wastewater treatment (*21*). Thus, an adequate survey of proper catalyst materials is to be formulated for better oxygen reduction, active electron transfer, and outstanding yield (*22*).

Furthermore, genetic engineering techniques were also used to enhance the microbe's activity to be used as a biocatalyst in BES. These systems are considered to be a hybrid of microbiology and electrochemistry and are capable of producing energy output and value added products by the action of microbes catalyzing oxidation and reduction reactions (*23*). They also facilitate improvement in production efficiency by refitting of current designs and configurations. In this chapter, we discuss a brief theory of BES followed by various catalysts materials, which are explored for different BES for numerous applications. These catalysts were found to benefit overall performance through enhancing efficiency. The working principles of BES and various catalyst materials explored in MFCs, EFCs, MECs, MDCs, and MSCs are comprehensively discussed in the following sections.

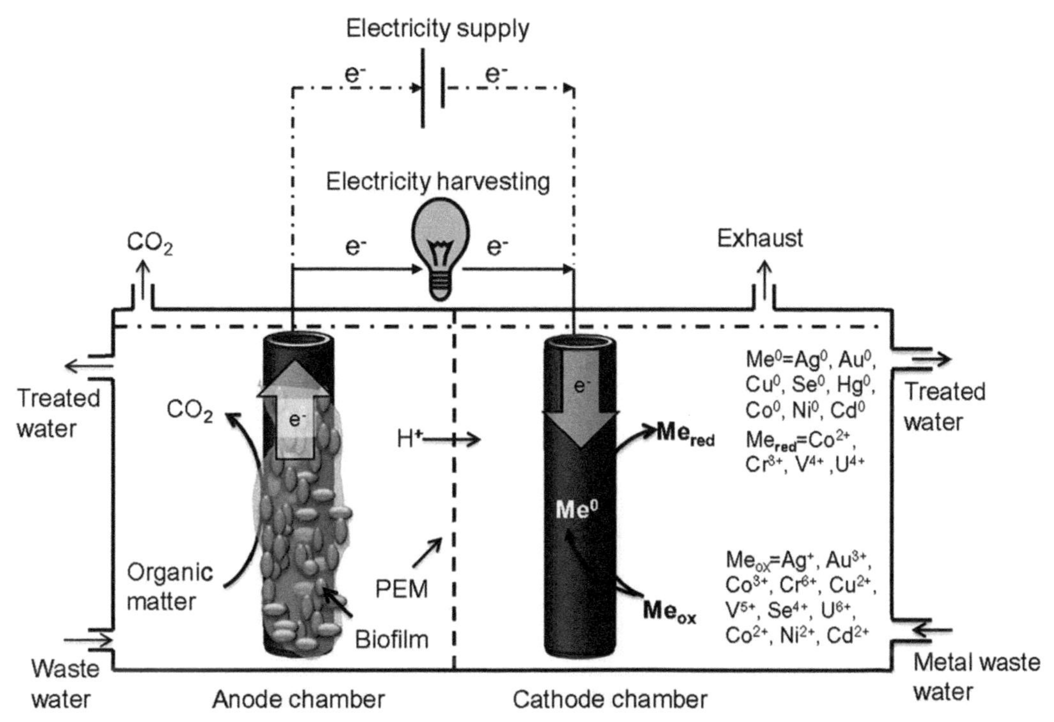

Figure 1. Schematic illustration of a BES for metal recovery and removal with a proton exchange membrane (PEM). Adapted with permission from reference (8). Copyright 2015 Elsevier.

Working Principle of BES

BES are found to oxidize the substrates on the anode by the development of the biofilm and transport of protons and electrons for the reduction reaction to the cathode by connecting to an external circuit (*24*). At the anode surface, the oxidation process results in electron production, and at the cathode surface, the electrons reduce terminal electron acceptors and lead to completion of the cathodic half-cell reaction (*25*). The cathode catalysts were, in fact, found to have the characteristic features for improvement in cathode performance and reduction in cost. An example of a redox reaction occurring in the anode and cathode chamber is shown in the equations below.

Anodic reaction:

$$CH_3COO^- + 2H_2O \rightarrow 2CO_2 + 7H^+ + 8e^- \quad (1)$$

$$C_6H_{12}O_6 + 6H_2O \rightarrow 6CO_2 + 24H^+ + 24e^- \quad (2)$$

Equations 1 and 2 show the oxidation reaction of acetic acid ions and glucose, which corresponds to a particular electrode potential.

Cathodic reaction:

$$O_2 + 4H^+ + 4e^- \xrightarrow{Pt} 2H_2O \quad (3)$$

$$O_2 + 2H^+ + 2e^- \xrightarrow{Pt} H_2O_2 \quad (4)$$

$$4H^+ + 4e^- \xrightarrow{Pt} 2H_2 \quad (5)$$

Equation 3 represents the four-electron ORR when a catalyst such as Pt is taking part in the chemical reactions (*26*). Equation 4 shows the production of hydrogen peroxide, which can act as disinfectant for contamination-free cathodes and membranes (*27*). Equation 5 represents hydrogen gas evolution, which can be explored as an energy source (*28*). Figure 2 shows a Pt-catalyzed hydrogen evolving reaction possessing molybdenum as a coating material. The hydrogen is converted from chemical-to-chemical energy and is easily accessible for fuel purposes.

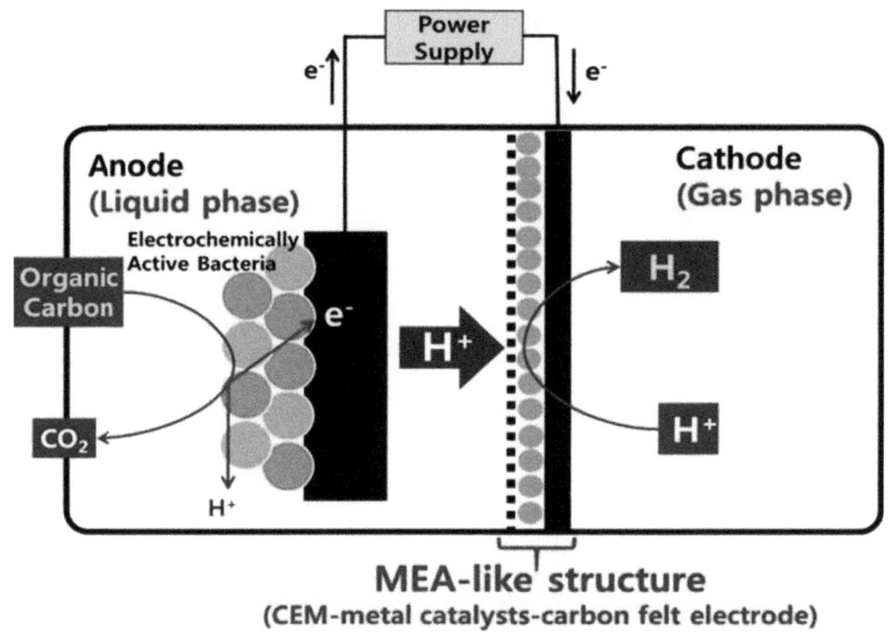

Figure 2. A schematic illustration of hydrogen gas production using a platinum/carbon (Pt/C) and nickel/carbon (Ni/C) catalyst by membrane electrode assembly (MEA) in a bioelectrochemical system. MEA was prepared by spraying the catalyst-coated cathode on carbon felt, and the membrane was hot pressed for 5 minutes at 140°C. Adapted with permission from reference (29). Copyright 2019 Elsevier.

Advances in Catalyst Materials in MFCs

MFCs are capable of generating bio-energy and power by bacterial interactions in the absence of oxygen (*30*). To aid in better performance of the cathode, several developments in electrocatalysts have been going on for decades (*14*). Since 50% of the total capital cost accounts for Pt in small-scale MFCs, Pt-based cathodes are not economically feasible for large-scale applications (*18*). The base material also helps in the inter- and intraparticular electron transfer mechanism for ORR (*31*). In this direction, improvements in the energy output of MFCs were made by exploring various substrates, base materials such as electrodes along with their modification, and the catalyst materials (*32*). The use of different catalysts with their power output in various BES is described in Table 1.

A wide range of substrates have been exploited, which serve as efficient carbon and nitrogen sources for any BES. Some of them are reported to be acetate, glucose, phenol, propionate, glucuronic acid, sodium fumarate, xylitol, and different wastewaters from various sources (*32*). For the electron transfer in the catalyst, the electrode material in the MFC plays a vital role in ORR because of its versatility, cost-effectiveness, and large surface area. Nanomaterials offer large surface area to volume ratios and have been previously used for various applications ranging from drug delivery to bioremediation (*33–35*). Carbon (C) and its derivatives are explored as excellent electrode materials (*36, 37*). For conventional support material for cathode catalysts, activated carbon was normally used (*38*). The synthesis of sugarcane-bagasse–derived activated carbon was used as a cathode catalyst in MFCs and achieved a power density of 110 ± 6.58 mW/m^2 and a peak current density of 0.40 mA/m^2, which was quite comparable with that of Pt catalysts (*39*). Xerogel, cryogel, aerogel, and a polymerized gel form of carbon are also extensively used as support materials because of their meso- and microporous structures (*40*). For upgraded performance of MFCs, xerogel-based catalysts have been studied due to their good stability and variable pore size. This catalyst was doped with nitrogen (N) and iron (Fe) and further modified with graphene oxide (GO) at a high temperature (800°C). The power density obtained was 26.8% higher (176.5 ± 6 mW/m^2) than that of a plain graphite electrode. The effect of pH is also significant, and when pH of the catholyte was increased to 12, power density was increased to 48.6% (207 ± 4 mW/m^2) (*41*). The xerogel-based materials were proven to be efficient support materials for both metals and nonmetals (*42, 43*). The xerogel support material loaded with reduced graphene oxide (RGO) as an electrocatalyst has also demonstrated enhanced performance and efficient wastewater treatment in MFCs (*44*). Synthesis of carbide-derived carbon (CDC) from the spent silicon carbide (SiC) heating rods were productively used in MFC application as a cost-effective catalyst support material. This material was reported to have large surface area and excellent electroconductivity. The highest power density obtained using this material was ~1570 ± 30 mW/m^2 (*45*). The same material has been found to be a potential adsorbent for Cr (VI) with an adsorption capacity of 95 ± 5 mg/g. In addition to this, a recent study has illustrated the efficacy of the CDC-supported and cobalt (Co)–impregnated iron phthalocyanine (FePc) for cathodes and catalyzing oxygen reduction for enhanced power generation as well as coulombic efficiency. The Co-FePC/CDC catalyst produced a peak power density of 1.57 W/m^2 using acetate-based synthetic wastewater. These results highlighted CDC-based materials to be efficient electrode materials as well as good adsorbents for environmental applications (*31*). In a separate study, nitrogen and phosphorus dual-doped catalysts for air cathodes in MFCs were derived by carbon from chitosan, which resulted in five times better output (1603.6 ± 80 mW/m^2) than that of other carbon-based air cathodes (*46, 47*), and they can be used as efficient, low-cost catalysts.

Pt-based materials were conventionally used as cathode catalysts. Using Pt-based catalysts, the bond dissociation energy of oxygen was reduced to 399 from 498 kJ/mole, respectively (*48–51*). In spite of having excellent capacity to decrease the ORR in the system, these catalysts have not been widely commercialized because of high cost, short shelf life, limited stability, poor performance, and many other drawbacks. A major difficulty associated with the decline of voltage in MFCs is the loss of activation overpotential due to bond dissociation energy of oxygen as shown in eq 3 (*52*). Thus, for the cost-effective and optimization of energy output, novel electrocatalysts with better oxygen reduction have a significant role in MFCs (*53*).

Nitrogen codoped nanocarbon and transition metal catalysts were made by the process of pyrolysis of multiwalled carbon nanotubes (CNTs) in the presence of a nitrogen source and cobalt or iron chloride. The maximum power densities achieved by Co-N-CNT and Fe-N-CNT cathode catalysts were found to be 5.1 W/m^3 and 6 W/m^3, respectively. The better output and enhanced ORR activity were due to formation of nitrogen-metal centers, so Co-N-CNT or Fe-N-CNT can be used as a substitute for expensive metal-based catalysts (*54*). Nickel (Ni) and nickel oxide–based catalysts have been used to reduce ORR in MFCs for the replacement of noble metal catalysts

(*19*, *55*–*59*). These catalysts are inexpensive and offer better electrochemical conductivity, although only few studies have been reported for Ni-based catalysts due to their less catalytic surface area. To overcome this difficulty, Ni nanostructures have been developed focusing on enhanced output productivity (*59*). By increasing the surface area, better efficiency with respect to performance was reported. One such study involved the increase in surface area of carbon felt electrodes to 3.1 m^2/g from 0.9 m^2/g after coating with PtNi/C catalyst material (*29*). Supporting materials for such operations were reported to overcome the aggregation effect offered by metal catalysts. Aggregation in certain cases may have advantages, but it often leads to depletion of catalyst reactivity, speciation and impacts the polymerization process between amide and imide groups (*60*). Sodium cobalt oxide was reported as another alternative to Pt catalysts in a separate study using synthetic wastewater for electricity production. This catalyst was found to be low cost and an easily available substitute to be used as a potential cathode catalyst because of its high thermoelectric properties and low resistivity. It exhibited the highest power density of 0.6 W/m^2 (*61*). In another report, α-MnO_2 nanowires along with α-MnO_2–supported carbon vulcan were synthesized by the hydrothermal method and applied in air-cathode MFCs, which substituted Pt catalysts (*20*).

Bimetallic materials were also investigated as catalysts in MFCs. Integration of $MnFe_2O_4$ (manganese iron oxide) catalysts with peroxymonosulphate in MFCs revealed simultaneous azo dye degradation and removal along with electricity generation (*62*). The maximum power density of 206.2 ± 3.1 mW/m^2 was observed during the dye degradation process. Synthesis of bimetallic iron ferrite was used as an excellent catalyst material for the replacement of expensive metal catalysts. This material as a cathode was proven to result in better power recovery than conventional Pt/C catalysts, and the maximum power density was 176.9 ± 4.2 mW/m^2. This process can be efficiently used for scaling up in large-scale applications (*63*). Hybrid metal oxide nanorods of Mn and Co_3O_4 (MCON) exhibited improved oxygen reduction kinetics compared to single metal oxide (Co_3O_4) nanorods (CON) and also attained a higher power density of 587 mW/m^2 as compared to CON (454 mW/m^2) (*64*). In a similar vein, an antifouling membrane accompanied with MnO_2 catalyst, polyvinylidenefluoride (PVDF), and RGO was developed by the combination of MFCs and membrane bioreactors (MBRs) for the treatment of high strength wastewater and better power density. This conductive membrane along with the catalyst was set on the surface of carbon fiber cloth from the oxidation of RGO, which binds GO using $KMnO_4$ to decrease oxygen reduction overpotential. Thus, the developed system performed dual role of membrane filter and cathode. The maximum power density obtained using the quartz sand chamber was reported to be 228 mW/m^3, and it can be used as an alternative to the expensive proton exchange membrane (PEM) (*65*). Another report of bimetallic nickel-cobalt (Ni-Co) supported on sulphonated polyaniline exhibited a better yield (~659.79 mW/m^2) than traditional high cost Pt catalyst (~483.48 mW/m^2) and provided long-term stability for ORR. This study reveals the potential of using conducting polymers as supporting material with Ni-Co catalysts as opposed to Pt/C catalysts and provides an understanding for development of bimetallic nanocatalysts, which can be used as an alternative for Pt/C catalysts (*53*).

Iron-based catalyst materials have also been widely used in these systems. Cathodic iron-nitrogen-carbon (Fe-N-C) catalysts synthesized by the modification of the sacrificial support method were known to enhance the ORR kinetics, and a peak power density of 218 mW/cm^2 was obtained, which was 10% higher than with the commercial Pt/C catalysts (*66*, *67*). In another study, the maximum power output obtained using Fe and aminoantipyrine as starting materials was reported to be 167 ± 6 μW/cm². This material was also found to be stable when waste treatment technology was applied (*68*). Using iron-mebendazole catalysts, the highest power density of 0.33 W/cm² was achieved. This material significantly contributes to the overall performance of the BES because of excellent pore connectivity (*67, 69, 70*). Research has also been carried out on enhanced power generation using a Fe-N-C cathode catalysts in a super-capacitive MFC stack (*71*). Iron-manganese (Fe-Mn) bimetallic nanocatalysts were hydrothermally synthesized for better performance and resulted in good power output. MnO_2 has been found to be much better with improved power generation when compared with conventional Pt catalysts (*72*). Figure 3 shows the possibility of using Fe-Mn catalysts to enhance the ORR mechanism in air-cathode MFCs, and its better performance is depicted by the catalytic mechanism.

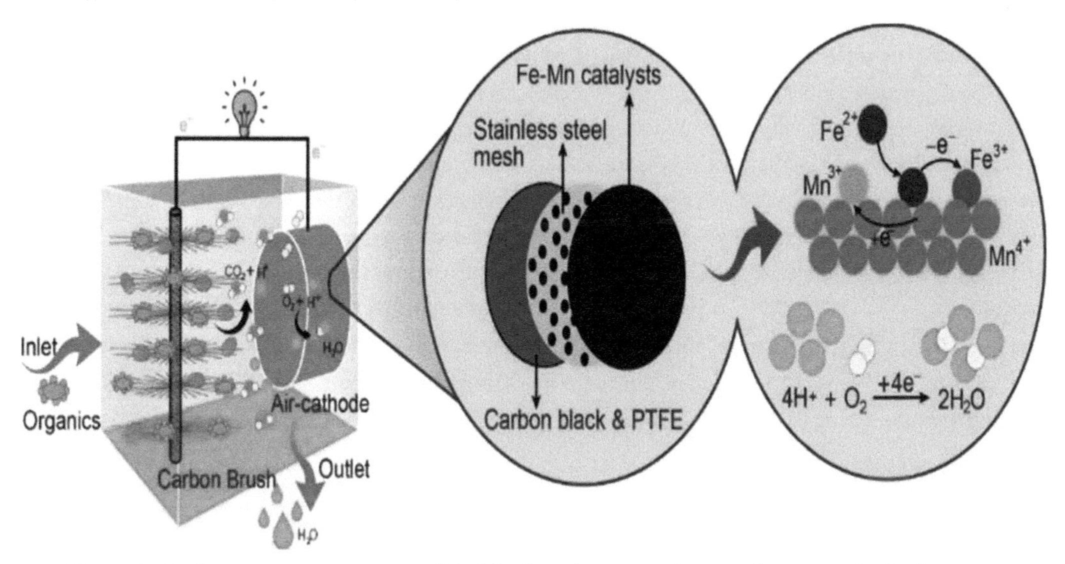

Figure 3. A schematic representation of Fe-Mn–based nanocatalysts used in air-cathodes for power generation in MFCs and their ORR mechanism C. The air-cathode was made of a stainless steel mesh current collector, a catalyst layer and a diffusion layer (consisted of carbon black and polytetrafluoroethylene [PTFE]). Adapted with permission from reference (72). Copyright 2019 Elsevier.

Another metal, vanadium, has been extensively used in MFCs because of its availability, economy, and nanosized structure in the form of nanowires, nanotubes, and nanorods. A study has shown the possibility of using vanadium pentoxide (V_2O_5) nanorods prepared by the hydrothermal method in air-cathode single chambered MFCs as an effective cathode catalyst. V_2O_5 has also been used previously for various sensors, supercapacitors, fuel cells, and lithium ion batteries because of its smaller ion diffusion distance and large surface area (*73–75*). The first report of V_2O_5/polyaniline nanocomposite used as a cathode catalyst accompanied by Nafion as PEM was reported and compared with respect to their performance with conventional Pt catalysts (*76*). The current density obtained with V_2O_5 as a catalyst was 0.0069 A/cm² and quite comparable with Pt/C, which was found to be 0.0107 A/cm². Although the current density of V_2O_5 was found to be lower than the conventional Pt/C, the ORR trend was found to be relatively analogous to the Pt/C catalyst.

These materials have led to a new insight for the conversion of waste to value-added products in the upcoming years. Therefore, the major focus of the research should relate to non-Pt–based catalysts for the reduction of ORR (*77–81*). Many studies were done for development of novel catalyst materials, and for this purpose, cheaper transition metals and their oxides such as cobalt-doped carbon, xerogel-based catalyst materials, and polyaniline-supported iron are reported to have enhanced power density when compared to costly Pt-based catalysts with better ORR activity (*57, 82–84*). Still, appropriate optimization of parameters are required for the further enhancement of output productivity. It is therefore necessary to improve the ORR catalytic activity of these catalysts, which may also contribute to the mechanism of power enhancement strategies.

Advances in Catalyst Materials in EFCs

EFCs are the bioelectronic devices that use enzymes as electrocatalysts for catalyzing redox reactions. This ultimately leads to the conversion of energy to electricity. Various substrates such as glucose, sucrose, fructose, methanol, ethanol, and glycerol have been widely exploited for bioelectricity generation in EFCs (*85, 86*). The enzymes that are frequently used in these fuel cells are glucose oxidase, glucose dehydrogenase, cellulose dehydrogenase, and fructose dehydrogenase. These enzymes are usually fuel-oxidizing enzymes. Oxygen-reducing enzymes such as laccase, bilirubin oxidase, and ascorbate oxidase are also used (*87*). Glucose dehydrogenase has an advantage over glucose oxidase since oxygen is not its natural electron acceptor. Also, cellobiose (CDH) is advantageous over enzymes such as glucose oxidase and glucose dehydrogenase, as it has low sensitivity of oxygen content. It also possesses the capacity of direct electron transfer (DET) and does not require any cofactors (*88*). A pictorial representation of an enzymatic fuel cell has been depicted in Figure 4. This figure illustrates the exploration of mediated bioanodes, which typically give better output than DET-based biocathodes, although mediated electron transfer (MET) mechanisms have their limitations owing to their less stable properties.

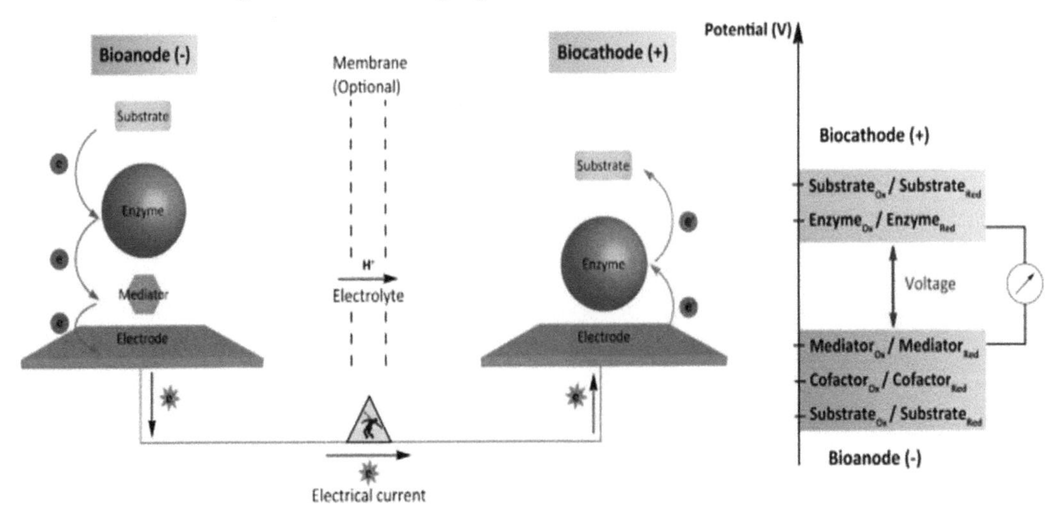

Figure 4. Enzymatic pathway in EFCs for bioelectricity generation using a mediator in the anodic chamber and direct electron transfer mechanism employed in the cathodic chamber C. Adapted with permission from reference (89). Copyright 2016 Elsevier.

Cofactors such as Nicotinamide Adenine Dinucleotide (NAD), Flavin Adenine Dinucleotide, and heme are commonly used for enzymatic activity. Moreover, oxygen, NAD, and quinone have been widely used as electron acceptors in this system, although hydrogen peroxide is also reported to have been used in such studies (*86*). Several redox active species are integrated into the BES that are known as mediators and have been used for increasing the electron transfer mechanism between electrodes and enzymes. These mediators are ferrocene, benzoquinone, osmium, and tetrathiafulvalene. Ferrocene and its derivatives are considered to be most efficient because of their solubility with various solvents and because they are nontoxic to the human body (*85*, *86*, *90*).

A wide range of electrodes such as laccase bioelectrodes have gained a lot of attention for research because of their efficiency in catalytic activity (*89*). Also, modification of electrodes for use as anodes utilizing NAD-dependent dehydrogenases for the enhancement of Nicotinamide Adenine Dinucleotide Hydrogen (NADH) oxidation in the BES was studied (*91*). Certain shortcomings of electrodes are associated with the performance of EFCs such as leaching of mediators and enzymes and the films showing low conductivity in spite of having hands-on applications (*92–94*). In general, stabilization and immobilization of the electrodes are critical for advancement in EFCs. Bioelectrocatalysis is a good stabilizing technique, although enzymes may leach out during its use in the EFCs (*95*). Crosslinking the enzymes into hydrogel and entrapping them allows efficient transportation of products through the membrane and provides good stability for the overall process (*96*). Using glucose oxidase as a biocatalyst, studies have been done in osmium hydrogels for efficient mediator-transfer mechanisms (*97*, *98*).

To improve the shelf life of the electrodes, Nafion, a commonly used PEM, is often used. Ferrocene-Nafion films have been found to show promising results (*99*, *100*). To improve the properties of these films, CNTs are used because the signal produced between enzymes and electrodes provides outstanding biocompatibility (*101–104*).

Research has been done for the use of enzymes; pyranose-2-oxidase is an anode biocatalyst that showed about 25% higher power output (8.45 ± 1.6 μW/cm^2) when compared with glucose oxidase and can be used for the enhancement of the power generation in small electronic devices (*105*, *106*). A study demonstrated the oxidation of xylose, a commonly used substrate for bioelectricity production, reassembling in vitro xylose through a utilization pathway. It generated a higher power density (0.36 mW/cm^2) than other xylose-fed BES (*107*). Another report for the application of in vitro enzymatic pathways by the co-utilization of mixed sugars resulted in a power density of 0.80–1.08 mW/m^2 (*108*). This serves as an excellent strategy for extracting energy output from a low-cost source. Using a membraneless configuration, a power density of 480 μW/cm^2 was obtained with laccase-based cathode catalysts (*109*). Many studies have been carried out using a mediator-based system, which used glucose dehydrogenase as a biocatalyst and immobilized 2-methyl-1,4-napthoquinone and potassium ferricyanide as anodic and cathodic mediators. CNTs and azine dye mediators have also been reported using glucose dehydrogenase as a biocatalyst, and the highest power densities of 52 μW/cm^2 and 1.45 ± 0.24 mW/cm^2 were achieved (*110*, *111*). Both DET and MET electrode design along with glucose oxidase as a biocatalyst were reported (*112*). Studies have been carried out utilizing carbon nanoparticles along with the covalent and noncovalent modifications for DET-based biofuel cells (*113*). In the case of MET, the potential developed is often close to that of the fuel and therefore increases power output. Ethanol production by mixed cultures through the MET mechanism was reported (*114*). The use of horseradish peroxidase (HRP) as a catalyst was reported, which reduced the production of hydrogen peroxide (H_2O_2). H_2O_2 often interferes with the stability and shelf life of bioanodes. It is formed as a result of using glucose oxidase in the system (*115*, *116*). A maximum of power density achieved using horseradish peroxidase as a catalyst material was reported to be 4.7 μW/cm^2 as compared to sole glucose oxidase–based EFCs (*117*).

Although enzymatic fuel cells are promisingly excellent in terms of power generation, they can rarely compete against microorganisms with respect to fuel utilization and long-term stability (*86*). EFCs have efficiency issues and depletion of lifetime stability, but they employ non-Pt–based catalysts that add to its cost-effectiveness and nontoxicity; therefore, they can be used as implantable power supplies for sensors in vivo by overcoming engineering and scientific obstacles (*87*).

Advances in Catalyst Materials in MECs

A similar technology related to MFCs is the MEC, which usually produces methane or hydrogen from fermentation end products such as ethanol, lactic acid, and acetate by the application of electrical current (*118–120*). These systems are also known to simultaneously remove various organic matter and salts from wastewater (*121*). Unlike MFCs, MECs require additional power for the production of hydrogen (*122*). One significant point in MEC is that the cathode chamber should be anaerobic, resulting in no oxygen being present and no power output generation. Graphene- and graphite–based carbon electrodes are studied, as they are inexpensive and offer better conductive properties in MECs (*123, 124*). Although MECs are an efficient technology for hydrogen production when acetate is used as a substrate or carbon source, when wastewater is used as substrate, the hydrogen recovery and power yield is decreased in the system (*11*). This may be due to the presence of various organics in the wastewater, which may often interfere with the hydrogen productivity. Certain activated carbons for ORR mechanisms and metal alloys and molybdenum disulphide catalysts have been exploited for hydrogen production (*77, 125–127*). In a separate strategy, dark fermentation was combined with MEC to produce hydrogen and methane, respectively, from the effluent of a palm oil mill (*128*).

Ni-based catalysts have been widely explored in MEC, giving high yield compared to conventional Pt catalysts. A study has been carried out for MECs containing Ni powder catalysts in cathode chambers, and only a slight difference in terms of hydrogen production between Ni catalysts (1.3 $m^3/d/m^3$) and Pt catalysts (1.6 $m^3/d/m^3$) was observed (*129*). In one such study, metals such as Ni- and copper (Cu)–loaded carbon catalysts were compared with Pt catalysts using MECs. Although current density was 75.5% higher in Pt/C catalyst when compared to PtNi/C, PtNi/C performed better than single-metal catalysts (Ni/C or Cu/C). The current density obtained using PtNi/C catalysts was found to be 3.90 mA/cm^2. The hydrogen production rate of Ni/C catalysts was analogous to Pt/C catalysts (0.068 $m^3/d/m^3$); however, it was slightly lower in case of Cu/C (0.053 $m^3/d/m^3$). This study shows the advantage of using metal-based catalysts such as Ni and Cu over the expensive Pt for improved hydrogen productivity, catalyst stability, and cost-effectiveness (*29*). MECs with Pt catalyzed cathodes were compared with Ni-alloy–based catalysts in MECs, and the hydrogen production rate was slightly lower than the former (2.0 $m^3/d/m^3$) together with current density (12 A/m^2) (*130*). Also, by using nickel alloys such as Ni-Mo catalysts (prepared through electrodeposition on carbon cloth), cathodes have shown only a 13% lower hydrogen production rate as compared with Pt catalysts in another study (*131*). This material therefore shows pioneering advancement in lowering the internal resistance developed in MECs as well as enhancing the hydrogen productivity along with power generation (*131*). Stainless steel alloys were used as catalysts for enhanced hydrogen production owing to their low activation overpotential, stability, and low cost. The hydrogen recovery using stainless steel was 61% compared to Pt (47%), and a maximum hydrogen production rate of 1.5 $m^3/d/m^3$ was achieved (*130*). Tungsten carbide was also reported to be used as a cathode catalyst because of its low price in MECs. Although it lacks stability to a certain extent and may also lead to corrosion, it may improve the anodic metabolite oxidation efficiency (*132*). In another report, enhancement of hydrogen recovery and electron transfer rate was obtained by a three-dimensional biphasic nanosheet matrix of nickel, which was coupled to a bioanode and showed 1.5 times better results than a nickel-foam cathode. The hydrogen production rate achieved was 9.78 $\pm$ 0.38 mL/d/cm^2 and, hence, provided a better performance along with long-term stability due to proper adherence of microbes on the matrix (*133*). A Cu sheet coated with Ni, Fe, and molybdenum (Mo) alloy catalysts displayed a good hydrogen production rate (50 $m^3/d/m^3$) and higher catalytic activity compared to Ni-coated cathode catalysts (*126*). In addition to this, catalysts consisting of graphite carbon and iron-based composite nanorods have been used for hydrogen production and long-term stability and therefore have proven to be better than nonmodified cathodes and CNTs. This catalyst did not reach the significant level of power yield against Pt catalysts, yet proved to be a low-cost material that accounts for a 5% less power cost than conventional Pt/C catalysts for large-scale production in MECs (*134*). Thus, these catalysts have proven to be an efficient means of hydrogen production. Further research is required for accurate optimization of the systems, which includes consistent effort and considering a potential candidate for a wide range of applications.

Advances in Catalyst Materials in MDCs

MDCs are the modification of MFCs facilitating in-situ desalination of saline water with a better power recovery. This technology was used to improve the shelf life of reverse osmosis membranes and the treatment of saline water (*135, 136*). This process provides an advantage over reverse osmosis, nanofiltration, or any other desalination system because of cost-effectiveness and certain environmental factors associated with BES (*137*). MDCs differ from MFCs in the way that they do not require a mediator for the power generation and instead, depend on charged components of the sludge (*138*). Some of the applications of MDCs are seawater desalination, brackish water desalination, and groundwater denitrification (*139–141*), and additionally, wastewater treatment (*142*). Studies were conducted to reduce ORR overpotential developed using Pt as a catalyst. Around 43–67% of desalination efficiency was obtained using Pt catalysts from saline water (*143*). But an alternative to expensive catalysts such as Pt is desirable and of primary concern. Cathode catalysts consisting of silver-tin dioxide (Ag-SnO_2) have been explored, and a desalination efficiency of 72.6 $\pm$ 3% and a power density of 1.47 W/m^3 were achieved when compared with a system without catalysts (0.88 W/m^3). This report shows the better catalytic activity of Ag-SnO_2 and a prospective cathode catalyst (*121*). An iron-nitrogen-carbon (Fe-N-C)–based cathode catalyst prepared by nicarbazin was compared with Pt catalysts and activated carbon cathode catalysts and has been shown to have enhanced power generation. The Fe-N-C–based catalyst showed a decline in its performance efficiency (15%) after a certain period of time (4 cycles) but showed better activity than Pt and activated carbon catalysts and achieved a maximum power density of 49 $\pm$ 2 μW cm^2 (*144*).

Several studies have been carried out using microbes as biocatalysts (*145*). An aerobic biocathode MDC using bacterial (mixed culture from anaerobic reactor) catalysts and carbon felt was studied. Due to desalination, the power generation (0.96 W/m^2) was found to be higher when compared with an air-cathode MDC. Chemical oxygen demand (COD) removal in MDCs was also reported, and a maximum of 62.6 $\pm$ 4.4% removal efficiency was obtained. Using aerobic biocathode obtained a maximum voltage of 609 mV (136 mV higher than air-cathode MDCs) along with a desalination rate of 2.83 mg/h and hence proved to be an efficient approach (*142*). Using anmmox bacteria as biocathodes or biocatalysts, simultaneous removal of nitrogen and carbon compounds (although not in very high concentrations) from wastewater in addition to power generation were reported. Ammonium removal efficiency in this study was found to be 90% from the 0.04 mg/L initial concentration to 100 mg/L, and a maximum power density of 0.092 W/m^3 has been attained (*146*). Some of the anammox bacteria commonly found in wastewater were *Candidatus Scalindua brodae* and *Candidatus Brocadia fulgida* (*147*). These catalysts deal to substitute expensive Pt-based material with the possibility of simultaneous electricity production along with enhanced desalination efficiency.

Advances in Catalyst Materials in MSCs

MSCs, also known as photosynthetic MFCs (PMFCs), are considered to generate electricity by the use of sunlight without the addition of any fuel or oxidants. MSCs are also known to be a sustainable and reasonable technique for sequestering carbon (*148*). MSCs are basically made by the combination of electroactive and photosynthetic microbes and waste constituents to generate various compounds such as methane, hydrogen, hydrogen peroxide, and ethanol (*42, 114, 148, 149*). Since MSCs entirely rely upon sunlight for the energy production, this technique has gained a wide range of interest among researchers. MSCs generally exploit live plants (PMFCs) where the roots of the entire plant aid in powering the electroactive species in the anode by the excretion process of rhizodeposits (*150–156*). Out of all MSCs exploited, *Spartina anglica*, a cord-glass PMFC, is considered to be the best in terms of power generation. The most commonly used bacteria is *Desulfobulbus* (*156*). But it is not well established whether electroactive species are actively participating in the entire process or not (*157*). Modified electrodes (usually biofilms), photosynthetic microorganisms such as bacteria, algae and mixed culture were utilized for improved power production. A maximum power density obtained was found to be 110 mW/m^2 using ferricyanide as a catalyst (*158, 159*). Characterization of the electrochemical property was carried out using both nonphotosynthetic and photosynthetic components of the biocathode where a positive current response was evaluated. These nonphotosynthetic catalysts are capable of integrating inorganic carbon for cathodic oxygen reduction for various BES (*160*). Another report stated that solar energy was converted to electrical current using a phototrophic biofilm on the anode of an MSC. This included the utilization of a mixed culture, though pure cultures have also been reported (*161*). Several chemical catalysts have also been employed apart from these biocatalysts, which are a part of biofilm consortia in MSCs for the production of hydrogen. However, they are found to be toxic and not self-sustainable (*157*) to be used in future. Since MSCs mostly use reproducing organisms or microbes, there is less chance of lethal contamination to the surroundings and they add no significant level of pollution.

Figure 5 represents the entire process of power generation in MSCs. The process involves the transportation of organic material by the oxidation reaction occurring in the anodic chamber aided by the microbes and the reduction reaction for the conversion of oxygen to water occurring in the cathodic chamber followed by photosynthesis.

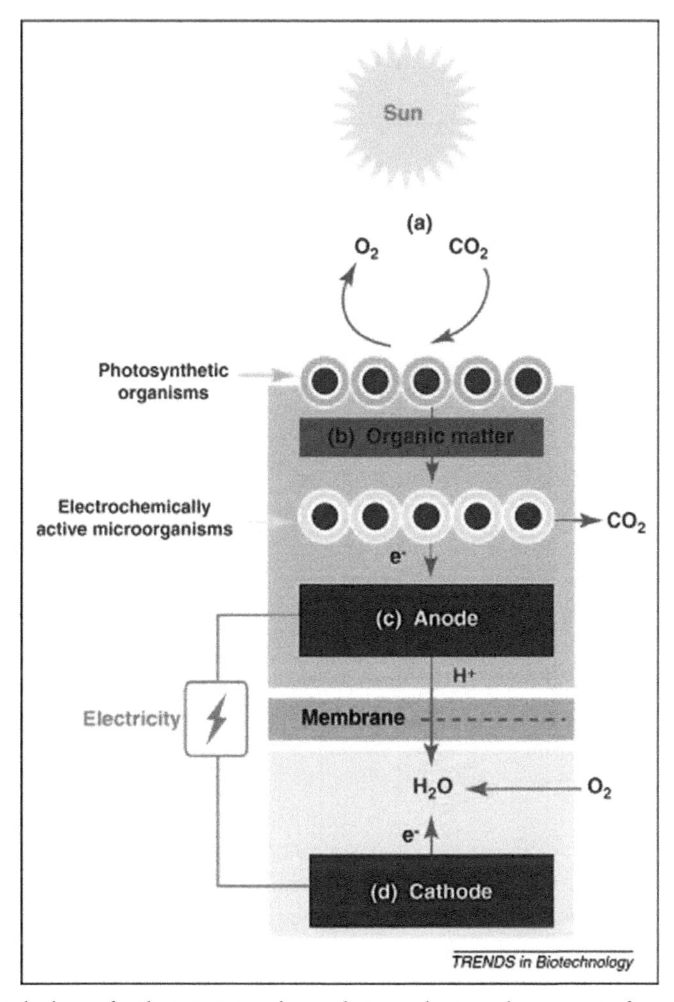

Figure 5. A pictorial scheme for the MSCs involving photosynthesis and transport of organic components to the anode followed by anodic oxidation by electroactive microbes and cathodic reduction of oxygen to water. Adapted with permission from reference (157). Copyright 2011 Elsevier.

Table 1. Various Catalysts Used for Different BES and Their Respective Power Yields

S.N.	*BES*	*Catalysts*	*Power Output*	*Comments*	*Ref.*
1	MFCs	Vanadium pentoxide (V_2O_5) cathode catalyst	1073 ± 18 mW/m^2	Double loaded MFCs are more efficient than single and triple loaded MFCs	*(162)*
2		Manganese iron oxide ($MnFe_2O_4$) cathode catalyst	206.2 ± 3.1 mW/m^2 and 574.4 mA/m^2	Azo dye degradation and catalyst regeneration	*(62)*
3		Cobalt-Iron (Co-Fe) phthalocyanine supported on carbide ORR catalyst	1.57 W/m^2	Enhanced ORR kinetics and COD removal efficiency	*(31)*
4		Bimetallic nanocomposite cathode catalyst supported on sulfonated polyaniline	659.79 mW/m^2	Conducting polymer supported matrix and improved efficiency when integrated with Ni-Co catalyst system	*(53)*
5		Nickel (Ni) nanoparticles cathode catalyst	560 and 540 mW/m^2	Anaerobic fermentation and pretreatment of macroalgae improves performance	*(59)*
6		Iron-manganese (Fe-Mn) bimetallic nanocatalysts as cathode catalyst	1940 ± 31 mW/m^2 and 19.4 A/m^2	24% and 37% higher electrochemical output than platinum on carbon (Pt/C) catalyst	*(72)*
7		Iron-nitrogen-carbon (Fe-N-C) cathode catalyst	36.9 W/m^3	Power decreased with increase in pulse time	*(71)*
8		α-MnO_2 nanowires supported on carbon vulcan as ORR catalyst	180 mW/m^2	Alternative for Pt-based catalyst and provides good stability	*(20)*
9		Nitrogen-Phosphorus dual-doped carbon cathode catalyst	1603.6 ± 80 mW/m^2	Presence of nitrogen, phosphorus and carbon oxygen bonds enhanced the cathode performance	*(46)*
10		Cobalt-Nitrogen-CNT (Co-N-CNT) and Iron Nitrogen-CNT (Fe-N-CNT) cathode catalyst	5.1 W/m^3 and 6 W/m^3	Due to formation of nitrogen-metal centers, power output and ORR kinetics improved	*(54)*
11		Sodium cobalt oxide ($NaCo_2O_4$) cathode catalyst	0.6 W/m^2 3 A/m^2	Cathode preparation method were optimized for enhanced performance	*(61)*
12		Bimetallic iron ferrite $Co_{0.5}Zn_{0.5}Fe_2O_4$ ORR catalyst	1.3 ± 0.5 W/m^3 and 176.9 ± 4.2 mW/m^2	Power recovered were 4 times higher than Pt/C and good alternative for platinum	*(63)*
13		Xerogel-based cathode catalyst	176.5 ± 6 mW/m^2	When pH of catholyte was increased, power output also enhanced	*(41)*

Table 1. (Continued). Various Catalysts Used for Different BES and Their Respective Power Yields

S.N.	*BES*	*Catalysts*	*Power Output*	*Comments*	*Ref.*
14	MBR + MFC	Manganese dioxide (MnO_2) cathode catalyst	228 mW/m^3 and 9.59 mA/m^3	Alternative to Proton Exchange Membrane (PEM) using carbon-based conductive membrane	(*65*)
15	MDC	Silver-tin dioxide (Ag-SnO_2) cathode catalyst	1.47 W/m^3	Good catalytic activity of the catalyst for large scale application	(*121*)
16	MSC	Nonphotosynthetic biocathode catalyst	0.035 A/m^2 and 0.045 A/m^2	The two organisms where one is active photosynthetically and the other catalyses oxygen reduction rate	(*163*)
17	MEC	Transition metal/carbon nanoparticle cathode catalyst	1856.0 mA/m^2	Hydrogen production and power generation	(*29*)
18	EFC	Nonimmobilized enzymes as anode catalysts	0.80-1.08 mW/cm^2	An alternative for energy extraction from mixed sugars and power generation	(*108*)
19		Glucose oxidase and laccase as anode and cathode catalysts	100 nW/cm^2	Enzymatic activity may be lowered due to presence of various metabolites in wastewater	(*164*)
20		Pyranose-2-oxidase anode catalyst	$8.45 \pm 1.6\ \mu W/cm^2$	Gas diffusion cathode and carbon paper electrode were used to improve the output	(*106*)

Conclusions

BES are one of the most promising technologies for their wide range of applications such as electricity generation, hydrogen production, methane production, desalination from salt or brine water, wastewater treatment, removal of heavy metals, and many more. But it is not finding large-scale application and is not well established for commercialization for several reasons. Optimization of certain parameters is required in the cathodic chamber. In this case, cathode catalysts often play a significant role in overall performance and diminish the costs associated. Platinum has been a more frequently used catalyst although appropriate and satisfactory power output results were not achieved. Moreover, platinum (Pt) accounts for the greatest cost of the entire system. So, other alternatives such as transition metals, metal oxides, biocatalysts such as microbes, and enzymes were investigated and incorporated with Pt or used individually. Carbon-based nanoparticles and bimetallic materials have excellent potential for their use as catalyst materials not only because they are less expensive but also because they enhanced the overall power performance in the BES. MFCs are the most commonly used BES and attained satisfactory power output although industrial scale-ups have not been reported to date because of several critical factors. EFCs employ enzymes for catalysis and have also been found to be the best among all the BES, although power yield was found to be quite low compared to other configurations of BES. EFCs are suited for low power applications such as biosensors and other biomedical devices. Other configurations of BES such as MECs and MSCs are used for hydrogen production and MDCs for the desalination process. It can be expected that with these advanced improvements in catalyst materials along with small-scale studies, the next big step would be for economically sustainable energy production with agreeable outcomes. A combination of different configurations of BES may be investigated for a particular application to achieve a better performance. A tunable system can be developed, which can be used as different configurations of BES by simply modifying the electrode catalysts. Priority for practical applications and the development of low-cost and highly durable materials should be in focus for future studies in a comprehensive manner. This study may aid in the further expansion of the development of novel catalyst materials improving various BES schemes and strategies in the future.

References

1. Sharma, S.; Datta, P.; Kumar, B.; Tiwari, P.; Pandey, L. M. Production of Novel Rhamnolipids via Biodegradation of Waste Cooking Oil Using *Pseudomonas aeruginosa* MTCC7815. *Biodegredation* **2019**, 1–12.
2. Pandey, L. M. Enhanced Adsorption Capacity of Designed Bentonite and Alginate Beads for the Effective Removal of Methylene Blue. *Appl. Clay Sci.* **2019**, *169*, 102–111.
3. Sharma, S.; Hasan, A.; Kumar, N.; Pandey, L. M. Removal of Methylene Blue Dye from Aqueous Solution Using Immobilized *Agrobacterium fabrum* Biomass Along with Iron Oxide Nanoparticles as Biosorbent. *Environ. Sci. Pollut. Res.* **2018**, *25* (22), 21605–21615.
4. Sharma, S.; Tiwari, S.; Hasan, A.; Saxena, V.; Pandey, L. M. Recent Advances in Conventional and Contemporary Methods for Remediation of Heavy Metal-Contaminated Soils. *3 Biotech.* **2018**, *8* (4), 216.
5. Jamil, F.; Ala'a, H.; Myint, M. T. Z.; Al-Hinai, M.; Al-Haj, L.; Baawain, M.; Al-Abri, M.; Kumar, G.; Atabani, A. Biodiesel Production by Valorizing Waste *Phoenix dactylifera* L. Kernel Oil in the Presence of Synthesized Heterogeneous Metallic Oxide Catalyst (Mn@ MgO-ZrO 2). *Energy Convers. Manage.* **2018**, *155*, 128–137.

6. Tiwari, S.; Hasan, A.; Pandey, L. M. A Novel Bio-Sorbent Comprising Encapsulated *Agrobacterium fabrum* (SLAJ731) and Iron Oxide Nanoparticles for Removal of Crude Oil Co-Contaminant, Lead Pb (II). *J. Environ. Chem. Eng.* **2017**, *5* (1), 442–452.
7. Chang, I. S.; Jang, J. K.; Gil, G. C.; Kim, M.; Kim, H. J.; Cho, B. W.; Kim, B. H. Continuous Determination of Biochemical Oxygen Demand Using Microbial Fuel Cell Type Biosensor. *Biosens. Bioelectron.* **2004**, *19* (6), 607–613.
8. Nancharaiah, Y.; Mohan, S. V.; Lens, P. Metals Removal and Recovery in Bioelectrochemical Systems: A Review. *Bioresour. Technol.* **2015**, *195*, 102–114.
9. Mook, W.; Chakrabarti, M.; Aroua, M.; Khan, G.; Ali, B.; Islam, M.; Hassan, M. A. Removal of Total Ammonia Nitrogen (TAN), Nitrate and Total Organic Carbon (TOC) from Aquaculture Wastewater Using Electrochemical Technology: A Review. *Desal.* **2012**, *285*, 1–13.
10. Borole, A. P.; Mielenz, J. R. Estimating Hydrogen Production Potential in Biorefineries Using Microbial Electrolysis Cell Technology. *Int. J. Hydrogen Energy* **2011**, *36* (22), 14787–14795.
11. Pant, D.; Singh, A.; Van Bogaert, G.; Olsen, S. I.; Nigam, P. S.; Diels, L.; Vanbroekhoven, K. Bioelectrochemical Systems (BES) for Sustainable Energy Production and Product Recovery from Organic Wastes and Industrial Wastewaters. *Rsc Adv.* **2012**, *2* (4), 1248–1263.
12. Sevda, S.; Pandey, L.; Singh, S.; Garlapati, V. K.; Sharma, S. Sustainability Assessment of Microbial Fuel Cells. *Waste to Sustainable Energy: MFCs–Prospects through Prognosis*; CRC Press: Boca Raton, FL, 2019; pp 313–330.
13. He, Z.; Angenent, L. T. Application of Bacterial Biocathodes in Microbial Fuel Cells. *Electroanalysis (N.Y.N.Y.)* **2006**, *18* (19–20), 2009–2015.
14. Yuan, H.; Hou, Y.; Abu-Reesh, I. M.; Chen, J.; He, Z. Oxygen Reduction Reaction Catalysts Used in Microbial Fuel Cells for Energy-Efficient Wastewater Treatment: A Review. *Mater. Horiz.* **2016**, 3 (5), 382–401.
15. Gupta, R.; Bekele, W.; Ghatak, A. Harvesting Energy of Interaction Between Bacteria and Bacteriophage in a Membrane-Less Fuel Cell. *Bioresour. Technol.* **2013**, *147*, 654–657.
16. Hu, H.; Fan, Y.; Liu, H. Hydrogen Production Using Single-Chamber Membrane-Free Microbial Electrolysis Cells. *Water Res.* **2008**, *42* (15), 4172–4178.
17. Chae, K.-J.; Choi, M.-J.; Kim, K.-Y.; Ajayi, F. F.; Chang, I.-S.; Kim, I. S. A Solar-Powered Microbial Electrolysis Cell with a Platinum Catalyst-Free Cathode to Produce Hydrogen. *Environ. Sci. Technol.* **2009**, *43* (24), 9525–9530.
18. Rozendal, R. A.; Hamelers, H. V.; Rabaey, K.; Keller, J.; Buisman, C. J. Towards Practical Implementation of Bioelectrochemical Wastewater Treatment. *Trends Biotechnol.* **2008**, *26* (8), 450–459.
19. Huang, J.; Zhu, N.; Yang, T.; Zhang, T.; Wu, P.; Dang, Z. Nickel Oxide and Carbon Nanotube Composite (NiO/CNT) as a Novel Cathode Non-Precious Metal Catalyst in Microbial Fuel Cells. *Biosens. Bioelectro.* **2015**, *72*, 332–339.
20. Majidi, M. R.; Farahani, F. S.; Hosseini, M.; Ahadzadeh, I. Low-Cost Nanowired α-MnO2/C as an ORR Catalyst in Air-Cathode Microbial Fuel Cell. *Bioelectrochemistry* **2019**, *125*, 38–45.
21. Zhang, Q.; Hu, J.; Lee, D.-J. Microbial Fuel Cells as Pollutant Treatment Units: Research Updates. *Bioresour. Technol.* **2016**, *217*, 121–128.

22. Lv, C.; Liang, B.; Zhong, M.; Li, K.; Qi, Y. Activated Carbon-Supported Multi-Doped Graphene as High-Efficient Catalyst to Modify Air Cathode in Microbial Fuel Cells. *Electrochim. Acta* **2019**, *304*, 360–369.
23. Li, S.; Chen, G. Factors Affecting the Effectiveness of Bioelectrochemical System Applications: Data Synthesis and Meta-Analysis. *Batteries* **2018**, *4* (3), 34.
24. Wang, H.; Park, J.-D.; Ren, Z. J. Practical Energy Harvesting for Microbial Fuel Cells: A Review. *Environ. Sci. Technol.* **2015**, *49* (6), 3267–3277.
25. Rismani-Yazdi, H.; Carver, S. M.; Christy, A. D.; Tuovinen, O. H. Cathodic Limitations in Microbial Fuel Cells: An Overview. *J. Power Sources* **2008**, *180* (2), 683–694.
26. Ayyaru, S.; Mahalingam, S.; Ahn, Y.-H. A Non-Noble V2O5 Nanorods as an Alternative Cathode Catalyst for Microbial Fuel Cell Applications. *Int. J. Hydrogen Energy* **2019**, *44* (10), 4974–4984.
27. Logan, B. E.; Hamelers, B.; Rozendal, R.; Schröder, U.; Keller, J.; Freguia, S.; Aelterman, P.; Verstraete, W.; Rabaey, K. Microbial Fuel Cells: Methodology and Technology. *Environ. Sci. Technol.* **2006**, *40* (17), 5181–5192.
28. Yuvaraj, A.; Santhanaraj, D. A Systematic Study on Electrolytic Production of Hydrogen Gas by Using Graphite as Electrode. *Mater. Res.* **2014**, *17* (1), 83–87.
29. Choi, M.-J.; Yang, E.; Yu, H.-W.; Kim, I. S.; Oh, S.-E.; Chae, K.-J. Transition Metal/Carbon Nanoparticle Composite Catalysts as Platinum Substitutes for Bioelectrochemical Hydrogen Production Using Microbial Electrolysis Cells. *Int. J. Hydrogen Energy* **2019**, *44* (4), 2258–2265.
30. Lv, Z.; Chen, Y.; Wei, H.; Li, F.; Hu, Y.; Wei, C.; Feng, C. One-Step Electrosynthesis of Polypyrrole/Graphene Oxide Composites for Microbial Fuel Cell Application. *Electrochim. Acta* **2013**, *111*, 366–373.
31. Noori, M. T.; Verma, N. Cobalt-Iron Phthalocyanine Supported on Carbide-Derived Carbon as an Excellent Oxygen Reduction Reaction Catalyst for Microbial Fuel Cells. *Electrochim. Acta* **2019**, *298*, 70–79.
32. Pant, D.; Van Bogaert, G.; Diels, L.; Vanbroekhoven, K. A Review of the Substrates Used in Microbial Fuel Cells (MFCs) for Sustainable Energy Production. *Bioresour. Technol.* **2010**, *101* (6), 1533–1543.
33. Saxena, V.; Sharma, S.; Pandey, L. M. Fe (III) Doped ZnO Nano-Assembly as a Potential Heterogeneous Nano-Catalyst for the Production of Biodiesel. *Mater. Lett.* **2019**, *237*, 232–235.
34. Karmakar, S.; Saxena, V.; Chandra, P.; Pandey, L. M. Novel Therapeutics and Diagnostics Strategies Based on Engineered Nanobiomaterials. *Nanotechnology in Modern Animal Biotechnology*; Springer Nature: Singapore, 2019; pp 1–27.
35. Saxena, V.; Chandra, P.; Pandey, L. M. Design and Characterization of Novel Al-Doped ZnO Nanoassembly as an Effective Nanoantibiotic. *Appl. Nanosci.* **2018**, *8* (8), 1925–1941.
36. Trogadas, P.; Fuller, T. F.; Strasser, P. Carbon as Catalyst and Support for Electrochemical Energy Conversion. *Carbon* **2014**, 75, 5–42.
37. Asset, T.; Job, N.; Busby, Y.; Crisci, A.; Martin, V.; Stergiopoulos, V.; Bonnaud, C. L.; Serov, A.; Atanassov, P.; Chattot, R. L. Porous Hollow PtNi/C Electrocatalysts: Carbon Support

Considerations to Meet Performance and Stability Requirements. *ACS Catal.* **2018**, *8* (2), 893–903.

38. Janicek, A.; Gao, N.; Fan, Y.; Liu, H. High Performance Activated Carbon/Carbon Cloth Cathodes for Microbial Fuel Cells. *Fuel Cells* **2015**, *15* (6), 855–861.
39. Bose, D.; Sridharan, S.; Dhawan, H.; Vijay, P.; Gopinath, M. Biomass Derived Activated Carbon Cathode Performance for Sustainable Power Generation from Microbial Fuel Cells. *Fuel* **2019**, *236*, 325–337.
40. Czakkel, O.; Marthi, K.; Geissler, E.; László, K. Influence of Drying on the Morphology of Resorcinol–Formaldehyde-Based Carbon Gels. *Micropor. Mesopor. Mat.* **2005**, *86* (1–3), 124–133.
41. Thapa, B. S.; Seetharaman, S.; Chetty, R.; Chandra, T. Xerogel Based Catalyst for Improved Cathode Performance in Microbial Fuel Cells. *Enzyme Microb. Technol.* **2019**, *124*, 1–8.
42. Rozendal, R. A.; Hamelers, H. V.; Euverink, G. J.; Metz, S. J.; Buisman, C. J. Principle and Perspectives of Hydrogen Production through Biocatalyzed Electrolysis. *Int. J. Hydrogen Energy* **2006**, *31* (12), 1632–1640.
43. Wu, J.; Yang, Z.; Sun, Q.; Li, X.; Strasser, P.; Yang, R. Synthesis and Electrocatalytic Activity of Phosphorus-Doped Carbon Xerogel for Oxygen Reduction. *Electrochim. Acta* **2014**, *127*, 53–60.
44. Li, M.; Zhou, S.; Xu, M. Graphene Oxide Supported Magnesium Oxide as an Efficient Cathode Catalyst for Power Generation and Wastewater Treatment in Single Chamber Microbial Fuel Cells. *Chem. Eng. J.* **2017**, *328*, 106–116.
45. Gupta, S.; Yadav, A.; Singh, S.; Verma, N. Synthesis of Silicon Carbide-Derived Carbon as an Electrode of a Microbial Fuel Cell and an Adsorbent of Aqueous Cr (VI). *Ind. Eng. Chem. Res.* **2017**, *56* (5), 1233–1244.
46. Liang, B.; Li, K.; Liu, Y.; Kang, X. Nitrogen and Phosphorus Dual-Doped Carbon Derived from Chitosan: An Excellent Cathode Catalyst in Microbial Fuel Cell. *Chem. Eng. J.* **2019**, *358*, 1002–1011.
47. Gao, N.; Qu, B.; Xing, Z.; Ji, X.; Zhang, E.; Liu, H. Development of Novel Polyethylene Air-Cathode Material for Microbial Fuel Cells. *Energy* **2018**, *155*, 763–771.
48. Xu, Y.; Ruban, A. V.; Mavrikakis, M. Adsorption and Dissociation of O_2 on Pt–Co and Pt–Fe Alloys. *J. Am. Chem. Soc.* **2004**, *126* (14), 4717–4725.
49. Cheng, S.; Liu, H.; Logan, B. E. Power Densities Using Different Cathode Catalysts (Pt and CoTMPP) and Polymer Binders (Nafion and PTFE) in Single Chamber Microbial Fuel Cells. *Environ. Sci. Technol.* **2006**, *40* (1), 364–369.
50. Jacob, T. The Mechanism of Forming H_2O from H_2 and O_2 over a Pt Catalyst via Direct Oxygen Reduction. *Fuel Cells* **2006**, *6* (3–4), 159–181.
51. Wang, S.; Chu, X.; Zhang, X.; Zhang, Y.; Mao, J.; Yang, Z. A First-Principles Study of O_2 Dissociation on Platinum Modified Titanium Carbide: A Possible Efficient Catalyst for the Oxygen Reduction Reaction. *J. Phys. Chem. C* **2017**, *121* (39), 21333–21342.
52. Kim, J.; Gewirth, A. A. Mechanism of Oxygen Electroreduction on Gold Surfaces in Basic Media. *J. Phys.Chem. B* **2006**, *110* (6), 2565–2571.
53. Papiya, F.; Pattanayak, P.; Kumar, P.; Kumar, V.; Kundu, P. P. Development of Highly Efficient Bimetallic Nanocomposite Cathode Catalyst, Composed of Ni: Co Supported

Sulfonated Polyaniline for Application in Microbial Fuel Cells. *Electrochim. Acta* **2018**, *282*, 931–945.

54. Türk, K.; Kruusenberg, I.; Kibena-Põldsepp, E.; Bhowmick, G.; Kook, M.; Tammeveski, K.; Matisen, L.; Merisalu, M.; Sammelselg, V.; Ghangrekar, M. Novel Multi Walled Carbon Nanotube Based Nitrogen Impregnated Co and Fe Cathode Catalysts for Improved Microbial Fuel Cell Performance. *Int. J. Hydrogen Energy* **2018**, *43* (51), 23027–23035.
55. Nie, Y.; Li, L.; Wei, Z. Recent Advancements in Pt and Pt-Free Catalysts for Oxygen Reduction Reaction. *Chem. Soc. Rev.* **2015**, *44* (8), 2168–2201.
56. Singh, S.; Modi, A.; Verma, N. Enhanced Power Generation Using a Novel Polymer-Coated Nanoparticles Dispersed-Carbon Micro-Nanofibers-Based Air-Cathode in a Membrane-Less Single Chamber Microbial Fuel Cell. *Int. J. Hydrogen Energy* **2016**, *41* (2), 1237–1247.
57. Modi, A.; Singh, S.; Verma, N. In Situ Nitrogen-Doping of Nickel Nanoparticle-Dispersed Carbon Nanofiber-Based Electrodes: Its Positive Effects on the Performance of a Microbial Fuel Cell. *Electrochim. Acta* **2016**, *190*, 620–627.
58. Ghasemi, M.; Daud, W. R. W.; Rahimnejad, M.; Rezayi, M.; Fatemi, A.; Jafari, Y.; Somalu, M. R.; Manzour, A. Copper-Phthalocyanine and Nickel Nanoparticles as Novel Cathode Catalysts in Microbial Fuel Cells. *Int. J. Hydrogen Energy* **2013**, *38* (22), 9533–9540.
59. Gebresemati, M.; Das, G.; Park, B. J.; Yoon, H. H. Electricity Production from Macroalgae by a Microbial Fuel Cell Using Nickel Nanoparticles as Cathode Catalysts. *Int. J. Hydrogen Energy* **2017**, *42* (50), 29874–29880.
60. Aluthge, D.; Ahn, J.; Mehrkhodavandi, P. Overcoming Aggregation in Indium Salen Catalysts for Isoselective Lactide Polymerization. *Chem. Sci.* **2015**, *6* (9), 5284–5292.
61. Hirooka, K.; Ichihashi, O.; Takeguchi, T. Sodium Cobalt Oxide as a Non-Platinum Cathode Catalyst for Microbial Fuel Cells. *Sustain. Environ. Res.* **2018**, *28* (6), 322–325.
62. Xu, H.; Quan, X.; Chen, L. A Novel Combination of Bioelectrochemical System with Peroxymonosulfate Oxidation for Enhanced Azo Dye Degradation and MnFe2O4 Catalyst Regeneration. *Chemosphere* **2019**, *217*, 800–807.
63. Das, I.; Noori, M. T.; Bhowmick, G. D.; Ghangrekar, M. Synthesis of Bimetallic Iron Ferrite Co0.5Zn0.5Fe2O4 as a Superior Catalyst for Oxygen Reduction Reaction to Replace Noble Metal Catalysts in Microbial Fuel Cell. *Int. J. Hydrogen Energy* **2018**, *43* (41), 19196–19205.
64. Kumar, R.; Singh, L.; Ab Wahid, Z.; Mahapatra, D. M.; Liu, H. Novel Mesoporous MnCo2O4 Nanorods as Oxygen Reduction Catalyst at Neutral pH in Microbial Fuel Cells. *Bioresour. Technol.* **2018**, *254*, 1–6.
65. Gao, C.; Liu, L.; Yu, T.; Yang, F. Development of a Novel Carbon-Based Conductive Membrane with in-situ Formed MnO2 Catalyst for Wastewater Treatment in Bio-Electrochemical System (BES). *J. Membr. Sci.* **2018**, *549*, 533–542.
66. Serov, A.; Robson, M. H.; Smolnik, M.; Atanassov, P. Tri-Metallic Transition Metal–Nitrogen–Carbon Catalysts Derived by Sacrificial Support Method Synthesis. *Electrochim. Acta* **2013**, *109*, 433–439.
67. Hossen, M. M.; Artyushkova, K.; Atanassov, P.; Serov, A. Synthesis and Characterization of High Performing Fe-NC Catalyst for Oxygen Reduction Reaction (ORR) in Alkaline Exchange Membrane Fuel Cells. *J. Power Sources* **2018**, *375*, 214–221.

68. Robson, M. H.; Serov, A.; Artyushkova, K.; Atanassov, P. A Mechanistic Study of 4-Aminoantipyrine and Iron Derived Non-Platinum Group Metal Catalyst on the Oxygen Reduction Reaction. *Electrochim. Acta* **2013**, *90*, 656–665.
69. Stariha, S.; Artyushkova, K.; Workman, M. J.; Serov, A.; Mckinney, S.; Halevi, B.; Atanassov, P. PGM-Free Fe-NC Catalysts for Oxygen Reduction Reaction: Catalyst Layer Design. *J. Power Sources* **2016**, *326*, 43–49.
70. Santoro, C.; Serov, A.; Villarrubia, C. W. N.; Stariha, S.; Babanova, S.; Artyushkova, K.; Schuler, A. J.; Atanassov, P. High Catalytic Activity and Pollutants Resistivity Using Fe-AAPyr Cathode Catalyst for Microbial Fuel Cell Application. *Sci. Rep.* **2015**, *5*, 16596.
71. Santoro, C.; Kodali, M.; Shamoon, N.; Serov, A.; Soavi, F.; Merino-Jimenez, I.; Gajda, I.; Greenman, J.; Ieropoulos, I.; Atanassov, P. Increased Power Generation in Supercapacitive Microbial Fuel Cell Stack Using FeNC Cathode Catalyst. *J. Power Sources* **2019**, *412*, 416–424.
72. Guo, X.; Jia, J.; Dong, H.; Wang, Q.; Xu, T.; Fu, B.; Ran, R.; Liang, P.; Huang, X.; Zhang, X. Hydrothermal Synthesis of FeMn Bimetallic Nanocatalysts as High-Efficiency Cathode Catalysts for Microbial Fuel Cells. *J. Power Sources* **2019**, *414*, 444–452.
73. Huang, T.; Mao, S.; Zhou, G.; Wen, Z.; Huang, X.; Ci, S.; Chen, J. Hydrothermal Synthesis of Vanadium Nitride and Modulation of Its Catalytic Performance for Oxygen Reduction Reaction. *Nanoscale* **2014**, *6* (16), 9608–9613.
74. Noori, M. T.; Ghangrekar, M.; Mukherjee, C. V2O5 Microflower Decorated Cathode for Enhancing Power Generation in Air-Cathode Microbial Fuel Cell Treating Fish Market Wastewater. *Int. J. Hydrogen Energy* **2016**, *41* (5), 3638–3645.
75. Pan, A.; Wu, H. B.; Yu, L.; Zhu, T.; Lou, X. W. Synthesis of Hierarchical Three-Dimensional Vanadium Oxide Microstructures as High-Capacity Cathode Materials for Lithium-Ion Batteries. *ACS Appl. Mater. Interfaces* **2012**, *4* (8), 3874–3879.
76. Ghoreishi, K. B.; Ghasemi, M.; Rahimnejad, M.; Yarmo, M. A.; Daud, W. R. W.; Asim, N.; Ismail, M. Development and Application of Vanadium Oxide/Polyaniline Composite as a Novel Cathode Catalyst in Microbial Fuel Cell. *Int. J. Energy Res.* **2014**, *38* (1), 70–77.
77. Zhang, F.; Cheng, S.; Pant, D.; Van Bogaert, G.; Logan, B. E. Power Generation Using an Activated Carbon and Metal Mesh Cathode in a Microbial Fuel Cell. *Electrochem. Commun.* **2009**, *11* (11), 2177–2179.
78. Liew, K. B.; Daud, W. R. W.; Ghasemi, M.; Leong, J. X.; Lim, S. S.; Ismail, M. Non-Pt Catalyst as Oxygen Reduction Reaction in Microbial Fuel Cells: A Review. *Int. J. Hydrogen Energy* **2014**, *39* (10), 4870–4883.
79. Noori, M. T.; Bhowmick, G. D.; Tiwari, B. R.; Ghangrekar, M.; Mukhrejee, C. Application of Low-Cost Cu–Sn Bimetal Alloy as Oxygen Reduction Reaction Catalyst for Improving Performance of the Microbial Fuel Cell. *MRS Adv.* **2018**, *3* (13), 663–668.
80. Noori, M. T.; Bhowmick, G.; Tiwari, B.; Ghangrekar, O.; Ghangrekar, M.; Mukherjee, C. Carbon Supported Cu-Sn Bimetallic Alloy as an Excellent Low-Cost Cathode Catalyst for Enhancing Oxygen Reduction Reaction in Microbial Fuel Cell. *J. Electrochem. Soc.* **2018**, *165* (9), F621–F628.
81. Li, M.; Zhang, H.; Xiao, T.; Wang, S.; Zhang, B.; Chen, D.; Su, M.; Tang, J. Low-Cost Biochar Derived from Corncob as Oxygen Reduction Catalyst in Air Cathode Microbial Fuel Cells. *Electrochim. Acta* **2018**, *283*, 780–788.

82. Singh, S.; Verma, N. Graphitic Carbon Micronanofibers Asymmetrically Dispersed with Alumina-Nickel Nanoparticles: A Novel Electrode for Mediatorless Microbial Fuel Cells. *Int. J. Hydrogen Energy* **2015**, *40* (17), 5928–5938.
83. Tang, X.; Ng, H. Y. Cobalt and Nitrogen-Doped Carbon Catalysts for Enhanced Oxygen Reduction and Power Production in Microbial Fuel Cells. *Electrochim. Acta* **2017**, *247*, 193–199.
84. Tang, X.; Li, H.; Du, Z.; Ng, H. Y. Polyaniline and Iron Based Catalysts as Air Cathodes for Enhanced Oxygen Reduction in Microbial Fuel Cells. *RSC Adv.* **2015**, *5* (97), 79348–79354.
85. Harkness, J. K.; Murphy, O. J.; Hitchens, G. D. Enzyme Electrodes Based on Ionomer Films Coated on Electrodes. *J. Electroanal. Chem.* **1993**, *357* (1–2), 261–272.
86. Ivanov, I.; Vidaković-Koch, T.; Sundmacher, K. Recent Advances in Enzymatic Fuel Cells: Experiments and Modeling. *Energies* **2010**, *3* (4), 803–846.
87. Atanassov, P.; Apblett, C.; Banta, S.; Brozik, S.; Barton, S. C.; Cooney, M.; Liaw, B. Y.; Mukerjee, S.; Minteer, S. D. Enzymatic Biofuel Cells. *Interface-Electrochem. Soc.* **2007**, *16* (2), 28–31.
88. Stoica, L.; Dimcheva, N.; Ackermann, Y.; Karnicka, K.; Guschin, D.; Kulesza, P.; Rogalski, J.; Haltrich, D.; Ludwig, R.; Gorton, L. Membrane-Less Biofuel Cell Based on Cellobiose Dehydrogenase (Anode)/Laccase (Cathode) Wired via Specific Os-Redox Polymers. *Fuel Cells* **2009**, *9* (1), 53–62.
89. Rasmussen, M.; Abdellaoui, S.; Minteer, S. D. Enzymatic Biofuel Cells: 30 Years of Critical Advancements. *Biosens. Bioelectron.* **2016**, *76*, 91–102.
90. Stepnicka, P. *Ferrocenes: Ligands, Materials and Biomolecules*; John Wiley & Sons: Hoboken, NJ, 2008.
91. Gorton, L.; Torstensson, A.; Jaegfeldt, H.; Johansson, G. Electrocatalytic Oxidation of Reduced Nicotinamide Coenzymes by Graphite Electrodes Modified with an Adsorbed Phenoxazinium Salt, Meldola Blue. *J. Electroanal. Chem. Interfacial Electrochem.* **1984**, *161* (1), 103–120.
92. Chen, M.; Wei, X.; Qian, H.; Diao, G. Fabrication of GNPs/CDSH-Fc/nafion Modified Electrode for the Detection of Dopamine in the Presence of Ascorbic Acid. *Mater.Sci. Eng: C* **2011**, *31* (7), 1271–1277.
93. Chinnadayyala, S. R.; Kakoti, A.; Santhosh, M.; Goswami, P. A Novel Amperometric Alcohol Biosensor Developed in a 3rd Generation Bioelectrode Platform Using Peroxidase Coupled Ferrocene Activated Alcohol Oxidase as Biorecognition System. *Biosens. Bioelectron.* **2014**, *55*, 120–126.
94. Ghosh, T.; Sarkar, P.; Turner, A. P. A Novel Third Generation Uric Acid Biosensor Using Uricase Electro-Activated with Ferrocene on a Nafion Coated Glassy Carbon Electrode. *Bioelectrochemistry* **2015**, *102*, 1–9.
95. Armstrong, F. A.; Wilson, G. S. Recent Developments in Faradaic Bioelectrochemistry. *Electrochim. Acta* **2000**, *45* (15–16), 2623–2645.
96. Barton, S. C.; Pickard, M.; Vazquez-Duhalt, R.; Heller, A. Electroreduction of O_2 to Water at 0.6 V (SHE) at pH 7 on the 'Wired' *Pleurotus ostreatus* Laccase Cathode. *Biosens. Bioelectron.* **2002**, *17* (11–12), 1071–1074.

97. Heller, A. Electrical Connection of Enzyme Redox Centers to Electrodes. *J. Phys. Chem.* **1992,** *96* (9), 3579–3587.
98. Schuhmann, W.; Ohara, T. J.; Schmidt, H. L.; Heller, A. Electron Transfer Between Glucose Oxidase and Electrodes via Redox Mediators Bound with Flexible Chains to the Enzyme Surface. *J. Am. Chem. Soc.* **1991,** *113* (4), 1394–1397.
99. Dong, S.; Wang, B.; Liu, B. Amperometric Glucose Sensor with Ferrocene as an Electron Transfer Mediator. *Biosens. Bioelectron.* **1992,** *7* (3), 215–222.
100. Vaillancourt, M.; Wei Chen, J.; Fortier, G.; Bélanger, D. Electrochemical and Enzymatic Studies of Electron Transfer Mediation by Ferrocene Derivatives with Nafion-Glucose Oxidase Electrodes. *Electroanalysis* **1999,** *11* (1), 23–31.
101. Dai, H. Carbon Nanotubes: Synthesis, Integration, and Properties. *Acc. Chem. Res.* **2002,** *35* (12), 1035–1044.
102. Meredith, M. T.; Kao, D.-Y.; Hickey, D.; Schmidtke, D. W.; Glatzhofer, D. T. High Current Density Ferrocene-Modified Linear Poly (Ethylenimine) Bioanodes and Their Use in Biofuel Cells. *J. Electrochem. Soc.* **2011,** *158* (2), B166–B174.
103. Smart, S.; Cassady, A.; Lu, G.; Martin, D. The Biocompatibility of Carbon Nanotubes. *Carbon* **2006,** *44* (6), 1034–1047.
104. Tran, T. O.; Lammert, E. G.; Chen, J.; Merchant, S. A.; Brunski, D. B.; Keay, J. C.; Johnson, M. B.; Glatzhofer, D. T.; Schmidtke, D. W. Incorporation of Single-Walled Carbon Nanotubes into Ferrocene-Modified Linear Polyethylenimine Redox Polymer Films. *Langmuir* **2011,** *27* (10), 6201–6210.
105. Yu, E. H.; Wang, X.; Krewer, U.; Li, L.; Scott, K. Direct Oxidation Alkaline Fuel Cells: from Materials to Systems. *Energy Environ. Sci.* **2012,** *5* (2), 5668–5680.
106. Şahin, S.; Wongnate, T.; Chuaboon, L.; Chaiyen, P.; Yu, E. H. Enzymatic Fuel Cells with an Oxygen Resistant Variant Of Pyranose-2-oxidase as Anode Biocatalyst. *Biosens. Bioelectron.* **2018,** *107*, 17–25.
107. Wu, R.; Ma, C.; Zhang, Y. H. P.; Zhu, Z. Complete Oxidation of Xylose for Bioelectricity Generation by Reconstructing a Bacterial Xylose Utilization Pathway in vitro. *ChemCatChem* **2018,** *10* (9), 2030–2035.
108. Zhu, Z.; Ma, C.; Zhang, Y.-H. P. Co-Utilization of Mixed Sugars in an Enzymatic Fuel Cell Based on an in Vitro Enzymatic Pathway. *Electrochim. Acta* **2018,** *263*, 184–191.
109. Mano, N.; Mao, F.; Heller, A. A Miniature Membrane-Less Biofuel Cell Operating at +0.60 V Under Physiological Conditions. *ChemBioChem* **2004,** *5* (12), 1703–1705.
110. Sakai, H.; Nakagawa, T.; Tokita, Y.; Hatazawa, T.; Ikeda, T.; Tsujimura, S.; Kano, K. A High-Power Glucose/Oxygen Biofuel Cell Operating under Quiescent Conditions. *Energy Environ. Sci.* **2009,** *2* (1), 133–138.
111. Li, X.; Zhang, L.; Su, L.; Ohsaka, T.; Mao, L. A Miniature Glucose/O_2 Biofuel Cell with a High Tolerance Against Ascorbic Acid. *Fuel Cells* **2009,** *9* (1), 85–91.
112. Ramanavicius, A.; Ramanaviciene, A. Hemoproteins in Design of Biofuel Cells. *Fuel Cells* **2009,** *9* (1), 25–36.
113. Giroud, F.; Milton, R. D.; Tan, B.-X.; Minteer, S. D. Simplifying Enzymatic Biofuel Cells: Immobilized Naphthoquinone as a Biocathodic Orientational Moiety and Bioanodic Electron Mediator. *ACS Catal.* **2015,** *5* (2), 1240–1244.

114. Steinbusch, K. J.; Hamelers, H. V.; Schaap, J. D.; Kampman, C.; Buisman, C. J. Bioelectrochemical Ethanol Production through Mediated Acetate Reduction by Mixed Cultures. *Environ. Sci. Technol.* **2009**, *44* (1), 513–517.

115. Jia, W.; Jin, C.; Xia, W.; Muhler, M.; Schuhmann, W.; Stoica, L. Glucose Oxidase/ Horseradish Peroxidase Co-Immobilized at a CNT-Modified Graphite Electrode: Towards Potentially Implantable Biocathodes. *Chem: Eur. J.* **2012**, *18* (10), 2783–2786.

116. Elouarzaki, K.; Bourourou, M.; Holzinger, M.; Le Goff, A.; Marks, R. S.; Cosnier, S. Freestanding HRP–GOx Redox Buckypaper as an Oxygen-Reducing Biocathode for Biofuel Cell Applications. *Energy Environ. Sci.* **2015**, *8* (7), 2069–2074.

117. Ramanavicius, A.; Kausaite-Minkstimiene, A.; Morkvenaite-Vilkonciene, I.; Genys, P.; Mikhailova, R.; Semashko, T.; Voronovic, J.; Ramanaviciene, A. Biofuel Cell Based on Glucose Oxidase from *Penicillium funiculosum* 46.1 and Horseradish Peroxidase. *Chem. Eng. J.* **2015**, *264*, 165–173.

118. Cardeña, R.; Cercado, B.; Buitrón, G. Microbial Electrolysis Cell for Biohydrogen Production. *Biohydrogen*; Elsevier: Amsterdam, 2019; pp 159–185.

119. Badwal, S. P.; Giddey, S. S.; Munnings, C.; Bhatt, A. I.; Hollenkamp, A. F. Emerging Electrochemical Energy Conversion and Storage Technologies. *Front. Chem.* **2014**, *2*, 79.

120. Wang, L.; Singh, L.; Liu, H. Revealing the Impact of Hydrogen Production-Consumption Loop Against Efficient Hydrogen Recovery in Single Chamber Microbial Electrolysis Cells (MECs). *Int. J. Hydrogen Energy* **2018**, *43* (29), 13064–13071.

121. Anusha, G.; Noori, M. T.; Ghangrekar, M. Application of Silver-Tin Dioxide Composite Cathode Catalyst for Enhancing Performance of Microbial Desalination Cell. *Mater. Sci. Energy Technol.* **2018**, *1* (2), 188–195.

122. Logan, B. E.; Call, D.; Cheng, S.; Hamelers, H. V.; Sleutels, T. H.; Jeremiasse, A. W.; Rozendal, R. A. Microbial Electrolysis Cells for High Yield Hydrogen Gas Production from Organic Matter. *Environ. Sci. Technol.* **2008**, *42* (23), 8630–8640.

123. Pasupuleti, S. B.; Srikanth, S.; Mohan, S. V.; Pant, D. Development of Exoelectrogenic Bioanode and Study on Feasibility of Hydrogen Production Using Abiotic VITO-CoRE™ and VITO-CASE™ Electrodes in a Single Chamber Microbial Electrolysis Cell (MEC) at Low Current Densities. *Bioresour. Technol.* **2015**, *195*, 131–138.

124. Walcarius, A.; Minteer, S. D.; Wang, J.; Lin, Y.; Merkoçi, A. Nanomaterials for Bio-Functionalized Electrodes: Recent Trends. *J. Mater. Chem. B* **2013**, *1* (38), 4878–4908.

125. Zhang, X.; Pant, D.; Zhang, F.; Liu, J.; He, W.; Logan, B. E. Long-Term Performance of Chemically and Physically Modified Activated Carbons in Air Cathodes of Microbial Fuel Cells. *ChemElectroChem* **2014**, *1* (11), 1859–1866.

126. Jeremiasse, A. W.; Bergsma, J.; Kleijn, J. M.; Saakes, M.; Buisman, C. J.; Stuart, M. C.; Hamelers, H. V. Performance of Metal Alloys as Hydrogen Evolution Reaction Catalysts in a Microbial Electrolysis Cell. *Int. J. Hydrogen Energy* **2011**, *36* (17), 10482–10489.

127. Tokash, J.; Logan, B. Electrochemical Evaluation of a Molybdenum Disulfide Catalyst for the Hydrogen Evolution Reaction under Solution Conditions Applicable to Microbial Electrolysis Cells. *Int. J. Hydrogen Energy* **2010**, *36*, 9439.

128. Krishnan, S.; Din, M. F. M.; Taib, S. M.; Nasrullah, M.; Sakinah, M.; Wahid, Z. A.; Kamyab, H.; Chelliapan, S.; Rezania, S.; Singh, L. Accelerated Two-Stage Bioprocess for Hydrogen and

Methane Production from Palm Oil Mill Effluent Using Continuous Stirred Tank Reactor and Microbial Electrolysis Cell. *J. Cleaner Prod.* **2019**, *229*, 84–93.
129. Selembo, P. A.; Merrill, M. D.; Logan, B. E. Hydrogen Production with Nickel Powder Cathode Catalysts in Microbial Electrolysis Cells. *Int. J. Hydrogen Energy* **2010**, *35* (2), 428–437.
130. Selembo, P. A.; Merrill, M. D.; Logan, B. E. The Use of Stainless Steel and Nickel Alloys as Low-Cost Cathodes in Microbial Electrolysis Cells. *J. Power Sources* **2009**, *190* (2), 271–278.
131. Hu, H.; Fan, Y.; Liu, H. Hydrogen Production in Single-Chamber Tubular Microbial Electrolysis Cells Using Non-Precious-Metal Catalysts. *Int. J. Hydrogen Energy* **2009**, *34* (20), 8535–8542.
132. Harnisch, F.; Schröder, U.; Quaas, M.; Scholz, F. Electrocatalytic and Corrosion Behaviour of Tungsten Carbide in Near-Neutral pH Electrolytes. *Appl. Catal. B: Environ.* **2009**, *87* (1–2), 63–69.
133. Cai, W.; Liu, W.; Sun, H.; Li, J.; Yang, L.; Liu, M.; Zhao, S.; Wang, A. Ni 5 P 4-NiP 2 Nanosheet Matrix Enhances Electron-Transfer Kinetics for Hydrogen Recovery in Microbial Electrolysis Cells. *Appl. Energy* **2018**, *209*, 56–64.
134. Xiao, L.; Wen, Z.; Ci, S.; Chen, J.; He, Z. Carbon/Iron-Based Nanorod Catalysts for Hydrogen Production in Microbial Electrolysis Cells. *Nano Energy* **2012**, *1* (5), 751–756.
135. Pradhan, H.; Ghangrekar, M. Multi-Chamber Microbial Desalination Cell for Improved Organic Matter and Dissolved Solids Removal from Wastewater. *Water Sci. Technol.* **2014**, *70* (12), 1948–1954.
136. Utami, T. S.; Arbianti, R.; Manaf, B. N. Sea Water Desalination Using *Debaryomyces hansenii* with Microbial Desalination Cell Technology. *Int. J. Technol.* **2015**, *6* (7), 1094–1100.
137. Sevda, S.; Yuan, H.; He, Z.; Abu-Reesh, I. M. Microbial Desalination Cells as a Versatile Technology: Functions, Optimization and Prospective. *Desalination* **2015**, *371*, 9–17.
138. Sophia, A. C.; Bhalambaal, V.; Lima, E. C.; Thirunavoukkarasu, M. Microbial Desalination Cell Technology: Contribution to Sustainable Waste Water Treatment Process, Current Status and Future Applications. *J. Environ. Chem. Eng.* **2016**, *4* (3), 3468–3478.
139. Zhang, B.; He, Z. Energy Production, Use and Saving in a Bioelectrochemical Desalination System. *RSC Adv.* **2012**, *2* (28), 10673–10679.
140. Morel, A.; Zuo, K.; Xia, X.; Wei, J.; Luo, X.; Liang, P.; Huang, X. Microbial Desalination Cells Packed with Ion-Exchange Resin to Enhance Water Desalination Rate. *Bioresour. Technol.* **2012**, *118*, 43–48.
141. Zhang, Y.; Angelidaki, I. A New Method for in situ Nitrate Removal from Groundwater Using Submerged Microbial Desalination–Denitrification Cell (SMDDC). *Water Res.* **2013**, *47* (5), 1827–1836.
142. Wen, Q.; Zhang, H.; Chen, Z.; Li, Y.; Nan, J.; Feng, Y. Using Bacterial Catalyst in the Cathode of Microbial Desalination Cell to Improve Wastewater Treatment and Desalination. *Bioresour. Technol.* **2012**, *125*, 108–113.
143. Mehanna, M.; Saito, T.; Yan, J.; Hickner, M.; Cao, X.; Huang, X.; Logan, B. E. Using Microbial Desalination Cells to Reduce Water Salinity Prior to Reverse Osmosis. *Energy Environ. Sci.* **2010**, *3* (8), 1114–1120.

144. Santoro, C.; Talarposhti, M. R.; Kodali, M.; Gokhale, R.; Serov, A.; Merino-Jimenez, I.; Ieropoulos, I.; Atanassov, P. Microbial Desalination Cells with Efficient Platinum-Group-Metal-Free Cathode Catalysts. *ChemElectroChem* **2017**, *4* (12), 3322–3330.
145. Al-Mamun, A.; Baawain, M. S.; Dhar, B. R.; Kim, I. S. Improved Recovery of Bioenergy and Osmotic Water in an Osmotic Microbial Fuel Cell Using Micro-Diffuser Assisted Marine Aerobic Biofilm on Cathode. *Biochem. Eng. J.* **2017**, *128*, 235–242.
146. Kokabian, B.; Gude, V. G.; Smith, R.; Brooks, J. P. Evaluation of Anammox Biocathode in Microbial Desalination and Wastewater Treatment. *Chem. Eng. J.* **2018**, *342*, 410–419.
147. Musabyimana, M. *Deammonification Process Kinetics and Inhibition Evaluation*. Ph.D. Dissertation, Virginia Tech, Blacksburg, VA, 2008.
148. Gude, V.; Kokabian, B.; Gadhamshetty, V. Beneficial Bioelectrochemical Systems for Energy, Water, and Biomass Production. *J. Microb. Biochem. Technol.* **2013**, *6*, 2.
149. Hamelers, H. V.; Ter Heijne, A.; Sleutels, T. H.; Jeremiasse, A. W.; Strik, D. P.; Buisman, C. J. New Applications and Performance of Bioelectrochemical Systems. *Appl. Microbiol. Biotechnol.* **2010**, *85* (6), 1673–1685.
150. Strik, D. P.; Hamelers, H.; Snel, J. F.; Buisman, C. J. Green Electricity Production with Living Plants and Bacteria in a Fuel Cell. *Int. J. Energy Res.* **2008**, *32* (9), 870–876.
151. Schamphelaire, L. D.; Bossche, L. V. D.; Dang, H. S.; Höfte, M.; Boon, N.; Rabaey, K.; Verstraete, W. Microbial Fuel Cells Generating Electricity from Rhizodeposits of Rice Plants. *Environ. Sci. Technol.* **2008**, *42* (8), 3053–3058.
152. Kaku, N.; Yonezawa, N.; Kodama, Y.; Watanabe, K. Plant/Microbe Cooperation for Electricity Generation in a Rice Paddy Field. *Appl. Microbiol. Biotechnol.* **2008**, *79* (1), 43–49.
153. Timmers, R. A.; Strik, D. P.; Hamelers, H. V.; Buisman, C. J. Long-Term Performance of a Plant Microbial Fuel Cell with *Spartina anglica*. *Appl. Microbiol. Biotechnol.* **2010**, *86* (3), 973–981.
154. Helder, M.; Strik, D.; Hamelers, H.; Kuhn, A.; Blok, C.; Buisman, C. Concurrent Bio-Electricity and Biomass Production in Three Plant-Microbial Fuel Cells Using *Spartina anglica, Arundinella anomala* and *Arundo donax*. *Bioresour. Technol.* **2010**, *101* (10), 3541–3547.
155. Takanezawa, K.; Nishio, K.; Kato, S.; Hashimoto, K.; Watanabe, K. Factors Affecting Electric Output from Rice-Paddy Microbial Fuel Cells. *Biosci. Biotechnol. Biochem.* **2010**, *74* (6), 1271–1273.
156. De Schamphelaire, L.; Cabezas, A.; Marzorati, M.; Friedrich, M. W.; Boon, N.; Verstraete, W. Microbial Community Analysis of Anodes from Sediment Microbial Fuel Cells Powered by Rhizodeposits of Living Rice Plants. *Appl. Environ. Microbiol.* **2010**, *76* (6), 2002–2008.
157. Strik, D. P.; Timmers, R. A.; Helder, M.; Steinbusch, K. J.; Hamelers, H. V.; Buisman, C. J. Microbial Solar Cells: Applying Photosynthetic and Electrochemically Active Organisms. *Trends Biotechnol.* **2011**, *29* (1), 41–49.
158. Malik, S.; Drott, E.; Grisdela, P.; Lee, J.; Lee, C.; Lowy, D. A.; Gray, S.; Tender, L. M. A Self-Assembling Self-Repairing Microbial Photoelectrochemical Solar Cell. *Energy Environ. Sci.* **2009**, *2* (3), 292–298.
159. Berk, R. S.; Canfield, J. H. Bioelectrochemical Energy Conversion. *Appl. Microbiol.* **1964**, *12* (1), 10–12.

160. Strycharz-Glaven, S. M.; Glaven, R. H.; Wang, Z.; Zhou, J.; Vora, G. J.; Tender, L. M. Electrochemical Investigation of a Microbial Solar Cell Reveals a Non-Photosynthetic Biocathode Catalyst. *Appl. Environ. Microbiol.* **2013**, *79* (13), 3933–3942.
161. Zou, Y.; Pisciotta, J.; Billmyre, R. B.; Baskakov, I. V. Photosynthetic Microbial Fuel Cells with Positive Light Response. *Biotechnol. Bioeng.* **2009**, *104* (5), 939–946.
162. Ayyaru, S.; Mahalingam, S.; Ahn, Y.-H. A Non-Noble V2O5 Nanorods as an Alternative Cathode Catalyst for Microbial Fuel Cell Applications. *Int. J. Hydrogen Energy* **2019**, *44* (10), 4974–4984.
163. Strycharz-Glaven, S. M.; Glaven, R. H.; Wang, Z.; Zhou, J.; Vora, G. J.; Tender, L. M. Electrochemical Investigation of a Microbial Solar Cell Reveals a Nonphotosynthetic Biocathode Catalyst. *Appl. Environ. Microbiol.* **2013**, *79* (13), 3933–3942.
164. Kiliç, M.; Korkut, S.; Hazer, B. Enzymatic Fuel Cells for Electric Power Generation from Domestic Wastewater. *WIT Transactions Ecol. Environ.* **2014**, *181*, 213–223.

Chapter 4

Oxygen Reduction Reaction Electrocatalysts for Microbial Fuel Cells

Miao Gao, Jia-Yuan Lu, and Wen-Wei Li*

CAS Key Laboratory of Urban Pollutant Conversion, Department of Applied Chemistry, University of Science & Technology of China, Hefei 230026, China
USTC-CityU Joint Advanced Research Center, Suzhou 215123, China

***E-mail: wwli@ustc.edu.cn.**

Microbial fuel cells (MFC) are considered to be a promising technology for sustainable wastewater treatment and environmental remediation, but the poor cathode performance usually restricts their application. This chapter summarizes recent advances in the fabrication and use of oxygen reduction reaction (ORR) electrocatalysts for an MFC cathode. The factors governing the activity, selectivity, and stability of various ORR electrocatalysts, ranging from metal-based materials and metal-free carbon materials to biocatalysts, are introduced, and the corresponding regulation strategies are discussed. The current knowledge gaps on the catalyst structure–function correlations are highlighted. Lastly, the future challenges and efforts needed for development of efficient, robust, and low-cost ORR electrocatalysts to suit MFC applications are discussed.

Introduction

Microbial fuel cells (MFC) are an emerging ecofriendly biotechnology that directly couples wastewater treatment with electricity generation (*1*, *2*). A typical MFC consists of a microbe-colonized anode for anaerobic oxidation of organic matter in wastewater and a cathode for reducing various electron acceptors, mostly oxygen due to its high redox potential and abundant availability. The oxygen reduction reaction (ORR) involves multistep electron transfer processes (eqs 1–6) that can be divided into four-electron and two-electron pathways.

Two-electron pathway:

$$O_{2(g)} + \left[H^+ + e^-\right] \rightarrow {}^*OOH_{(aq)} \tag{1}$$

$${}^*OOH_{(aq)} + \left[H^+ + e^-\right] \rightarrow H_2O_{2(aq)} \tag{2}$$

© 2020 American Chemical Society

Four-electron pathway:

$$O_{2(g)} + \left[H^+ + e^-\right] \rightarrow {}^*OOH_{(aq)} \quad (3)$$

$$^*OOH_{(aq)} + \left[H^+ + e^-\right] \rightarrow {}^*O_{(aq)} + H_2O_{(aq)} \quad (4)$$

$$^*O_{(aq)} + \left[H^+ + e^-\right] \rightarrow {}^*OH_{(aq)} \quad (5)$$

$$^*OH_{(aq)} + \left[H^+ + e^-\right] \rightarrow {}^*H_2O_{(aq)} \quad (6)$$

The cathodic ORR involves complicated interfacial transfer of mass and electrons and typically takes place under ambient pH and temperature conditions to suit MFC applications. This generally results in sluggish reaction kinetics and a high cathodic overpotential (3), which constrains the MFC electrochemical performance. Another challenge is the use of noble metal catalysts, which are costly and unstable during long-term operation in a complicated wastewater matrix. The need to address these challenges has spurred intensive research interest in development of more efficient, robust, and cost-effective ORR electrocatalysts and cathode materials over the past few years.

In this chapter, the fundamentals and criteria for selecting and designing ORR electrocatalysts for MFC applications are introduced, and the remaining challenges and the future research needed to advance this technology are discussed. Future prospects for MFC electrocatalyst development are analyzed in terms of catalyst structure, morphology, and electrode configuration.

Fundamentals of ORR Electrocatalysts

Desired Features of ORR Electrocatalysts

An electrocatalyst should ideally possess high conductivity and high specific surface area, exhibit excellent durability, and be economically affordable. There are extra requirements for ORR electrocatalysts to be used in MFCs, especially those with a single-chamber configuration where the cathode and a bioanode coexist in the same electrolyte. First, the ORR catalyst should be electrochemically active in a neutral pH environment, which is essential to sustaining the growth and metabolic activities of the anodic microorganisms. Moreover, it should be resistant to electrode deactivation, because many water contaminants such as sulfide and chloride may deposit on the electrode surface and decrease the catalytic activity during long-term operation. All these factors need to be considered in designing suitable ORR electrocatalysts for MFC applications.

In a typical MFC, the cathodic reaction is dominated by ORR in a four-electron pathway. To improve the oxygen utilization efficiency, cathode materials that are conducive to oxygen adsorption and favor lower dissociation energy of O–O bonds are typically used (4). Compared with four-electron ORR, the two-electron pathway is less efficient in terms of power output by MFC, but it produces H_2O_2 as a valuable byproduct. Such electrochemically generated H_2O_2 can be used in situ for oxidative degradation of pollutants by creating an MFC-electro-Fenton system (Figure 1), with much lower power consumption and operation cost than a traditional electro-Fenton system (5). However, the usually low yield of H_2O_2, due to kinetic and thermodynamic restrictions and poor reaction selectivity of the electrocatalysts, is the main hurdle for its practical application.

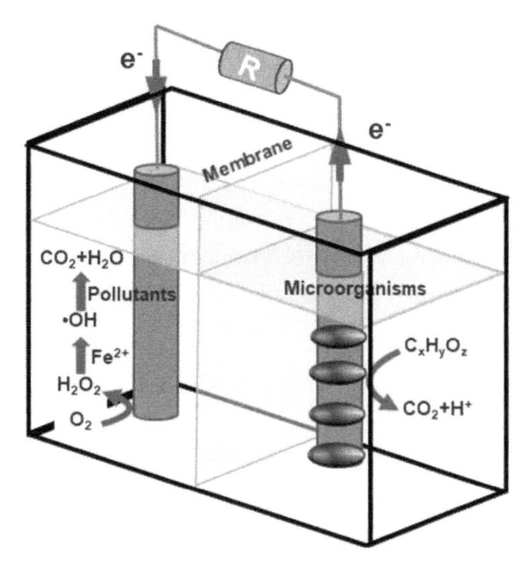

Figure 1. Schematic of MFC-electro-Fenton system that combines a bioanode and a H_2O_2-producing ORR cathode. Reproduced with permission from reference (5). Copyright 2018 Elsevier.

Electron Transfer Number of Cathodic ORR

In ORR, the electron transfer number per O_2 molecule reduced can be measured by a rotating disk electrode (RDE). With this technique, the ORR kinetic rates can be estimated based on the Koutecky–Levich equations (*6*):

$$\frac{1}{J} = \frac{1}{J_K} + \frac{1}{Bw^{\frac{1}{2}}} \tag{7}$$

$$B = 0.62nFv^{-\frac{1}{6}}C_{O_2}D_{O_2}^{\frac{2}{3}} \tag{8}$$

where J and J_K are the measured current density and kinetic current density (mA cm^{-2}), respectively, w is the angular velocity (rad s^{-1}), n is the electron transfer number per O_2 molecule, F is the Faraday constant (C mol^{-1}), v is the kinematic viscosity (cm^2 s^{-1}), C_{O2} is the bulk concentration of O_2 (mol cm^{-3}), and D_{O2} is the diffusion coefficient of O_2 in the electrolyte (cm^2 s^{-1}). Based on the RDE measurement, the electron transfer number for a specific electrocatalyst during ORR can be calculated. The H_2O_2 selectivity during ORR can also be evaluated by the rotating ring disk electrode (RRDE) method according to the following equation (*7*):

$$H_2O_2\% = \frac{200I_r}{I_r + NI_d} \tag{9}$$

where I_r is the ring current, I_d is the disk current, and N is the collection efficiency (0.24–0.5). Tuning the two-electron or four-electron selectivity of ORR electrocatalysts is important for realizing different applications of MFC.

ORR Selectivity

Noble Metal Materials

The ORR selectivity is a vital factor to be considered in the selection and design of an MFC cathodic catalyst. Noble metal catalysts were commonly used in earlier MFC studies. Many noble metals (e.g., Pt, Pd, and Au) have superior electron-donating ability to favor the dissolution of O–O bonds, thereby preferring four-electron ORR. Notably, the reaction selectivity can be tuned by introducing species that are catalytically less active. For example, the presence of an inactive Hg heteroatom was found to isolate the active sites in noble metals and optimize the binding energy of intermediate *OOH, resulting in high selectivity towards H_2O_2 generation even up to 100% (*8*, *9*). Choi et al. (*10*) suggest that the H_2O_2 selectivity is highly associated with the type of O_2 molecule adsorption on electrocatalysts, that is, adsorption by "side-on" or "end-on" configurations (Figure 2). In general, the "end-on" configuration favors the two-electron pathway, whereas the "side-on" configuration is thermodynamically more favorable for dissociation of HO–O and hence prefers the four-electron pathway. Therefore, strategies that reduce the "side-on" adsorption of the O_2 molecule (e.g., coating with an amorphous carbon layer) could be used to tune the ORR selectivity of Pt toward the two-electron pathway.

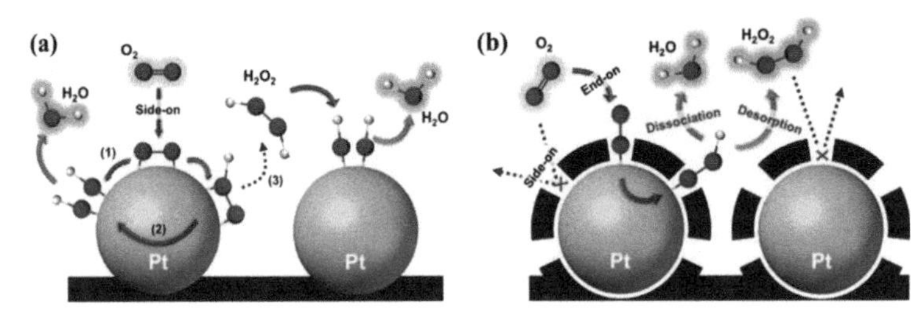

Figure 2. ORR pathway on (a) pristine Pt surface and (b) carbon-coated Pt surface. Reproduced with permission from reference (10). Copyright 2014 American Chemical Society.

Transition Metal Materials

Nanoscale transition metals are considered as promising low-cost alternatives to noble metal electrocatalysts. Many transition metal materials also possess high ORR activity, and the reaction selectivity can be readily tuned. For example, zero-valent ion nanoparticles (nZVI) and $Fe@Fe_2O_3$ nanowires show good four-electron ORR selectivity (*11*), with an electron transfer number approaching 4. However, the formation of an iron phosphate coating (*12*) significantly suppressed the four-electron pathway and resulted in an electron transfer number of only 1.13 (Figure 3), indicating the dominance of one- and two-electron pathways (*13*).

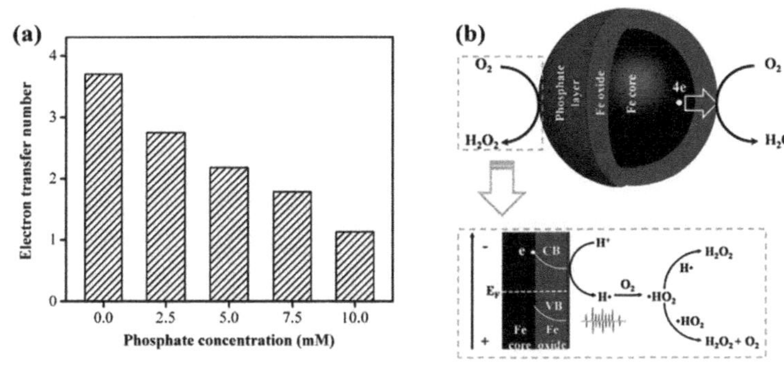

Figure 3. (a) ORR electron transfer number of Fe@Fe_2O_3 in the presence of different phosphate concentrations; (b) schematic illustration for enhanced H_2O_2 generation in the presence of phosphate. Reproduced with permission from reference (13). Copyright 2017 American Chemical Society.

Carbon Materials

With the rapid advances in carbon nanomaterials, many metal-free materials also exhibit attractive ORR activity. Unlike metal-based materials, most carbon materials tend to catalyze H_2O_2 generation. For example, hierarchically porous carbon materials with sp^3-C and defects showed up to 95.0% H_2O_2 selectivity (*14*). In addition, the H_2O_2 selectivity of various carbon-based ORR catalysts was also strongly influenced by the zeta potential and nitrogen doping: a positive zeta potential and low nitrogen doping of the electrode surface were beneficial for H_2O_2 production due to easy protonation, whereas a high nitrogen content was detrimental for H_2O_2 generation due to increased active surface site density (*15*). The oxygen functional groups of carbon materials critically determine their ORR activity and selectivity, because the C atoms adjacent to these functional groups (e.g., the –COOH in the armchair edge and the C–O–C in the basal plane of the graphene) typically serve as the active sites (*7*). The ORR selectivity is also affected by the electrocatalyst loading amount. A smaller loading is more preferred for H_2O_2 generation because the low density of active sites restricts further reduction of H_2O_2 to H_2O (*16*).

Metal-Based Electrocatalysts

In accordance with the above selection criteria and catalyst design principles, various ORR electrocatalysts for MFC applications have been developed (*17*). In particular, metal-based electrocatalysts have found widespread application in MFCs due to generally higher activity and application maturity than nonmetal materials.

Noble Metal Based Electrocatalysts

Pt is the earliest adopted and still prevailing ORR electrocatalyst so far. A Pt-loaded carbon electrode has been used as MFC cathode to efficiently treat acid elutriation effluents of piggery waste with simultaneous electric power generation (Figure 4) of up to 1553 mW cm^{-2} (*18*). However, despite the high catalytic activity, the high cost and low abundance of Pt restrict its practical applications (*19*). Combining Pt with non-noble metals (e.g., V, Cr, Mn, Fe, Co, Cu, and Ni) offers an effective strategy to decrease the amount of Pt used without sacrificing the catalytic performance. Such bimetal or multimetal alloys give the non-noble metals dissimilar electronic character, enable a synergy between the different metals that significantly improves the electron and mass transfer (*20–25*). For example, alloying Pt with Ni could decrease the particle size and create more active sites for oxygen adsorption and reduction, thus enabling much higher power output from the MFC (0.637 W m^{-2}) than with a Pt cathode (0.180 W m^{-2}) (*26*). Apart from the electrocatalyst components, the anions in supporting electrolytes also affect the ORR catalytic performance. This explains the anion-dependent ORR activities of different Pt-based catalysts: Pt_3Ni > Pt_3Co > Pt in H_2SO_4 solution and Pt skin > Pt_3Co > Pt_3Ni > Pt in $HClO_4$, where the coordination environment was altered by different anions (*27*).

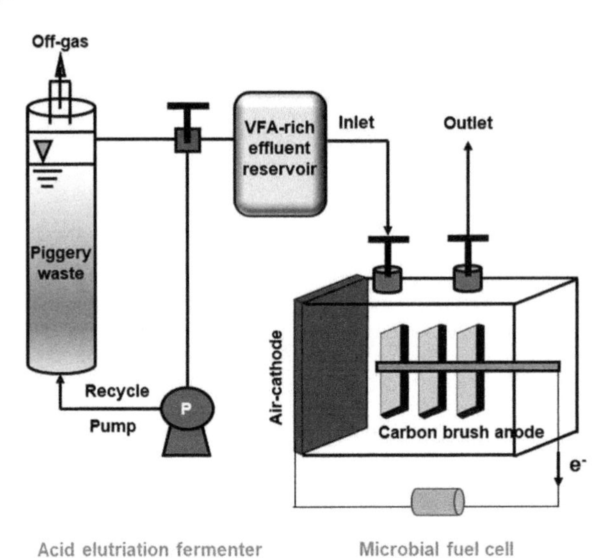

Figure 4. Schematic diagram of an MFC coupled with an acid elutriation fermenter. Reproduced with permission from reference (18). Copyright 2017 Elsevier.

Other noble metal electrocatalysts such as Au, Ag, and Pd can also serve as effective ORR catalysts. A superior activity and four-electron ORR selectivity of ultrasmall gold nanomolecules (AuNMs) (<2 nm diameter and consisting of 36 atoms of Au) have been recently reported, ascribed to the high surface density of Au with low coordination number that facilitates interfacial interactions (*28*). The highly active Ag and Pd are also promising electrocatalysts for MFC (*29–31*). However, the ORR performance of these elemental noble metals is restricted by the initial step of O–O bond cleavage being thermodynamically unfavorable. Such a drawback can be overcome by alloying them with other metals (*32, 33*). For example, by introducing tungsten carbide to strengthen the dissociation of the O–O bond, a silver–tungsten carbide nanohybrid catalyst exhibited ORR activity comparable to commercial Pt/C in MFC applications (*33*). Similar synergy has also been shown by a Cu-Au electrode (*34*).

Non-Noble Metal Electrocatalysts

With increasing recognition of the potential of non-noble metal materials, many researchers have started to seek options that fully abandon the use of noble metals (*35, 36*). Plentiful evidence has shown that properly designed non-noble metal materials can achieve ORR activity and long-term durability comparable to, or even better than, Pt in MFC applications. In this respect, many transition metal oxides such as MnO_2 (*37, 38*), Mn_3O_4 (*39*), Co_3O_4 (*40, 41*), V_2O_5 (*42*), Fe_3O_4 (*43*), and

CeO_2 (*44*) have been widely used. Manganese oxides (MnO_x) are promising ORR catalysts for MFC due to their environmental benignity, low cost, and high chemical stability, but the relatively low activity presents a major restriction. Recent studies suggest that the catalytic activity of MnO_x can be substantially boosted by introducing oxygen vacancies or doping other metals, which alters the electronic structure and electron transfer behaviors. For example, oxygen-deficient $Cu_{1.5}Mn_{1.5}O_4$ showed significantly enhanced electron enrichment and transfer on the catalyst surface and exhibited a superior catalytic activity exceeding commercial Pt/C electrocatalyst in MFC operation (*45*). Similar synergy has also been shown by bimetal catalysts such as $MnCo_2O_4$/C and Co_3O_4@$MnCo_2O_{4.5}$ (*46, 47*). Interestingly, a simple physical mixture of several active ingredients such as graphite, γ-MnO_2, and MoS_2 could also to a certain extent boost the catalytic performance of such non-noble catalysts (*48*).

Another attractive feature of transition metals is the ability to activate H_2O_2 in situ and generate hydroxyl radicals, a class of highly oxidative species that can destroy various organic pollutants (*49*). As a special case, nickel foam could act as a cathode electrocatalyst to produce H_2O_2 at a high rate of 0.65 mg L^{-1} min^{-1}. Nickel foam displayed better two-electron catalytic activity than carbon fiber due to its stronger ability to activate molecular oxygen to produce O_2^- and H_2O_2. The H_2O_2 could be subsequently decomposed to hydroxyl radicals in the presence of Ni(II) to favor efficient pollutant degradation (*50*).

Carbon-Supported Metal-Based Electrocatalysts

Despite the superior catalytic activity of metal-based materials, especially at the nanoscale, they are prone to aggregate and suffer from a decrease in activity during repeated use. A common solution is to support such metal nanoparticles on porous-structured carbon materials (*51*), which not only favors catalyst dispersion and exposure of more active sites but also facilitates electrolyte diffusion and charge transfer (*52*). For instance, spinel nano-Co_3O_4 doped activated carbon exhibited much lower internal resistance than a commercial Co_3O_4 electrode, accomplishing 41.2% higher power density output (1500 mW m^{-2}) in MFC (*53*). A nitrogen-doped activated carbon support has also been used to improve the ORR catalytic activity of spinel $CoFe_2O_4$, with a maximum power density of 1770.8 mW m^{-2} achieved in MFC (*54*). In addition, a carbon nanotube (CNT)/Pt composite exhibited up to 32.2% higher ORR activity than pristine Pt (*55*), and the electron transfer on the surface of MnO_x could be significantly accelerated by CNTs (*56, 57*). The Co_3O_4 grown on Fe-encapsulated graphite layers with rich C_2N functional groups also showed catalytic activity comparable to Pt/C (*58*).

Besides supporting metal oxides on porous carbon materials, metal–carbon composites can also be directly derived from metal–organic frameworks (MOF) through carbonization. Cobalt nanoparticle/nitrogen-doped CNTs were prepared from MOF precursors and used as an efficient ORR electrocatalyst in MFC, achieving a maximum power output of 2252 mW m^{-2} (*59*). Furthermore, Ni, Co, and N codoped carbon materials prepared from Co-MOF also exhibited a remarkably high ORR activity and excellent stability, enabling a power density of 4335.6 mW m^{-2} when serving as an MFC cathode (*60*). Fe-MOF (e.g., ZIF-8) has also been used to obtain Fe, N, and S codoped carbon matrix–CNT composites, which exhibited a half-wave potential of 0.91 V versus a standard hydrogen electrode for ORR under alkaline conditions, comparable to commercial Pt/C (*61*).

Interestingly, a metal–carbon composite with metal nanoparticles dispersed in porous biochar can also be obtained through pyrolysis of waste biomass, making the fabrication process more sustainable. Nitrogen-doped, hierarchically porous carbon with dispersed Pt nanoparticles and a large, electrochemically active surface area was derived from pig bones and, when functioning as ORR catalyst, exhibited and outstanding catalytic performance comparable to Pt/carbon black (*62*). Bacterial cells with biogenic Pd nanoparticles have also been carbonized to obtain heteroatom-doped, Pd nanoparticle-embedded porous carbon (Figure 5), which as an ORR electrocatalyst displayed 2.2 times higher specific mass catalytic activity and better stability than Pt/C (*63*).

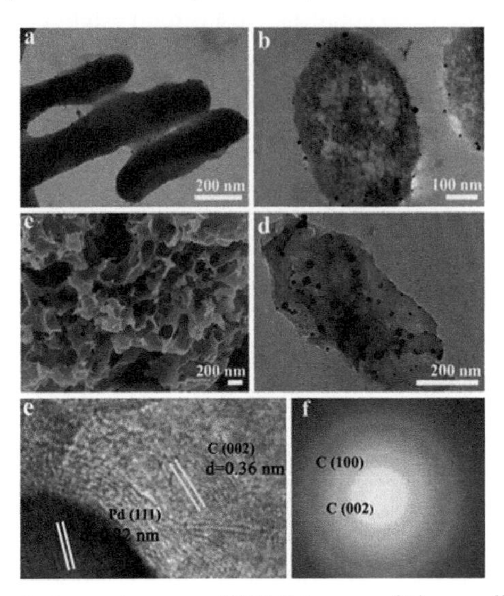

Figure 5. (a, b) Transmission electron microscopy (TEM) images of Shewanella oneidensis MR-1 cells with biogenic Pd; (c) scanning electron microscopy, (d) TEM, (e) HRTEM, and (f) Selected area electron diffraction images of the heteroatom-doped-carbon-Pd. Reproduced with permission from reference (63). Copyright 2015 Elsevier.

Carbon-supported metal-based materials can be used to generate and activate H_2O_2 in situ, enabling a heterogeneous electro-Fenton system for efficient degradation of pollutants. Bulk continuous iron-copper-carbon, with ultradispersed metallic Fe-Cu nanoparticles in the CO_2/N_2 activated carbon matrix, was used as an electro-Fenton cathode for efficient pollutant degradation over a wide pH range of 3–9 (*64*). Other Fe-containing carbon materials such as FeOOH/CNTs and Fe@Fe_2O_3/graphite have also been used as efficient cathode electrocatalysts in an MFC-electro-Fenton system (*65*, *66*). An iron–carbon composite synthesized from iron-rich sludge has been shown to efficiently catalyze H_2O_2 production and pollutant degradation over a broad pH range in electro-Fenton systems (*67*).

Metal-Nitrogen-Carbon (M-N-C) catalysts

In a conventional metal–carbon composite, metal components are typically loaded in a carbon matrix through simple physical interaction. Recently, a unique category of ORR electrocatalysts featuring coordination between metal ions and N-doped carbon through forming metal–nitrogen (M-N) bonds has emerged (commonly referred to as M-N-C materials). For example, with an M-

N-C cathode consisting of Co and Fe-coordinating N-doped carbon, an MFC produced two times higher power density than that with commercial Pt/C and exhibited good stability (i.e., the activity was not influenced by the crossover effect of organics) (*68*). Anchoring square-like cobalt oxide onto N-doped graphene (NG/Co-NS) also favored a more positive onset potential and significantly raised the limiting diffusion current, achieving 24.9% higher power density than Pt/C in MFC (*69*). The ORR pathways of M-N-C can be tuned by altering the composition and surface functional groups. A carbon-based material with abundant Co-N_x-C sites and oxygen functional groups exhibited excellent catalytic performance for electrosynthesis of H_2O_2, with H_2O_2 selectivity as high as 80% (*70*). Notably, although M-N-C materials exhibited high catalytic activity with M-N as active sites, the insufficient stability in acidic media might limit their actual use in electro-Fenton applications.

Metal-Free Electrocatalysts

One common challenge for most metal-based electrocatalysts is the leaching of metal ions during the electrochemical process, which results in activity loss and arouses environmental contamination concerns. These drawbacks can be circumvented by using metal-free electrocatalysts, including various carbon-based materials and even biocatalysts.

CNTs and Nanofibers

CNTs and carbon nanofibers (CNFs, whose aspect ratio is greater than 100) (*71*) are attractive electrocatalysts because of their superior electronic properties, large specific area, and high tensile strength (*72*). In particular, with simple treatment such as chemical activation by KOH to improve the specific surface area and porous structure, CNFs enabled a power density of 61.3 mW m^{-2} in MFC, and the catalyst cost was only 37.7% of the investment of Pt (*73*).

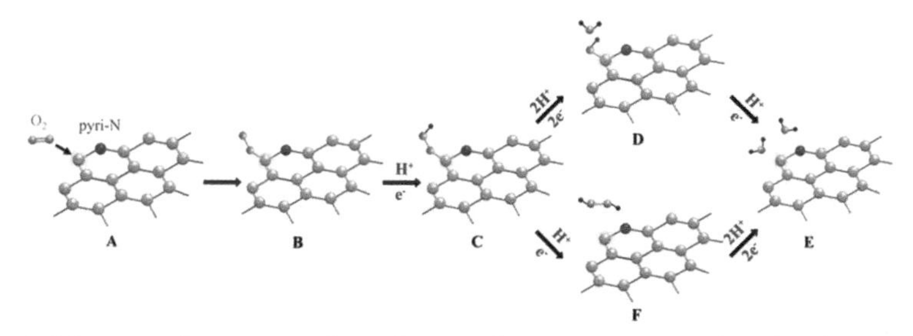

Figure 6. Schematic electron transfer pathways of ORR on nitrogen-doped carbon materials.

To further improve the ORR activity, CNTs and CNFs can be modified by heteroatom doping or combination with other active carbon materials (*74*). For example, vertically aligned nitrogen-doped CNTs exhibited a much higher ORR catalytic activity, long-term operation stability, and tolerance to crossover effect than Pt in an alkaline environment. This discovery has led to development of various N-doped carbon electrocatalysts (*75*), featuring high ORR activity at neutral to basic pH (*76*). In such electrocatalysts, the nitrogen doping alters the electronic structure and electron-donor property of carbon materials, making the C adjacent to N atoms more positively charged to favor adsorption and dissolution of O_2 molecules (Figure 6) (*17, 77*). Generally, four types of nitrogen species have been identified in N-doped carbon materials: pyridinic-N, pyrrolic-N, graphitic-N, and pyridinic-N-oxide (*78*). Although tremendous experimental and computational efforts have been devoted to identifying the active sites of such N-doped carbon materials, the exact mechanisms underlying the improved ORR activity are still controversial. CNTs and CNFs with surface modification can be used as the H_2O_2-producing cathode in MFC (*5, 79*). Strategies such as surface decoration through nitrate acid treatment to introduce more oxygen-containing groups have been demonstrated to favor a more positive onset potential and significantly improve the H_2O_2 selectivity (*7*).

Graphene

Graphene with superior crystallinity and electronic properties also shows remarkable ORR activity (*80*). The recent advances in preparation methods for low-cost mass production of reduced graphene oxide (GO), such as graphite exfoliation, have made their application in MFCs even more attractive (*81*). The catalytic properties of graphene depend strongly on its structure. Compared with activated carbon, graphene nanosheets typically show higher ORR activity but lower H_2O_2 selectivity in neutral media, enabling a power density output of up to 2.06 W m^{-2} in MFC (*82*). Similar to CNTs and CNFs, the ORR catalytic activity and durability of graphene can also be significantly strengthened via nitrogen doping (*83–86*) and achieve performance comparable to a Pt-based cathode in MFC applications (*87*). Doping other heteroatoms such as boron, silicon, phosphorous, and sulfur has also been shown to improve the electronic properties for O_2 adsorption and dissociation and to enhance the ORR activity (*78, 88*).

Graphene-based materials can also be used in an MFC-electro-Fenton system. In particular, oxidized graphene with a high oxygen atomic fraction of 5.7% in the form of C–O–C showed high catalytic activity for H_2O_2 production (Figure 7). With this cathode, the MFC-electro-Fenton system realized simultaneous wastewater treatment and power generation (*89*).

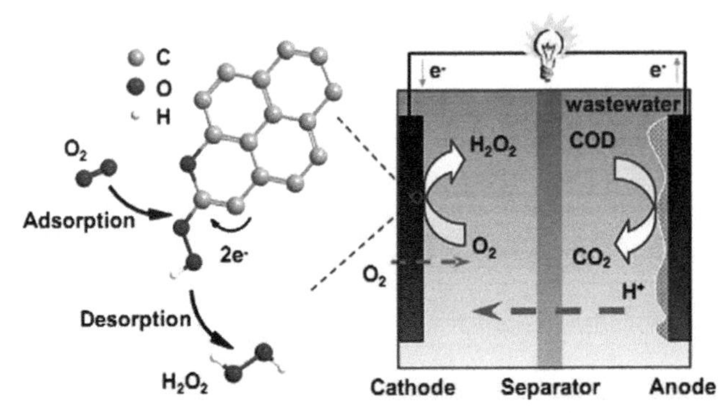

Figure 7. Possible mechanisms of ORR on the oxidized graphene air cathode in an MFC. Reproduced with permission from reference (89). Copyright 2018 Elsevier.

Graphite

Graphite has an sp^2 hybridized carbon structure with multiple layers that are bonded via van der Waals interaction. Bulk graphite can be used to produce graphene nanosheets through chemical and physical exfoliation. Although pristine graphite displays insufficient electrochemical activity, its performance can be improved by chemical activation or elemental doping. For example, treatment of graphite with H_3PO_4 significantly raised its ORR activity, achieving 2.4 times higher power density in MFC (7.9 W m^{-3}) than that with unactivated graphite (2.3 W m^{-3}) (*90*). A porous structure of graphite is also essential, to ensure a high specific surface area and to facilitate electrolyte diffusion and charge transfer (*91*). Like other carbon materials, heteroatom doping could also be applied to alter the surface charge distribution to improve ORR performance. A fine-tuned N-doped graphite with abundant pyridinic N as active sites even showed an ORR activity comparable to N-doped graphene materials (*17*).

The electrochemical performance of graphite can be further improved by adopting a three-dimensional (3D) electrode structure. Changing the MFC cathode from pure graphite rod to 3D graphite particle electrode was found to significantly raise the H_2O_2 generation from 78.85 mg L^{-1} to 196.50 mg L^{-1} within 12 h of operation (*92*). With this 3D electrode, 84% COD removal was achieved within 24 h in an MFC-electro-Fenton system (*93*). Besides, plasma, thermal, or chemical treatments could also be used to improve the wettability of graphite felt (*94*) and favor H_2O_2-selective ORR (*95*).

Organic Precursor–Derived Carbon Materials

Carbon materials with ORR activity can be directly derived from various organic precursors (*96, 97*). Such one-step fabrication approaches are attractive in that heteroatom-containing organic precursors can be directly converted into self-doped carbon materials through pyrolysis. The method has been adopted to prepared N and F codoped carbon black (BP-NF) from a mixture of polytetrafluoroethylene and BP-2000 under an ammonium atmosphere. The resulting material was applied in MFC, exhibiting a higher maximum power density (672 mA cm^{-2}) than that found for commercial Pt/C (572 mA cm^{-2}) (*97*). Likewise, N, P, and O codoped carbon materials with a hollow sphere structure have been produced from pyrolysis of polypyrrole-based precursors and exhibited superior ORR activity (electron transfer number of around 3.7) with excellent stability (*98*). MOF has also been used to prepare metal-free carbon materials (*99*). Liu et al. (*100*) fabricated hierarchically porous carbon from MOF-5 under H_2 atmosphere, where the metal component was lost during the carbonization. With a high content of sp^3-carbon and defects, large surface area, and favorable electrolyte diffusion properties, this material exhibited superior ORR activity and high H_2O_2 selectivity (70.2–95.0%), producing up to 222.6 mmol L^{-1} H_2O_2 after 2.5 h of reaction. The material was further applied in an electro-Fenton system for mineralization of perfluorooctanoate, outperforming most of the existing electron-Fenton cathodes (*14*).

Biomass-Derived Carbon Materials

Most carbon-based materials are still restricted by their relatively high cost and low yield. To favor MFC applications, biomass-derived carbon materials offer another promising option (*101*). The ORR selectivity of such materials is closely related to the physicochemical properties of the biomass. For example, porous carbon aerogels with high ORR activity and selectivity (electron transfer number of 3.7) were obtained from pyrolysis, hydrothermal carbonization, chemical activation, and ammoxidation of soy protein (*102*). In another study, a pomelo peel–derived nanoporous N/C electrocatalyst showed superior H_2O_2 selectivity (electron transfer number of 2) (*103*).

Biomass-derived carbon materials are usually synthesized from phytomass (carbohydrates, polysaccharides, and lignocellulose), showing different catalytic performances (*101, 104*). Lotus leaf–derived, hierarchically structured carbon materials exhibited high ORR activity and long-term durability for MFC operation (*105*). In addition, N-doped, partly graphitized carbon materials were fabricated from cornstalk and, when used as MFC cathode, exhibited better catalytic performance (maximum power density of 1122 mW m^{-2}) than Pt/C (*106*). Even sewage sludge can be made into carbon materials and function as an MFC cathode (Figure 8) (*107*). By addition of melamine to increase the nitrogen content, sludge-derived carbon materials exhibited excellent stability and better ORR activity than Pt/C in an alkaline medium (*108*). Bacterial cellulose, a typical industrially produced biomass with a 3D interconnected porous network, is also an attractive precursor for synthesis of functional CNFs (*109*). A nitrogen-doped CNF aerogel was obtained by carbonization of bacterial cellulose followed by NH_3 activation. This material possessed a high specific surface area of 916 m^2 g^{-1} and exhibited ORR activity comparable to Pt/C (*110*). In many zoomass materials such as feathers, blood, and bones, the inherently embodied heteroatoms such as N, P, and S can be inherited into the derived carbon materials to favor enhanced electrochemical properties without the need for external dosing. One good example is a hemoglobin-derived carbon electrocatalyst, which was used in an MFC cathode with reduced Fe(II) as the active sites to enable a high power density (up to 0.16 W cm^{-2}) (*111*).

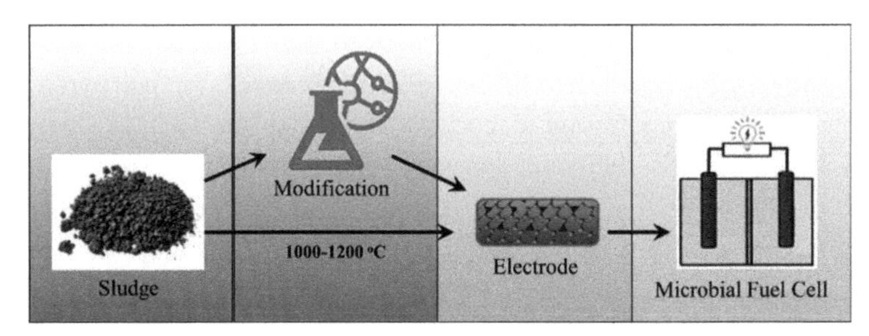

Figure 8. Schematic diagram for conversion of sewage sludge into MFC electrode material. Reproduced with permission from reference (107). Copyright 2019 Elsevier.

Analogous to other carbon materials, a biomass-derived carbon electrode may also be tuned to enable selective H_2O_2 production. A ramie-derived N-doped activated carbon was found to catalyze production of 18.1 μM H_2O_2 within 60 min in 0.05 M H_2SO_4 (*112*). The ORR activity and selectivity of such materials can be effectively tuned by strategies such as surface modification (*7*), heteroatom doping (*4*), introduction of porous or hollow structures, and optimum hybridization forms and defects (*14*).

Bioelectrocatalysts

Apart from chemically synthesized materials, enzymes or living microbial cells can also be utilized as cathodic ORR catalysts. For example, commercialized laccase could simultaneously catalyze ORR and dye decolorization at high efficiency (*113*). However, the high cost and easy deactivation of the enzymes restrict their practical application for wastewater treatment. Compared with chemical and enzymatic catalysts, living cells as bioelectrocatalysts have many advantages, such as low cost, self-regeneration, and environmental benignity. An efficient biocathode has been constituted by growing electroactive bacteria on a conductive, biocompatible electrode (typically carbon materials or stainless steel mesh) to support long-term sustained power generation in MFCs (*114*). However, the achievable power density in such a biocathode MFC (mostly <1 W m^{-2}) is an order of magnitude lower than those with chemical catalysts (*115*), mainly limited by the poor extracellular electron transfer ability and constrained oxygen diffusion across the cathode biofilm.

The conversion of organic carbon and nutrients in wastewater, which is essential for supporting microbial metabolism and growth, may also compete for oxygen and further impair the electrochemical ORR efficiency. Therefore, applications of biocatalysts in MFC cathodes are still scarce so far.

Future Perspectives

Although considerable advances have been made over the past decade in the fabrication and application of ORR electrocatalysts for MFC, some key hurdles to their practical application remain. Overcoming these challenges entails a better knowledge of and control over the structure and functions of various electrocatalysts, mainly in three aspects: crystallinity and phase, morphology, and electrode configuration.

Crystallinity and Phase

The activity and stability of ORR electrocatalysts are highly associated with the exposed crystal planes and phases. Taking Pt as an example, the (100) plane was less active than (111) for ORR in a H_2SO_4 environment due to different adsorption levels of sulfates (*116*). Pd could serve as the seed to direct the growth of Pt via reduction of K_2PtCl_4 by L-ascorbic acid in an aqueous solution, and ultimately form Pd-Pt bimetallic nanodendrites without the need for high temperature or organic solvent. The as-formed Pd-Pt bimetal with abundantly exposed (111) facets showed 2.5 times higher ORR activity than Pt/C (*117*). Even tuned semiconductors like TiO_2 can serve as an efficient ORR electrocatalyst. Pei et al. fabricated nanostructured TiO_2 with selectively exposed {001} facets (*118*), which exhibited a surprisingly high ORR activity and excellent durability. Moreover, by creating heterostructure and facets to boost charge transfer (*119, 120*), MOF-derived Pd@PdO-Co_3O_4 not only exhibited ORR activity comparable to commercial Pt/C but also higher stability, due to the synergy between PdO and Co_3O_4 (*121*).

To improve the acid tolerance of metal oxides for favorable application in acidic environments (e.g., in some electro-Fenton systems), transition metal chalcogenides and phosphides may be adopted as alternative cathodic electrocatalysts (*122*). Moreover, the covalent bond in these materials could provide better conductivity and more active sites than metal oxides alone (3). The codoping of iron sulfides ($Fe_{1-x}S$) and N and S into mesoporous graphitic carbon spheres significantly improved their ORR electrocatalytic activity to a level close to the state-of-the-art nonprecious metals (*123*). In addition, the platinum monophosphide (PtP) alloy also exhibited remarkable ORR activity and stability far exceeding plain Pt electrocatalysts (*124*).

Morphology

The physicochemical properties of electrocatalysts can also be tailored by regulating the morphology and catalyst dispersion (*125*). For example, Pt in the form of monodispersed nanocubes exhibited two times higher ORR activity than in bulk form (*126*), and 7 nm Pt nanocubes were four times more active than the 3 nm polyhedral form (*127*). Some favorable morphologies for ORR electrocatalysts, including porous one-dimensional (1D) nanostructure, hollow structure, and hierarchically porous structure, are summarized below.

The major porous 1D nanostructured materials for ORR cathodes include nanotubes, nanofibers, nanowires, nanoribbons, and nanorods. Such materials combine the merits of 1D and porous structures, such as large surface area, shortened electrolyte diffusion length, self-assembled interconnected networks to form free-standing electrodes, and porous regions as hosts for other species (*128*). These features make them attractive for ORR catalysis, although complicated fabrication procedures are usually needed (*98*, *129*). Hierarchically porous structured materials are especially beneficial for ORR in MFC. On one hand, the wide pore distribution from micropores to macropores and large specific surface area favor O_2 diffusion and adsorption (*130*). On the other hand, plentiful exposed active sites are ensured to decrease the overpotential and improve the catalytic efficiency (*14*, *46*, *62*). Such materials can be synthesized through in situ chlorination of carbides in ordered mesoporous titanium carbide–carbon composites (*130*), or by chemical activation (*105*) and soft template methods (*6*).

Electrode Configuration

The electrode configuration also significantly affects the catalytic efficiency of ORR electrocatalysts, because oxygen diffusion at the liquid–material interface is critically governed by the electrode configuration. Gas diffusion electrodes (GDE) with large active area and porous structure to favor O_2 diffusion are commonly adopted as MFC cathodes (*131*, *132*). A binder-free GDE with CO_3O_4 particles directly grown on a stainless steel mesh was recently reported, where the oxidized stainless steel mesh with plentiful oxygen-containing groups bond strongly with CO_3O_4 to form a highly active interface that favors O_2 molecule adsorption (*133*). This electrode displayed comparable catalytic activity, better durability, and lower cost compared with conventional Pt/C under pH-neutral electrolyte. Graphite felt is a good candidate for constructing GDE (*134*). attributed to its low cost, high specific surface area, and good mechanical integrity. Moreover, it allows efficient regeneration of Fe^{2+}, making it a promising cathode material for MFC-electro-Fenton systems (*94*). Various 3D electrode configurations have been frequently adopted for MFC applications because they allow greater electrocatalyst loading and provide more reactive surface area than two-dimensional ones (*135*). In particular, 3D electrodes of activated carbon or graphite particles generally possess high activity for H_2O_2 production (*93*).

Notably, although a number of high-performance ORR catalysts for MFC cathode applications have been developed, the specific mechanisms governing the ORR activity and selectivity and the correlations with catalyst physicochemical properties and environmental conditions are still poorly understood. In particular, the key factors determining the two- or four-electron ORR pathways are still under debate. These knowledge gaps have severely limited the development of electrocatalysts for MFC applications. Thus, future advances of ORR electrocatalysts for MFC may rely on a better understanding and control of the ORR properties based on knowledge from the chemical fuel cell field and the influences of MFC-specific operating environments. In addition, better fabrication methods to lower the cost and engineering strategies to further improve the catalytic performance in scaled-up systems are also needed to promote their practical applications.

Conclusions

The poor performance of ORR electrocatalysts is currently one important limiting factor for MFC application. Although they share many features in common with those for chemical fuel cell applications, MFC-targeted electrocatalysts have some extra requirements, such as operation at neutral pH and high robustness and stability, to suit the complicated wastewater environment. These needs have spurred intensive research and drastic advances in development of ORR electrocatalysts, broadly ranging from metal-based materials to metal-free carbon materials and biocatalysts. In addition, with better knowledge of the ORR mechanism, a number of strategies have been developed to tune the activity and selectivity. However, there is still a long way to go for practical applications of these electrocatalysts in MFC-based wastewater treatment systems, which warrant efforts to better understand catalyst structure–function correlations, finely control the catalyst properties, and explore effective strategies for production of cost-effective and scalable materials.

Acknowledgments

The authors thank the National Key Research and Development Program of China (2018YFA0901301), the National Natural Science Foundation of China (51778597, 51538012 and 51821006), and the Program for Changjiang Scholars for supporting this work.

References

1. Liu, X. W.; Li, W. W.; Yu, H. Q. Cathodic catalysts in bioelectrochemical systems for energy recovery from wastewater. *Chem. Soc. Rev.* **2014**, *43*, 7718–7745.
2. Sun, M.; Zhai, L. F.; Li, W. W.; Yu, H. Q. Harvest and utilization of chemical energy in wastes by microbial fuel cells. *Chem. Soc. Rev.* **2016**, *45*, 2847–2870.
3. Jiang, J.; Lu, S.; Gao, H.; Zhang, X.; Yu, H. Q. Ternary $FeNiS_2$ ultrathin nanosheets as an electrocatalyst for both oxygen evolution and reduction reactions. *Nano Energy* **2016**, *27*, 526–534.
4. Liu, J.; Song, P.; Ning, Z.; Xu, W. Recent advances in heteroatom-doped metal-free electrocatalysts for highly efficient oxygen reduction reaction. *Electrocatalysis* **2015**, *6*, 132–147.
5. Li, X.; Chen, S.; Angelidaki, I.; Zhang, Y. Bio-electro-Fenton processes for wastewater treatment: Advances and prospects. *Chem. Eng. J.* **2018**, *354*, 492–506.
6. Liu, Y. M.; Quan, X.; Fan, X. F.; Wang, H.; Chen, S. High-yield electrosynthesis of hydrogen peroxide from oxygen reduction by hierarchically porous carbon. *Angew. Chem. Int. Ed.* **2015**, *54*, 6837–6841.
7. Lu, Z.; Chen, G.; Siahrostami, S.; Chen, Z.; Liu, K.; Xie, J.; Liao, L.; Wu, T.; Lin, D.; Liu, Y.; Jaramillo, T. F.; Norskov, J. K.; Cui, Y. High-efficiency oxygen reduction to hydrogen peroxide catalysed by oxidized carbon materials. *Nat. Catal.* **2018**, *1*, 156–162.
8. Siahrostami, S.; Verdaguer-Casadevall, A.; Karamad, M.; Deiana, D.; Malacrida, P.; Wickman, B.; Escudero-Escribano, M.; Paoli, E. A.; Frydendal, R.; Hansen, T. W.; Chorkendorff, I.; Stephens, I. E. L.; Rossmeisl, J. Enabling direct H_2O_2 production through rational electrocatalyst design. *Nat. Mater.* **2013**, *12*, 1137–1143.
9. Verdaguer-Casadevall, A.; Deiana, D.; Karamad, M.; Siahrostami, S.; Malacrida, P.; Hansen, T. W.; Rossmeisl, J.; Chorkendorff, I.; Stephens, I. E. L. Trends in the electrochemical synthesis of H_2O_2: Enhancing activity and selectivity by electrocatalytic site engineering. *Nano Lett.* **2014**, *14*, 1603–1608.
10. Choi, C. H.; Kwon, H. C.; Yook, S.; Shin, H.; Kim, H.; Choi, M. Hydrogen peroxide synthesis via enhanced two-electron oxygen reduction pathway on carbon-coated Pt surface. *J. Phys. Chem. C* **2014**, *118*, 30063–30070.

11. Keenan, C. R.; Sedlak, D. L. Factors affecting the yield of oxidants from the reaction of manoparticulate zero-valent iron and oxygen. *Environ Sci. Technol.* **2008**, *42*, 1262–1267.
12. Wilfert, P.; Kumar, P. S.; Korving, L.; Witkamp, G.-J.; van Loosdrecht, M. C. M. The relevance of phosphorus and iron chemistry to the recovery of phosphorus from wastewater: A review. *Environ. Sci. Technol.* **2015**, *49*, 9400–9414.
13. Mu, Y.; Ai, Z.; Zhang, L. Phosphate shifted oxygen reduction pathway on $Fe@Fe_2O_3$ core-shell nanowires for enhanced reactive oxygen species generation and aerobic 4-chlorophenol degradation. *Environ. Sci. Technol.* **2017**, *51*, 8101–8109.
14. Liu, Y.; Quan, X.; Fan, X.; Wang, H.; Chen, S. High-yield electrosynthesis of hydrogen peroxide from oxygen reduction by hierarchically porous carbon. *Angew. Chem. Int. Ed.* **2015**, *54*, 6837–6841.
15. Sun, Y.; Sinev, I.; Ju, W.; Bergmann, A.; Dresp, S.; Kuehl, S.; Spoeri, C.; Schmies, H.; Wang, H.; Bernsmeier, D.; Paul, B.; Schmack, R.; Kraehnert, R.; Roldan Cuenya, B.; Strasser, P. Efficient electrochemical hydrogen peroxide production from molecular oxygen on nitrogen-doped mesoporous carbon catalysts. *ACS Catal.* **2018**, *8*, 2844–2856.
16. Bonakdarpour, A.; Dahn, T. R.; Atanasoski, R. T.; Debe, M. K.; Dahn, J. R. H_2O_2 release during oxygen reduction reaction on Pt nanoparticles. *Electrochem. Solid State Lett.* **2008**, *11*, B208–B211.
17. Guo, D.; Shibuya, R.; Akiba, C.; Saji, S.; Kondo, T.; Nakamura, J. Active sites of nitrogen-doped carbon materials for oxygen reduction reaction clarified using model catalysts. *Science* **2016**, *351*, 361–365.
18. Chandrasekhar, K.; Ahn, Y.-H. Effectiveness of piggery waste treatment using microbial fuel cells coupled with elutriated-phased acid fermentation. *Bioresour. Technol.* **2017**, *244*, 650–657.
19. Chandrasekhar, K. Effective and nonprecious cathode catalysts for oxygen reduction reaction in microbial fuel cells. In *Microbial Electrochemical Technology*; Venkata Mohan, S., Varjani, S., Pandey, A., Eds.; Elsevier: Netherlands, 2018; Chapter 3.5, pp 485–501.
20. Wang, Y.; Chen, K. S.; Mishler, J.; Cho, S. C.; Adroher, X. C. A review of polymer electrolyte membrane fuel cells: Technology, applications, and needs on fundamental research. *Appl. Energy* **2011**, *88*, 981–1007.
21. van der Vliet, D.; Wang, C.; Debe, M.; Atanasoski, R.; Markovic, N. M.; Stamenkovic, V. R. Platinum-alloy nanostructured thin film catalysts for the oxygen reduction reaction. *Electrochim. Acta* **2011**, *56*, 8695–8699.
22. Bing, Y.; Liu, H.; Zhang, L.; Ghosh, D.; Zhang, J. Nanostructured Pt-alloy electrocatalysts for PEM fuel cell oxygen reduction reaction. *Chem. Soc. Rev.* **2010**, *39*, 2184–2202.
23. Morozan, A.; Jousselme, B.; Palacin, S. Low-platinum and platinum-free catalysts for the oxygen reduction reaction at fuel cell cathodes. *Energy Environ. Sci.* **2011**, *4*, 1238–1254.
24. Wang, C.; Markovic, N. M.; Stamenkovic, V. R. Advanced platinum alloy electrocatalysts for the oxygen reduction reaction. *ACS Catal.* **2012**, *2*, 891–898.
25. Gupta, G.; Slanac, D. A.; Kumar, P.; Wiggins-Camacho, J. D.; Wang, X.; Swinnea, S.; More, K. L.; Dai, S.; Stevenson, K. J.; Johnston, K. P. Highly stable and active Pt-Cu oxygen reduction electrocatalysts based on mesoporous graphitic carbon supports. *Chem. Mater.* **2009**, *21*, 4515–4526.

26. Cetinkaya, A. Y.; Ozdemir, O. K.; Koroglu, E. O.; Hasimoglu, A.; Ozkaya, B. The development of catalytic performance by coating Pt-Ni on CMI7000 membrane as a cathode of a microbial fuel cell. *Bioresour. Technol.* **2015**, *195*, 188–193.
27. Stamenkovic, V.; Schmidt, T. J.; Ross, P. N.; Markovic, N. M. Surface composition effects in electrocatalysis: Kinetics of oxygen reduction on well-defined Pt_3Ni and Pt_3Co alloy surfaces. *J. Phys. Chem. B* **2002**, *106*, 11970–11979.
28. Sumner, L.; Sakthivel, N. A.; Schrock, H.; Artyushkova, K.; Dass, A.; Chakraborty, S. Electrocatalytic oxygen reduction activities of thiol-protected nanomolecules ranging in size from $Au_{28}(SR)_{20}$ to $Au_{279}(SR)_{84}$. *J. Phys. Chem. C* **2018**, *122*, 24809–24817.
29. Dai, H. Y.; Yang, H. M.; Jian, X.; Liu, X.; Liang, Z. H. Performance of Ag_2O/Ag electrode as cathodic electron acceptor in microbial fuel cell. *Acta Metall. Sin. (Engl. Lett.)* **2017**, *30*, 1243–1248.
30. Han, H. X.; Shi, C.; Yuan, L.; Sheng, G. P. Enhancement of methyl orange degradation and power generation in a photoelectrocatalytic microbial fuel cell. *Appl. Energy* **2017**, *204*, 382–389.
31. Pu, L.; Li, K.; Chen, Z.; Zhang, P.; Zhang, X.; Fu, Z. Silver electrodeposition on the activated carbon air cathode for performance improvement in microbial fuel cells. *J. Power Sources* **2014**, *268*, 476–481.
32. Jiang, J.; Gao, H.; Lu, S.; Zhang, X.; Wang, C. Y.; Wang, W. K.; Yu, H. Q. Ni-Pd core-shell nanoparticles with Pt-like oxygen reduction electrocatalytic performance in both acidic and alkaline electrolytes. *J. Mater. Chem. A* **2017**, *5*, 9233–9240.
33. Gong, X. B.; You, S. J.; Wang, X. H.; Gan, Y.; Zhang, R. N.; Ren, N. Q. Silver-tungsten carbide nanohybrid for efficient electrocatalysis of oxygen reduction reaction in microbial fuel cell. *J. Power Sources* **2013**, *225*, 330–337.
34. Kargi, F.; Eker, S. Electricity generation with simultaneous wastewater treatment by a microbial fuel cell (MFC) with Cu and Cu-Au electrodes. *J. Chem. Technol. Biotechnol.* **2007**, *82*, 658–662.
35. Jasinski, R. A new fuel cell cathode catalyst. *Nature* **1964**, *201*, 1212–1213.
36. Xiong, L.; Huang, Y. X.; Liu, X. W.; Sheng, G. P.; Li, W. W.; Yu, H. Q. Three-dimensional bimetallic Pd-Cu nanodendrites with superior electrochemical performance for oxygen reduction reaction. *Electrochim. Acta* **2013**, *89*, 24–28.
37. Valipour, A.; Hamnabard, N.; Meshkati, S. M. H.; Pakan, M.; Ahn, Y.-H. Effectiveness of phase- and morphology-controlled MnO_2 nanomaterials derived from flower-like delta-MnO_2 as alternative cathode catalyst in microbial fuel cells. *Dalton Trans.* **2019**, *48*, 5429–5443.
38. Zhang, S.; Su, W.; Wei, Y.; Liu, J.; Li, K. Mesoporous MnO_2 structured by ultrathin nanosheet as electrocatalyst for oxygen reduction reaction in air-cathode microbial fuel cell. *J. Power Sources* **2018**, *401*, 158–164.
39. Ma, A.; Chang, A.; Zhang, J.; Gui, J.; Liu, D.; Yu, Y.; Zhang, B. MnO_2-mediated synthesis of Mn_3O_4@$CaMn_7O_{12}$ core@shell nanorods for electrocatalytic oxygen reduction reaction. *ChemElectroChem* **2019**, *6*, 618–622.
40. Cheng, C.; Hu, Y.; Shao, S.; Yu, J.; Zhou, W.; Cheng, J.; Chen, Y.; Chen, S.; Chen, J.; Zhang, L. Simultaneous Cr(VI) reduction and electricity generation in plant-sediment microbial fuel

cells (p-smfcs): Synthesis of non-bonding Co_3O_4 nanowires onto cathodes. *Environ. Pollut.* **2019**, *247*, 647–657.
41. Zhang, S.; Su, W.; Li, K.; Liu, D.; Wang, J.; Tian, P. Metal organic framework-derived $Co_3O_4/NiCo_2O_4$ double-shelled nanocage modified activated carbon air-cathode for improving power generation in microbial fuel cell. *J. Power Sources* **2018**, *396*, 355–362.
42. Noori, M. T.; Mukherjee, C. K.; Ghangrekar, M. M. Enhancing performance of microbial fuel cell by using graphene supported V_2O_5-nanorod catalytic cathode. *Electrochim. Acta* **2017**, *228*, 513–521.
43. Zhu, H.; Zhang, S.; Huang, Y. X.; Wu, L.; Sun, S. Monodisperse $M_xFe_{3-x}O_4$ (M = Fe, Cu, Co, Mn) nanoparticles and their electrocatalysis for oxygen reduction reaction. *Nano Lett.* **2013**, *13*, 2947–2951.
44. Zhu, Y.; Liu, S.; Jin, C.; Bie, S.; Yang, R.; Wu, J. MnO_x decorated CeO_2 nanorods as cathode catalyst for rechargeable lithium-air batteries. *J. Mater. Chem. A* **2015**, 3, 13563–13567.
45. Wang, J.; Tian, P.; Li, K.; Ge, B.; Liu, D.; Liu, Y.; Yang, T.; Ren, R. The excellent performance of nest-like oxygen-deficient $Cu_{1.5}Mn_{1.5}O_4$ applied in activated carbon air-cathode microbial fuel cell. *Bioresour. Technol.* **2016**, *222*, 107–113.
46. Liu, Y.; Chi, X.; Han, Q.; Du, Y.; Huang, J.; Lin, X.; Liu, Y. Metal-organic framework-derived hierarchical Co_3O_4@$MnCo_2O_{4.5}$ nanocubes with enhanced electrocatalytic activity for Na-O_2 batteries. *Nanoscale* **2019**, *11*, 5285–5294.
47. Hu, D.; Zhang, G.; Wang, J.; Zhong, Q. Carbon-supported spinel nanoparticle $MnCo_2O_4$ as a cathode catalyst towards oxygen reduction reaction in dual-chamber microbial fuel cell. *Aust. J. Chem.* **2015**, *68*, 987–994.
48. Jiang, B.; Muddemann, T.; Kunz, U.; Silva e Silva, L. G.; Bormann, H.; Niedermeiser, M.; Haupt, D.; Schlaefer, O.; Sievers, M. Graphite/MnO_2 and MoS_2 composites used as catalysts in the oxygen reduction cathode of microbial fuel cells. *J. Electrochem. Soc.* **2017**, *164*, E519–E524.
49. Qian, X.; Ren, M.; Zhu, Y.; Yue, D.; Han, Y.; Jia, J.; Zhao, Y. Visible light assisted heterogeneous Fenton-like degradation of organic pollutant via alpha-FeOOH/mesoporous carbon composites. *Environ. Sci. Technol.* **2017**, *51*, 3993–4000.
50. Bocos, E.; Iglesias, O.; Pazos, M.; Angeles Sanroman, M. Nickel foam a suitable alternative to increase the generation of Fenton's reagents. *Process Saf. Environ. Prot.* **2016**, *101*, 34–44.
51. Deng, J.; Li, M.; Wang, Y. Biomass-derived carbon: Synthesis and applications in energy storage and conversion. *Green Chem.* **2016**, *18*, 4824–4854.
52. Antolini, E. Carbon supports for low-temperature fuel cell catalysts. *Appl. Catal., B* **2009**, *88*, 1–24.
53. Ge, B.; Li, K.; Fu, Z.; Pu, L.; Zhang, X. The addition of ortho-hexagon nano spinel Co_3O_4 to improve the performance of activated carbon air cathode microbial fuel cell. *Bioresour. Technol.* **2015**, *195*, 180–187.
54. Huang, Q.; Zhou, P.; Yang, H.; Zhu, L.; Wu, H. In situ generation of inverse spinel $CoFe_2O_4$ nanoparticles onto nitrogen-doped activated carbon for an effective cathode electrocatalyst of microbial fuel cells. *Chem. Eng. J.* **2017**, *325*, 466–473.

55. Ghasemi, M.; Ismail, M.; Kamarudin, S. K.; Saeedfar, K.; Daud, W. R. W.; Hassan, S. H. A.; Heng, L. Y.; Alam, J.; Oh, S.-E. Carbon nanotube as an alternative cathode support and catalyst for microbial fuel cells. *Appl. Energy* **2013**, *102*, 1050–1056.
56. Ben Liew, K.; Daud, W. R. W.; Ghasemi, M.; Loh, K. S.; Ismail, M.; Lim, S. S.; Leong, J. X. Manganese oxide/functionalised carbon nanotubes nanocomposite as catalyst for oxygen reduction reaction in microbial fuel cell. *Int. J. Hydrogen Energy* **2015**, *40*, 11625–11632.
57. Xu, N.; Nie, Q.; Luo, L.; Yao, C.; Gong, Q.; Liu, Y.; Zhou, X.-D.; Qiao, J. Controllable hortensia-like MnO_2 synergized with carbon nanotubes as an efficient electrocatalyst for long-term metal-air batteries. *ACS Appl. Mater. Interfaces* **2019**, *11*, 578–587.
58. Kim, J.; Gwon, O.; Kwon, O.; Mahmood, J.; Kim, C.; Yang, Y.; Lee, H.; Lee, J. H.; Jeong, H. Y.; Baek, J.-B.; Kim, G. Synergistic coupling derived cobalt oxide with nitrogenated holey two-dimensional matrix as an efficient bifunctional catalyst for metal-air batteries. *ACS Nano* **2019**, *13*, 5502–5512.
59. Zhang, S.; Su, W.; Wang, X.; Li, K.; Li, Y. Bimetallic metal-organic frameworks derived cobalt nanoparticles embedded in nitrogen-doped carbon nanotube nanopolyhedra as advanced electrocatalyst for high-performance of activated carbon air-cathode microbial fuel cell. *Biosens. Bioelectron.* **2019**, *127*, 181–187.
60. Tang, H.; Cai, S.; Xie, S.; Wang, Z.; Tong, Y.; Pan, M.; Lu, X. Metal-organic-framework-derived dual metal- and nitrogen-doped carbon as efficient and robust oxygen reduction reaction catalysts for microbial fuel cells. *Adv. Sci.* **2016**, 3, 1500265–1500272.
61. Jin, H.; Zhou, H.; He, D.; Wang, Z.; Wu, Q.; Liang, Q.; Liu, S.; Mu, S. MOF-derived 3D Fe-N-S co-doped carbon matrix/nanotube nanocomposites with advanced oxygen reduction activity and stability in both acidic and alkaline media. *Appl. Catal., B* **2019**, *250*, 143–149.
62. Liu, H.; Cao, Y.; Wang, F.; Zhang, W.; Huang, Y. Pig bone derived hierarchical porous carbon-supported platinum nanoparticles with superior electrocatalytic activity towards oxygen reduction reaction. *Electroanalysis* **2014**, *26*, 1831–1839.
63. Xiong, L.; Chen, J. J.; Huang, Y. X.; Li, W. W.; Xie, J. F.; Yu, H. Q. An oxygen reduction catalyst derived from a robust Pd-reducing bacterium. *Nano Energy* **2015**, *12*, 33–42.
64. Zhao, H.; Qian, L.; Guan, X.; Wu, D.; Zhao, G. Continuous bulk FeCuC aerogel with ultradispersed metal nanoparticles: An efficient 3D heterogeneous electro-Fenton cathode over a wide range of pH 3–9. *Environ. Sci. Technol.* **2016**, *50*, 5225–5233.
65. Feng, C. H.; Li, F. B.; Mai, H. J.; Li, X. Z. Bio-electro-Fenton process driven by microbial fuel cell for wastewater treatment. *Environ. Sci. Technol.* **2010**, *44*, 1875–1880.
66. Yong, X. Y.; Gu, D. Y.; Wu, Y. D.; Yan, Z. Y.; Zhou, J.; Wu, X. Y.; Wei, P.; Jia, H. H.; Zheng, T.; Yong, Y. C. Bio-electron-Fenton (BEF) process driven by microbial fuel cells for triphenyltin chloride (TPTC) degradation. *J. Hazard. Mater.* **2017**, *324*, 178–183.
67. Huang, B. C.; Jiang, J.; Wang, W. K.; Li, W. W.; Zhang, F.; Jiang, H.; Yu, H. Q. Electrochemically catalytic degradation of phenol with hydrogen peroxide in situ generated and activated by a municipal sludge-derived catalyst. *ACS Sustain. Chem. Eng.* **2018**, *6*, 5540–5546.
68. Zhao, Y.; Watanabe, K.; Hashimoto, K. Self-supporting oxygen reduction electrocatalysts made from a nitrogen-rich network polymer. *J. Am. Chem. Soc.* **2012**, *134*, 19528–19531.

69. Cao, C.; Wei, L.; Su, M.; Wang, G.; Shen, J. Enhanced power generation using nano cobalt oxide anchored nitrogen-decorated reduced graphene oxide as a high-performance air-cathode electrocatalyst in biofuel cells. *RSC Adv.* **2016**, *6*, 52556–52563.
70. Li, B. Q.; Zhao, C. X.; Liu, J. N.; Zhang, Q. Electrosynthesis of hydrogen peroxide synergistically catalyzed by atomic Co-N_x-C sites and oxygen functional groups in noble-metal-free electrocatalysts. *Adv. Mater.* **2019**, e1808173–e1808173.
71. Gautam, R. K.; Verma, A. Electrocatalyst materials for oxygen reduction reaction in microbial fuel cell. In *Microbial Electrochemical Technology*; Venkata Mohan, S., Varjani, S., Pandey, A., Eds.; Elsevier: Netherlands, 2018; Chapter 3.4, pp 451–483.
72. Baughman, R. H.; Zakhidov, A. A.; de Heer, W. A. Carbon nanotubes: The route toward applications. *Science* **2002**, *297*, 787–792.
73. Ghasemi, M.; Shahgaldi, S.; Ismail, M.; Kim, B. H.; Yaakob, Z.; Daud, W. R. W. Activated carbon nanofibers as an alternative cathode catalyst to platinum in a two-chamber microbial fuel cell. *Int. J. Hydrogen Energy* **2011**, *36*, 13746–13752.
74. Zhou, W.; Meng, X.; Gao, J.; Alshawabkeh, A. N. Hydrogen peroxide generation from O_2 electroreduction for environmental remediation: A state-of-the-art review. *Chemosphere* **2019**, *225*, 588–607.
75. Gong, K.; Du, F.; Xia, Z.; Durstock, M.; Dai, L. Nitrogen-doped carbon nanotube arrays with high electrocatalytic activity for oxygen reduction. *Science* **2009**, *323*, 760–764.
76. Maldonado, S.; Stevenson, K. J. Influence of nitrogen doping on oxygen reduction electrocatalysis at carbon nanofiber electrodes. *J. Phys. Chem. B* **2005**, *109*, 4707–4716.
77. Feng, L.; Chen, X.; Cao, Y.; Chen, Y.; Wang, F.; Chen, Y.; Liu, Y. Pyridinic and pyrrolic nitrogen-rich ordered mesoporous carbon for efficient oxygen reduction in microbial fuel cells. *RSC Adv.* **2017**, *7*, 14669–14677.
78. Wu, G.; Santandreu, A.; Kellogg, W.; Gupta, S.; Ogoke, O.; Zhang, H.; Wang, H.-L.; Dai, L. Carbon nanocomposite catalysts for oxygen reduction and evolution reactions: From nitrogen doping to transition-metal addition. *Nano Energy* **2016**, *29*, 83–110.
79. Zarei, M.; Salari, D.; Niaei, A.; Khataee, A. Peroxi-coagulation degradation of CI basic yellow 2 based on carbon-PTFE and carbon nanotube-PTFE electrodes as cathode. *Electrochim. Acta* **2009**, *54*, 6651–6660.
80. Geim, A. K.; Novoselov, K. S. The rise of graphene. *Nat. Mater.* **2007**, *6*, 183–191.
81. Wu, Z. S.; Zhou, G.; Yin, L. C.; Ren, W.; Li, F.; Cheng, H. M. Graphene/metal oxide composite electrode materials for energy storage. *Nano Energy* **2012**, *1*, 107–131.
82. Santoro, C.; Kodali, M.; Kabir, S.; Soavi, F.; Serov, A.; Atanassov, P. Three-dimensional graphene nanosheets as cathode catalysts in standard and supercapacitive microbial fuel cell. *J. Power Sources* **2017**, *356*, 371–380.
83. Qu, L.; Liu, Y.; Baek, J.-B.; Dai, L. Nitrogen-doped graphene as efficient metal-free electrocatalyst for oxygen reduction in fuel cells. *ACS Nano* **2010**, *4*, 1321–1326.
84. Hussain, S.; Erikson, H.; Kongi, N.; Treshchalov, A.; Rahn, M.; Kook, M.; Merisalu, M.; Matisen, L.; Sammelselg, V.; Tammeveski, K. Oxygen electroreduction on Pt nanoparticles deposited on reduced graphene oxide and N-doped reduced graphene oxide prepared by plasma-assisted synthesis in aqueous solution. *ChemElectroChem* **2018**, *5*, 2902–2911.

85. Buan, M. E. M.; Muthuswamy, N.; Walmsley, J. C.; Chen, D.; Ronning, M. Nitrogen-doped carbon nanofibers on expanded graphite as oxygen reduction electrocatalysts. *Carbon* **2016**, *101*, 191–202.
86. Kumar, S.; Gonen, S.; Friedman, A.; Elbaz, L.; Nessim, G. D. Doping and reduction of graphene oxide using chitosan-derived volatile N-heterocyclic compounds for metal-free oxygen reduction reaction. *Carbon* **2017**, *120*, 419–426.
87. Feng, L.; Chen, Y.; Chen, L. Easy-to-operate and low-temperature synthesis of gram-scale nitrogen-doped graphene and its application as cathode catalyst in microbial fuel cells. *ACS Nano* **2011**, *5*, 9611–9618.
88. Yu, S.; Zheng, W.; Wang, C.; Jiang, Q. Nitrogen/boron doping position dependence of the electronic properties of a triangular graphene. *ACS Nano* **2010**, *4*, 7619–7629.
89. Dong, H.; Liu, X.; Xu, T.; Wang, Q.; Chen, X.; Chen, S.; Zhang, H.; Liang, P.; Huang, X.; Zhang, X. Hydrogen peroxide generation in microbial fuel cells using graphene-based air-cathodes. *Bioresour. Technol.* **2018**, *247*, 684–689.
90. Zhang, L.; Lu, Z.; Li, D.; Ma, J.; Song, P.; Huang, G.; Liu, Y.; Cai, L. Chemically activated graphite enhanced oxygen reduction and power output in catalyst-free microbial fuel cells. *J. Clean. Prod.* **2016**, *115*, 332–336.
91. Xing, Z.; Gao, N.; Qi, Y.; Ji, X.; Liu, H. Influence of enhanced carbon crystallinity of nanoporous graphite on the cathode performance of microbial fuel cells. *Carbon* **2017**, *115*, 271–278.
92. Fu, L.; You, S. J.; Yang, F. L.; Gao, M. M.; Fang, X. H.; Zhang, G. Q. Synthesis of hydrogen peroxide in microbial fuel cell. *J. Chem. Technol. Biotechnol.* **2010**, *85*, 715–719.
93. Chen, J.-y.; Li, N.; Zhao, L. Three-dimensional electrode microbial fuel cell for hydrogen peroxide synthesis coupled to wastewater treatment. *J. Power Sources* **2014**, *254*, 316–322.
94. Zhou, L.; Zhou, M.; Hu, Z.; Bi, Z.; Serrano, K. G. Chemically modified graphite felt as an efficient cathode in electro-Fenton for p-nitrophenol degradation. *Electrochim. Acta* **2014**, *140*, 376–383.
95. Thi Xuan Huong, L.; Bechelany, M.; Cretin, M. Carbon felt based-electrodes for energy and environmental applications: A review. *Carbon* **2017**, *122*, 564–591.
96. Wang, J. G.; Liu, H.; Sun, H.; Hua, W.; Wang, H.; Liu, X.; Wei, B. One-pot synthesis of nitrogen-doped ordered mesoporous carbon spheres for high-rate and long-cycle life supercapacitors. *Carbon* **2018**, *127*, 85–92.
97. Meng, K.; Liu, Q.; Huang, Y.; Wang, Y. Facile synthesis of nitrogen and fluorine co-doped carbon materials as efficient electrocatalysts for oxygen reduction reactions in air-cathode microbial fuel cells. *J. Mater. Chem. A* **2015**, *3*, 6873–6877.
98. Huang, S.; Meng, Y.; Cao, Y.; He, S.; Li, X.; Tong, S.; Wu, M. N-, O- and P-doped hollow carbons: Metal-free bifunctional electrocatalysts for hydrogen evolution and oxygen reduction reactions. *Appl. Catal., B* **2019**, *248*, 239–248.
99. Lee, J.; Farha, O. K.; Roberts, J.; Scheidt, K. A.; Nguyen, S. T.; Hupp, J. T. Metal-organic framework materials as catalysts. *Chem. Soc. Rev.* **2009**, *38*, 1450–1459.
100. Liu, Y.; Chen, S.; Quan, X.; Yu, H.; Zhao, H.; Zhang, Y. Efficient mineralization of perfluorooctanoate by electro-Fenton with H_2O_2 electro-generated on hierarchically porous carbon. *Environ. Sci. Technol.* **2015**, *49*, 13528–13533.

101. Borghei, M.; Lehtonen, J.; Liu, L.; Rojas, O. J. Advanced biomass-derived electrocatalysts for the oxygen reduction reaction. *Adv. Mater.* **2018**, *30*, 1703691–1703717.
102. Alatalo, S.-M.; Qiu, K.; Preuss, K.; Marinovic, A.; Sevilla, M.; Sillanpaa, M.; Guo, X.; Titirici, M.-M. Soy protein directed hydrothermal synthesis of porous carbon aerogels for electrocatalytic oxygen reduction. *Carbon* **2016**, *96*, 622–630.
103. Yang, Y.; He, F.; Shen, Y.; Chen, X.; Mei, H.; Liu, S.; Zhang, Y. A biomass derived N/C-catalyst for the electrochemical production of hydrogen peroxide. *Chem. Commun.* **2017**, *53*, 9994–9997.
104. Yang, W.; Li, J.; Fu, Q.; Zhang, L.; Zhu, X.; Liao, Q. A simple method for preparing a binder-free paper-based air cathode for microbial fuel cells. *Bioresour. Technol.* **2017**, *241*, 325–331.
105. Ye, W.; Tang, J.; Wang, Y.; Cai, X.; Liu, H.; Lin, J.; Van der Bruggen, B.; Zhou, S. Hierarchically structured carbon materials derived from lotus leaves as efficient electrocatalyst for microbial energy harvesting. *Sci. Total Environ.* **2019**, *666*, 865–874.
106. Sun, Y.; Duan, Y.; Hao, L.; Xing, Z.; Dai, Y.; Li, R.; Zou, J. Cornstalk-derived nitrogen-doped partly graphitized carbon as efficient metal-free catalyst for oxygen reduction reaction in microbial fuel cells. *ACS Appl. Mater. Interfaces* **2016**, *8*, 25923–25932.
107. Mian, M. M.; Liu, G.; Fu, B. Conversion of sewage sludge into environmental catalyst and microbial fuel cell electrode material. *Sci. Total Environ.* **2019**, *666*, 525–539.
108. Liu, F.; Peng, H.; You, C.; Fu, Z.; Huang, P.; Song, H.; Liao, S. High-performance doped carbon catalyst derived from nori biomass with melamine promoter. *Electrochim. Acta* **2014**, *138*, 353–359.
109. Wu, Z. Y.; Liang, H. W.; Chen, L. F.; Hu, B. C.; Yu, S. H. Bacterial cellulose: A robust platform for design of three dimensional carbon-based functional nanomaterials. *Accounts Chem. Res.* **2016**, *49*, 96–105.
110. Liang, H. W.; Wu, Z. Y.; Chen, L. F.; Li, C.; Yu, S. H. Bacterial cellulose derived nitrogen-doped carbon nanofiber aerogel: An efficient metal-free oxygen reduction electrocatalyst for zinc-air battery. *Nano Energy* **2015**, *11*, 366–376.
111. Maruyama, J.; Okamura, J.; Miyazaki, K.; Abe, I. Two-step carbonization as a method of enhancing catalytic properties of hemoglobin at the fuel cell cathode. *J. Phys. Chem. C* **2007**, *111*, 6597–6600.
112. Xue, Y.; Du, X.; Cai, W.; Sun, Z.; Zhang, Y.; Zheng, S.; Zhang, Y.; Jin, W. Ramie biomass derived nitrogen-doped activated carbon for efficient electrocatalytic production of hydrogen peroxide. *J. Electrochem. Soc.* **2018**, *165*, E171–E176.
113. Savizi, I. S. P.; Kariminia, H.-R.; Bakhshian, S. Simultaneous decolorization and bioelectricity generation in a dual chamber microbial fuel cell using electropolymerized-enzymatic cathode. *Environ. Sci. Technol.* **2012**, *46*, 6584–6593.
114. De Schamphelaire, L.; Boeckx, P.; Verstraete, W. Evaluation of biocathodes in freshwater and brackish sediment microbial fuel cells. *Appl. Microbiol. Biot.* **2010**, *87*, 1675–1687.
115. Zhang, Y.; Sun, J.; Hu, Y.; Li, S.; Xu, Q. Carbon nanotube-coated stainless steel mesh for enhanced oxygen reduction in biocathode microbial fuel cells. *J. Power Sources* **2013**, *239*, 169–174.

116. Markovic, N. M.; Gasteiger, H. A.; Philip, N. Oxygen reduction on platinum low-index single-crystal surfaces in alkaline solution: Rotating ring disk(Pt(hkl)) studies. *J. Phys. Chem.* **1996**, *100*, 6715–6721.
117. Lim, B.; Jiang, M.; Camargo, P. H. C.; Cho, E. C.; Tao, J.; Lu, X.; Zhu, Y.; Xia, Y. Pd-Pt bimetallic nanodendrites with high activity for oxygen reduction. *Science* **2009**, *324*, 1302–1305.
118. Pei, D. N.; Gong, L.; Zhang, A. Y.; Zhang, X.; Chen, J. J.; Mu, Y.; Yu, H. Q. Defective titanium dioxide single crystals exposed by high-energy {001} facets for efficient oxygen reduction. *Nat. Commun.* **2015**, *6*, 8696–8705.
119. Zheng, Y.; Zhou, T.; Zhang, C.; Mao, J.; Liu, H.; Guo, Z. Boosted charge transfer in SnS/SnO_2 heterostructures: Toward high rate capability for sodium-ion batteries. *Angew. Chem. Int. Ed.* **2016**, *55*, 3408–3413.
120. Zhang, A. Y.; Wang, W.; Chen, J. J.; Liu, C.; Li, Q.; Zhang, X.; Li, W. W.; Si, Y.; Yu, H. Q. Epitaxial facet junction on TiO_2 single crystals for efficient photocatalytic water splitting. *Energy Environ. Sci.* **2018**, *11*, 1444–1448.
121. Li, H. C.; Zhang, Y. J.; Hu, X.; Liu, W. J.; Chen, J. J.; Yu, H. Q. Metal-organic framework templated Pd@PdO-Co_3O_4 nanocubes as an efficient bifunctional oxygen electrocatalyst. *Adv. Energy Mater.* **2018**, *8*, 1702734–1702743.
122. Han, A. L.; Jin, S.; Chen, H. L.; Ji, H. X.; Sun, Z. J.; Du, P. W. A robust hydrogen evolution catalyst based on crystalline nickel phosphide nanoflakes on three-dimensional graphene/nickel foam: High performance for electrocatalytic hydrogen production from pH 0–14. *J. Mater. Chem. A* **2015**, 3, 1941–1946.
123. Xiao, J.; Xia, Y.; Hu, C.; Xi, J.; Wang, S. Raisin bread-like iron sulfides/nitrogen and sulfur dual-doped mesoporous graphitic carbon spheres: A promising electrocatalyst for the oxygen reduction reaction in alkaline and acidic media. *J. Mater. Chem. A* **2017**, 5, 11114–11123.
124. Zhang, L.; Wei, M.; Wang, S.; Li, Z.; Ding, L. X.; Wang, H. Highly stable PtP alloy nanotube arrays as a catalyst for the oxygen reduction reaction in acidic medium. *Chem. Sci.* **2015**, *6*, 3211–3216.
125. Chen, J.; Lim, B.; Lee, E. P.; Xia, Y. Shape-controlled synthesis of platinum nanocrystals for catalytic and electrocatalytic applications. *Nano Today* **2009**, *4*, 81–95.
126. Wang, C.; Daimon, H.; Lee, Y.; Kim, J.; Sun, S. Synthesis of monodisperse Pt nanocubes and their enhanced catalysis for oxygen reduction. *J. Am. Chem. Soc.* **2007**, *129*, 6974–6975.
127. Wang, C.; Daimon, H.; Onodera, T.; Koda, T.; Sun, S. A general approach to the size- and shape-controlled synthesis of platinum nanoparticles and their catalytic reduction of oxygen. *Angew. Chem. Int. Ed.* **2008**, *47*, 3588–3591.
128. Wei, Q.; Xiong, F.; Tan, S.; Huang, L.; Lan, E. H.; Dunn, B.; Mai, L. Porous one-dimensional nanomaterials: Design, fabrication and applications in electrochemical energy storage. *Adv. Mater.* **2017**, *29*, 1602300–1602338.
129. Zhou, L.; Zhuang, Z.; Zhao, H.; Lin, M.; Zhao, D.; Mai, L. Intricate hollow structures: Controlled synthesis and applications in energy storage and conversion. *Adv. Mater.* **2017**, *29*, 1602914–1602942.

130. Liu, H. J.; Wang, J.; Wang, C. X.; Xia, Y. Y. Ordered hierarchical mesoporous/microporous carbon derived from mesoporous titanium-carbide/carbon composites and its electrochemical performance in supercapacitor. *Adv. Energy Mater.* **2011**, *1*, 1101–1108.
131. Luo, H.; Li, C.; Wu, C.; Zheng, W.; Dong, X. Electrochemical degradation of phenol by in situ electro-generated and electro-activated hydrogen peroxide using an improved gas diffusion cathode. *Electrochim. Acta* **2015**, *186*, 486–493.
132. Moreira, J.; Lima, V. B.; Goulart, L. A.; Lanza, M. R. V. Electrosynthesis of hydrogen peroxide using modified gas diffusion electrodes (MGDE) for environmental applications: Quinones and azo compounds employed as redox modifiers. *Appl. Catal., B* **2019**, *248*, 95–107.
133. Gong, X. B.; You, S.-J.; Wang, X. H.; Zhang, J. N.; Gan, Y.; Ren, N. Q. A novel stainless steel mesh/cobalt oxide hybrid electrode for efficient catalysis of oxygen reduction in a microbial fuel cell. *Biosens. Bioelectron.* **2014**, *55*, 237–241.
134. Zhao, Y.; Watanabe, K.; Hashimoto, K. Efficient oxygen reduction by a Fe/Co/C/N nano-porous catalyst in neutral media. *J. Mater. Chem. A* **2013**, *1*, 1450–1456.
135. Brown, C. J.; Pletcher, D.; Walsh, F. C.; Hammond, J. K.; Robinson, D. Studies of three-dimensional electrodes in the FM01-LC laboratory electrolyser. *J. Appl. Electrochem.* **1994**, *24*, 95–106.

Chapter 5

Bacterially Generated Nanocatalysts and Their Applications

Zhiyong Zheng,[1] Yong Xiao,[1,2] Feng Zhao,[2] Jens Ulstrup,[1] and Jingdong Zhang[*,1]

[1]Department of Chemistry, Technical University of Denmark, Kemitorvet, Building 207, DK-2800 Kongens Lyngby, Denmark

[2]CAS Key Laboratory of Urban Pollutant Conversion, Institute of Urban Environment, Chinese Academy of Sciences (CAS), 1799 Jimei Road, Xiamen 361021, China

[*]E-mail: jz@kemi.dtu.dk.

Nanocatalysts synthesized by bacteria, mainly *Shewanella oneidensis* MR-1, are reviewed. Mechanisms of nanocatalyst biosynthesis by *S. oneidensis* MR-1, including intracellular and extracellular biosynthesis, are also discussed. We present characterization techniques such as UV-Vis spectroscopy, scanning electron microscopy (SEM), transmission electron microscopy (TEM), atomic force microscopy, X-ray photoelectron spectroscopy (XPS), X-ray diffraction (XRD), and thermogravimetric analyses. The biosynthesis process and information about a variety of resulting biogenic nanoparticle (NP) catalysts, particularly metallic and non-metallic NPs, alloys, and metallic and non-metallic sulfide NPs, are discussed further. Representative applications in electrocatalysis, photocatalysis, and biocatalysis are discussed, and opportunities and challenges of biogenic nanocatalysts are summarized.

Introduction

Nanomaterials have been defined as "materials with any external dimension in the nanoscale or having internal structure or surface structure in the nanoscale." "Nanoscale" is defined as a "length range approximately from 1 nm to 100 nm" (*1*). Nanocatalysts, therefore, refer to catalysts in the nanoscale. Compared to bulk counterparts, nanocatalysts offer a wealth of advantages, such as high activity, economy in atomic consumption, high selectivity, and stability. Due to these unique properties, nanocatalysts are attracting wide attention, and numerous nanocatalysts for a range of applications in clean energy technology and chemical industry have emerged.

Different methods to produce nanomaterials have been developed. Physical methods include ion implantation (*2*), vapor deposition (*3*), pulsed laser deposition (*4*), and mechanical techniques (*5*). The nanomaterials produced using physical methods are morphologically well controlled and environmentally friendly, but the methods are also limited by high complexity, low yield, and low efficiency (*6*). Chemical methods, such as ion exchange (*7*), sol-gel deposition (*8*), hydrothermal

© 2020 American Chemical Society

reactions (*9*, *10*), and chemical co-precipitation (*11*), produce high yields and can synthesize uniformly dispersive and narrow-sized nanomaterials in solution, but are demanding regarding experimental conditions, fraught by side reactions, and potentially harmful to the environment. Recently, green synthesis using non-toxic chemicals such as starch has emerged for platinum (*12*) and gold nanoparticles (NPs) (*13*).

The biosynthesis methods are eco-friendly and non-toxic. Different biological reactive agents react with metal ions and form corresponding NPs intra- or extracellularly (*14*). The biosynthesized nanomaterials also benefit from excellent biocompatibility, mild experimental conditions, low cost, high yield, and good scalability. The biosynthesis processes are very attractive because they can be accomplished using bacteria, fungi, plant extracts, and DNA (*15*, *16*).

Due to fast growth, low cost, and established genetic manipulation, bacterial biosynthesis of nanomaterials has been widely investigated. Bacterial biosynthesis of nanomaterials is simple and straightforward. Bacteria are cultured in nutritionally rich media (e.g., Luria broth) to obtain sufficient numbers of bacterial cells. The bacterial cells are then washed with water or a 0.9% NaCl aqueous solution to remove secretion. The cell pellet is then resuspended in a deoxygenated defined medium, such as M9 medium (*17*, *18*), 3-(N-morpholino)propanesulfonic acid buffer (*19*, *20*), sodium 4-(2-hydroxyethyl)-1-piperazineethanesulfonic acid (HEPES) buffer (*21*–*24*), a 0.9% NaCl aqueous solution, or even distilled water (*25*, *26*). Phosphate buffer solution, which is widely used in other applications of bioelectrochemical systems (*27*), is rarely chosen because most metal ions precipitate with phosphate. The redox properties of the medium should also be considered. Maintaining neutral pH is not essential since many bacteria can survive in high concentrations of metal ions due to the extracellular electron transfer (EET) process (*28*); for example, *Lysinibacillus* sp. ZYM-1 can produce Se NPs over a range of pH 5 to 9 (*29*). Finally, depending on nanomaterial and synthesis conditions (i.e. bacteria or medium), the target nanomaterials are synthesized within a few minutes, several hours, or even days after introducing metal ions into the medium (mainly as an electron acceptor) of specific bacteria.

Moreover, the biosynthesis process of nanomaterials by bacteria can achieve environmental remediation and material recycling at the same time. For example, *Cupriavidus necator* and *Pseudomonas putida* can remove Pd(II) from waste materials and the recycled Pd(0) nanomaterials were found to be able to catalyze Suzuki–Miyaura and Mizoroki–Heck reactions (*30*). The C-C bond formation was also catalyzed using the Pd(0) NPs produced by *C. necator* and *Cupriavidus metallidurans* from an acidic leachate containing several heavy metals including Pd(II) (*31*). The remediation by bacteria can be very efficient. *C. necator, P. putida,* and *Paracoccus denitrificans* removed 80%, 100%, and 100% of Pd(II), respectively. The Pd(0) nanomaterials obtained catalyzes dihydrogen production from hypophosphite (*32*).

Depending on extracellular or intracellular biosynthesis, different separation methods (e.g., centrifugation, sonication, and freeze-thawing) can be employed to obtain the separated nanomaterials (*33*). Separation of nanomaterials from bacterial cells is not generally required. Nanomaterial-bacteria hybrids offer several advantages compared to the separated nanomaterials. Aggregation is avoided in biosynthesized nanomaterials. Aggregation of nanomaterials has been a problem in applications, and different surfactants or stabilizing agents (e.g., polyethylene glycol) are used to solve the problem. However, the addition of surfactants often contributes to the decreased performance of the nanomaterials. Bacterial cells act as a supporter and framework in nanomaterial-bacteria hybrids, in which the formation sites for nanomaterials are uniformly distributed on the bacterial cell surfaces (e.g., the outer cell membrane) to efficiently prevent aggregation in this way.

Additionally, yield is increased compared to the separated nanomaterials. For small-sized NP materials (e.g., quantum dots), it is a challenge to retain samples during concentration and purification without aggregation. In contrast, nanomaterials generated by bacteria are attached on or enclosed in the cells. Concentration and purification can thus be accomplished straightforwardly by centrifugation. Other functions are also introduced via chemical doping elements. The main elemental composition of bacterial cells includes carbon (C), hydrogen (H), oxygen (O), nitrogen (N), phosphorus (P), and sulfur (S), in which N, P, and S are essential doping elements for catalysts. These elements can be doped into the nanomaterials in subsequent processes (*17*). After the formation of the nanomaterials, obtained or separated nanomaterials from the nanomaterial-bacteria hybrids can be used as electrocatalysts (*34*) in biosensors (*35*) and in electrosynthesis (*36, 37*) as adsorbents (*24*) and as photothermal agents (*23*) without further treatment. In addition, further treatment (e.g., carbonization, hydrothermal processing, and microwave heating) can be implemented to enhance the electroconductivity and structure of the nanomaterials.

A variety of bacteria have been employed in the biosynthesis of nanomaterials, including *Saccharomyces cerevisiae* (*38*), *Bacillus subtilis* (*39*), *Shewanella* spp. (*17, 25, 34*), and *Escherichia coli* (*40*). As one of several models of electrochemically active bacteria (EAB) in bioelectrochemical systems, *Shewanella oneidensis* MR-1 has gained particular attention in the biosynthesis of various nanomaterials, including metal, metalloid, and inorganic compounds.

Shewanella was first classified in 1931 as belonging to the *Achromobacter* genus (*41*). Several reclassifications were conducted, and in 1985, the new name *Shewanella* was assigned to the genus to honor Dr. James M. Shewan's prolific contributions in fishery microbiology (*42*). *S. oneidensis* MR-1 was first discovered in Oneida Lake, New York in 1988 in relation to Mn^{4+} reduction and named as *Alteromonas putrefaciens* MR-1 (*43*). This bacterium was renamed as *Shewanella putrefaciens* MR-1 before finally being named *S. oneidensis* MR-1 after the lake where it was discovered (*44*). The term "MR" is the abbreviation for "manganese reducer".

S. oneidensis MR-1 is a facultatively aerobic Gram-negative bacterium (*45*), approximately 2–3 μm in length and 0.5 μm across (Figure 1). As a dissimilatory metal-reducing bacterium, *S. oneidensis* MR-1 can anaerobically reduce various metal ions such as Au(III) (*25*), Pd(II) (*18, 34*), Pt(IV) (*17*), and Ag(I) (*46*), to form the corresponding metallic nanomaterials. The metal ions are electron acceptors in bacterial biosynthesis. This process correlates with the EET process (Figure 2), in which EAB exchange electrons with external redox partners, electrochemical electrodes, or other EAB through one of three different pathways (*47, 48*). The first pathway is short-range direct EET, in which EAB exchange electrons with external redox partners via redox proteins in the outer cell membrane, such as OmcA and MtrC. The second pathway is long-range direct EET. *S. oneidensis* MR-1 develops conductive appendices (also termed as "pili" or "nanowires") to reach redox partners when nutrients are limited. The final pathway is mediated EET, in which *S. oneidensis* MR-1 secretes redox mediators (e.g., flavins) to shuttle EET. This pathway dominates the EET process of *S. oneidensis* MR-1 and contributes more than 70% of the EET (*49–52*). In all the pathways, the electrons are transferred by "hoping" in extracellular polymeric substances (EPSs), which envelop the bacteria (*49*).

Much attention has been given to the biosynthesis of nanomaterials, and several studies have summarized these developments (*53–59*). The present chapter provides a general view of metallic and non-metallic nanocatalysts synthesized by bacteria. In particular, the mechanism of bacterial NP biosynthesis with focus on *S. oneidensis* MR-1 is discussed. Characterization methods of the catalytic NPs are described, their applications are summarized, and new prospects and challenges of biosynthesized nanocatalysts are envisioned.

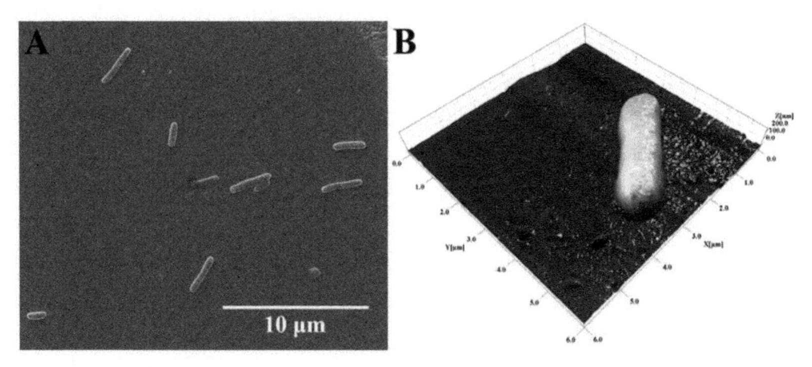

Figure 1. Morphology of S. oneidensis MR-1. (A) Scanning electron microscopy (SEM) and (B) atomic force microscopy (AFM) images of S. oneidensis MR-1.

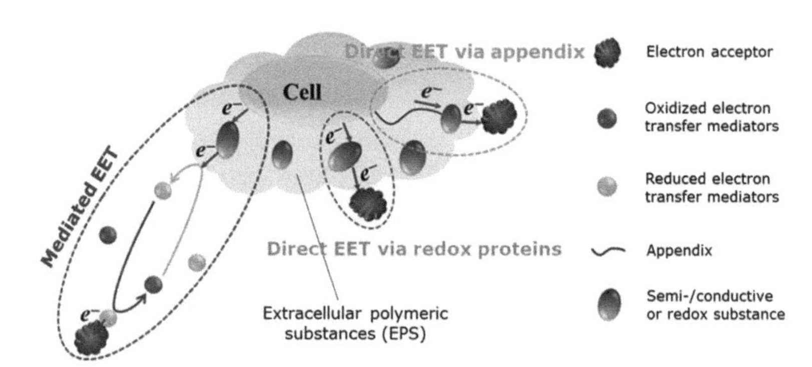

Figure 2. Three pathways of EET of S. oneidensis MR-1. Adapted with permission from reference (49). Copyright 2016 the American Association for the Advancement of Science.

Biosynthesis Mechanisms of Nanocatalysts

Metal ions and metal complexes are often positively charged while there are numerous negatively charged sites on the bacterial cell membrane due to the presence of glycoconjugates and surface groups such as carboxylate groups (*60, 61*). There are, however, also positively charged sites on the cell surface caused by positively charged proteins, which attract negatively charged metal complexes. As shown in Figure 3, target ions are captured from the environment by electrostatic interactions. The ions are then extracellularly converted into elemental metals by redox molecules on the outer membrane (e.g., cytochromes and enzymes). Alternatively, the ions are transported through the outer membrane, interact with redox molecules, and intracellularly transformed into nanocatalysts in the periplasm (*19, 25, 34, 60, 62*).

The exact routes for bacterial nanocatalyst biosynthesis are very complex and vary depending on the bacteria and ions. Presently, focus is on the mechanism of bacterial biosynthesis of nanocatalysts by the representative bacterium (*S. oneidensis* MR-1). This bacterium can reduce a variety of metal ions and EET of this particular species has been well studied.

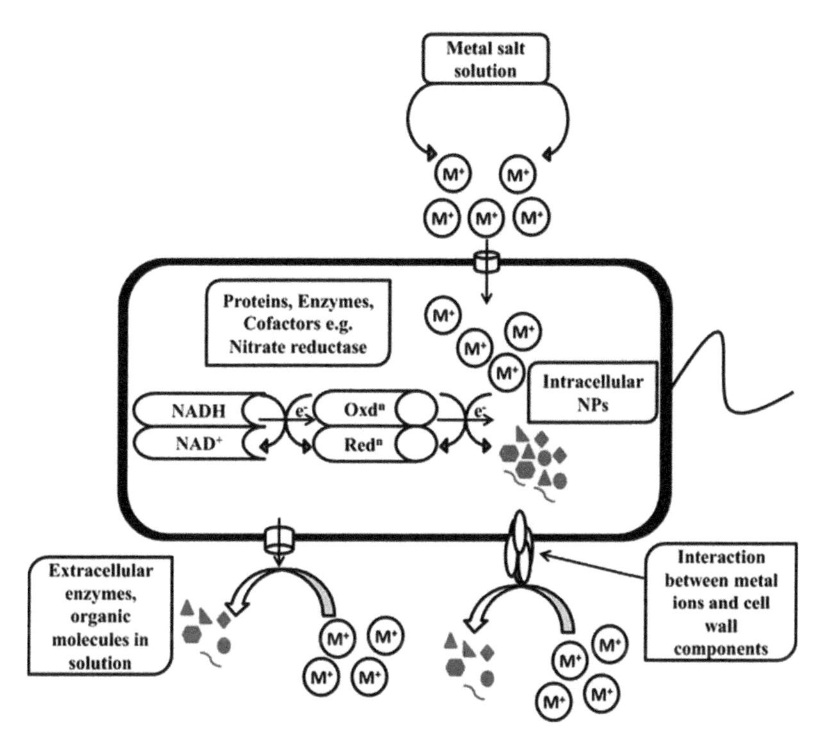

Figure 3. Microbial synthesis mechanism of nanocatalysts. Adapted with permission from reference (60). Copyright 2016 Springer Nature.

The cell membrane of *S. oneidensis* MR-1 is shown schematically in Figure 4. The formation of nanocatalysts correlates with the precipitation and bioreduction of metal ions by EAB (*63*). Metal ions precipitate on or within the bacterial cell because the metal concentration exceeds the stoichiometry per reactive sites on or in the cell. That means that a small amount of soluble metals can be associated with cell surface or inside the cell, while high metal concentration can cause precipitation on the reactive sites (*64*). Although three EET pathways for *S. oneidensis* MR-1 have been proposed, the short-range direct EET prevails in the biosynthesis of nanocatalysts compared to the other two pathways. Nanowires are unlikely to be formed in the biosynthesis, in which abundant metal ions are present as electron acceptors and are usually formed only when electron acceptors are limited (*65*). In the mediated EET pathway, *S. oneidensis* MR-1 excretes flavins, which mediate the EET between the cells and external electron acceptors (*52*). However, biosynthesis of nanocatalysts by flavins is hampered since their midpoint potentials are in the range -0.2 to -0.25 V compared to the standard hydrogen electrode (*52*). Different cytochromes *c* (e.g., MtrC, MtrA, and OmcA) are involved in the short-range direct EET and form the OmcA–MtrCAB pathway (*49, 66*). Electrons from oxidized quinol are first delivered to CymA, where the electrons are then carried to MtrA using FccA and a small tetraheme cytochrome (STC). This is followed by electron transport successively through MtrA, MtrB, and MtrC, which constitutes a complex penetrating the outer cell membrane. Finally, outer membrane cytochromes *c* MtrC and OmcA directly relay the electrons to metal ions (*67–69*). MtrA and OmcA are therefore important for the biosynthesis of metallic nanocatalysts and strongly affect the size of the nanocatalyst particles (*20*). Nonetheless, the presence of MtrA and OmcA is not indispensable for the formation of all nanomaterials since a mutant without MtrA and OmcA (*S. oneidensis* MR-1 *ΔomcA/mtrC*) can also synthesize nanocatalysts such as Ag (*20*), Se(IV) (*21*), and Au (*25*) NPs. The nanocatalysts synthesized by the mutant can differ in size and antibacterial activity from those synthesized by the wildtype, making controlling synthesis by gene technologies possible (*20*). The ability of *S. oneidensis* MR-1 *ΔomcA/mtrC* to synthesize extracellular nanocatalysts indicates that other reactive sites are present, but the details regarding these sites remain to be explored.

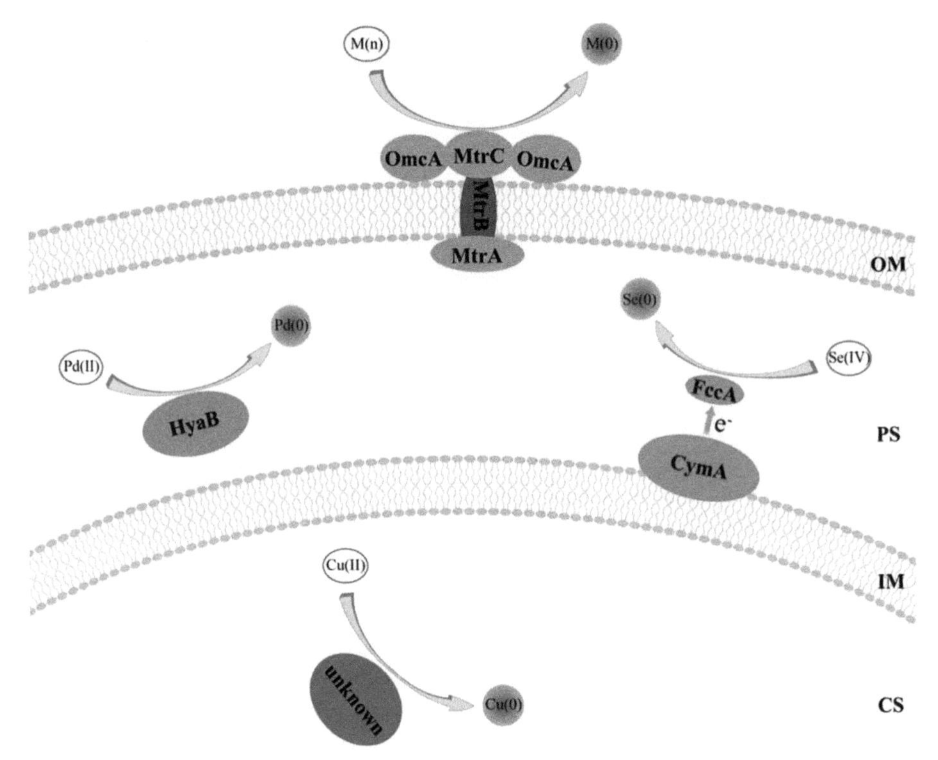

Figure 4. Schematic view of cell membrane structure of S. oneidensis MR-1 and main synthesis sites of metallic and metalloid nanomaterials. M(n) = Tc(VII), Ag(I), U(VI), and Au(III). OM: outer membrane; PS: periplasm; IM: inner membrane; CS: cytoplasm. Note that the indicated sites are the primary sites for the metals and metalloids, but other sites are also possible. For simplicity, the quinol and other structures are not presented.

Moreover, some studies show that MtrC and OmcA play an important role in the reduction of Fe(III), Cr(VI), Tc(VII), Ag(I), Au(III), and U(VI) (*20, 28, 70, 71*), while other studies imply that hydrogenases, specifically [NiFe]-hydrogenase HyaB, are responsible for the reduction of Pd(II) (*22*). Two main hydrogenases are present in the genome of *S. oneidensis* MR-1: HydA and HyaB (*72*). HydA is a periplasmic [Fe-Fe] hydrogenase, which is involved in dihydrogen formation, while HyaB is a bifunctional periplasmic [Ni-Fe] hydrogenase responsible for either the formation or oxidation of dihydrogen (*72*). The Pd(II) complex, normally $[PdCl_4]^{2-}$, has to penetrate the outer cell membrane in order to be reduced by HyaB, but only limited information about the mechanism of how $[PdCl_4]^{2-}$ is transported through the outer membrane is available. On the other hand, abundant Pd NPs were also found on the outer membrane (*34*).

The OmcA-MtrCAB pathway, nitrate and nitrite reductase, and hydrogenase are reported to have little effect on the reduction of SeO_3^{2-} to Se(0). In the periplasm, CymA relays electrons from quinol to fumarate reductase FccA, which further reduces SeO_3^{2-} to Se(0) (*21*). Different from

Au, Pd, or Pt NPs with uniform and small size, the Se NPs are apparently bigger and more widely dispersed in size.

Apart from the formation of nanocatalysts in the outer membrane and the periplasm of *S. oneidensis* MR-1, Cu particles are also found to be dispersed in the cytoplasm and periplasm. In the reduction of Cu(II), MtrC, OmcA, MtrF, MtrABCDEF, DmsE, S04360, and CctA did not play a key role. The reduction of Cu(II) could happen intracellularly, with possible unidentified reductases involved (*19*). Tellurium (Te) nanomaterials were also observed in the cytoplasm and periplasm (*73*, *74*), but detailed mechanisms are not clear.

Figure 5. Circular representation of the S. oneidensis MR-1 chromosome sequence related to EET process. The sequence originates from GenBank (accession numbers AE014299).

The genome of *S. oneidensis* MR-1 is composed of 4,969,803 base pairs (*45*), including 76 base pairs involving the EET process according to information from the National Center for Biotechnology Information database (Figure 5). Products of the 76 include cytochromes, hydrogenases, reductases, and flavodoxins, among others. It has been reported that there are 39 cythchromes *c* in *S. oneidensis* MR-1 including 8 decaheme cytochrome *c* (*45*). The vital genes in the OmcA–MtrCAB pathway are summarized in Table 1. Note that among the 39 cytochromes *c*, only 6 of them (MtrA, MtrC, OmcA, CymA, small tetraheme cytochrome, and FccA) have been well documented in the EET process, and even fewer have been fully investigated in the synthesis of nanomaterials.

Table 1. Important Genes in OmcA-MtrCAB Pathway

Locus tag	*No. of Haems*	*Location in the Cell*	*Reference for S. oneidensis MR-1*
SO_1776	-[a]	Outer membrane	MtrB
SO_1777	10	Periplasm	MtrA
SO_1778	10	Outer membrane	MtrC
SO_1779	10	Outer membrane	OmcA
SO_4591	4	Periplasm	CymA

[a] MtrB is not a cytochrome *c* and contains no haem.

In summary, various electron transport pathways in *S. oneidensis* MR-1 provide different biosynthesis mechanisms. Different enzymes and 42 cytochrome *c* species are present in the genome of the bacteria, but only some of them have been studied (e.g., MtrC, OmcA, MtrB, CymA). The three established EET pathways dominate the reduction of insoluble Mn(IV) and Fe(III) (hydro)oxides, but different EET processes operate for other soluble metal ions (Figure 4).

Characterization of Biogenic Nanocatalysts

In the bacterial synthesis of nanocatalysts, it is important to follow the formation process. The defined medium changes to pale yellow after bacterial cells (e.g., *S. oneidensis* MR-1) are resuspended into the medium. The medium can change to a specific color or the color of the metal ions fade as the nanocatalysts are formed, which means that the formation of some nanocatalysts can be followed directly with UV-Vis spectroscopy. For example, the absorption peak around the wavelength of maximum absorbance (λ_{max}) of 411 nm disappears when Pd nanocatalysts are formed (*34*). A peak at λ_{max} = 530 nm appears within 24 h in the biosynthesis of Au nanomaterial using *S. oneidensis* MR-1 (*28*). For Ag nanomaterials, a peak at λ_{max} = 418 nm emerges in the colorless $AgNO_3$ solution containing *S. oneidensis* MR-1 after 48 h (*46*). CuS NPs form simultaneously with the appearance of a peak at λ_{max} = 1100 nm (*23*). Moreover, since λ_{max} is strongly linked with the size or the structure of nanomaterials, the evolution of nanomaterials can be monitored by UV-Vis spectroscopy. For example, a spectral red shift indicates an increasing amount of CdS quantum dots (*38*).

Electron microscopy technologies are crucial for visualization of the precise size and morphology as well as the structure of the nanocatalyst NPs. The most common of these technologies are SEM and TEM. The SEM and TEM sample preparations are similar, and fixation (usually by glutaraldehyde) and dehydration (e.g., gradient ethanol dehydration) are required to maintain the original structure of the bacterial cells (*17, 34, 38, 75*). A relatively low voltage (5–20 kV) is applied in SEM of biological samples to avoid destroying the samples (*76*). SEM can image a large enough area for nanomaterial-bacteria hybrids, and the resolution can reach 1 nm. The diameter of bacterial cells can be several hundred nanometers, but slicing is not required if the inside structure of the cells is not addressed. Advanced SEM technologies have emerged and provided much more information. For example, "3D" images were constructed with serial block-face SEM, confirming the presence of Cu NPs inside the *S. oneidensis* MR-1 cells (*19*). TEM is very powerful for observing both the morphology and structure of the nanomaterials, but ultra-thin sections of the nanomaterial-bacterial hybrids must be sliced if the inside of the cells is to be mapped (*19*, *25*) since the electron beam of the TEM cannot penetrate samples thicker than 100 nm. The surface facets of the nanocatalysts are key factors for their performance, and the extraordinary nanometer-to-atomic TEM resolution makes the acquisition of such information possible, particularly the structure of the crystal lattice (*17*, *23*). Coupled with energy-dispersive X-ray spectroscopy (EDS), TEM can map not only the morphology of the nanocatalysts, but also the elemental composition, which is essential for the characterization of alloys and core-shell NPs. For example, the EDS mapping displayed in Figure 6 shows clearly overlapping Pd, Au, P, and S peaks, indicative of a doped alloy (*17*).

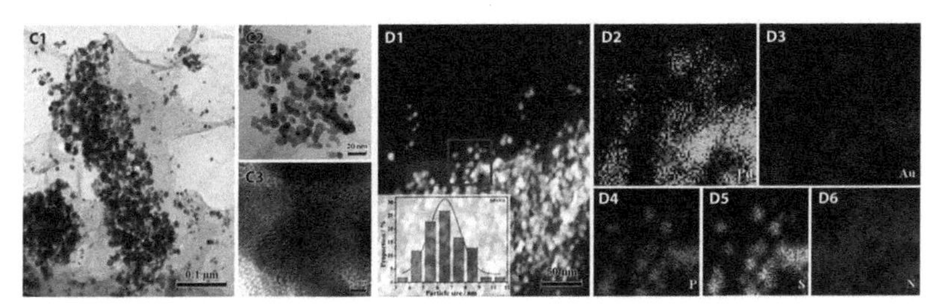

Figure 6. TEM and corresponding EDS mapping of PdAu alloy biosynthesized by S. oneidensis MR-1 with subsequent hydrothermal treatment. Reproduced with permission from reference (17). Copyright 2016 the American Association for the Advancement of Science.

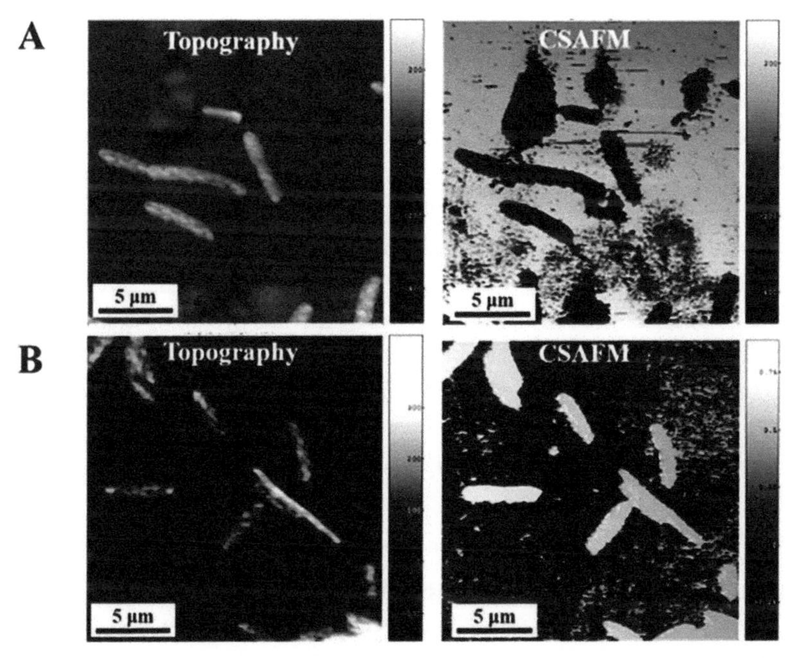

Figure 7. Representative topographic (left) and current-sensing AFM images (right) of (A) S. oneidensis MR-1 and (B) S. oneidensis MR-1 coated with biosynthesized Pd nanocatalysts. Reproduced with permission from reference (34). Copyright 2018 The Royal Society of Chemistry.

Another versatile technique is AFM. The resolution of AFM can also reach the nanometer level, and, AFM is therefore suitable for precise imaging of nanomaterials (*46*). The sample preparation of AFM is simple with a small number of samples immobilized and spread out on a flat surface. Fixation, dehydration, coating with metal layer, and slicing are not needed, which minimizes sample destruction. There is no need for a vacuum; AFM can be conducted in ambient atmosphere and even in solution. In addition, 3D images can be readily constructed from AFM measurements (Figure 1A). AFM can provide physical properties and information other than size and morphology. For instance, Figure 7 shows a comparison between the topographic and current-sensing AFM (CSAFM) images of *S. oneidensis* MR-1 cells coated with Pd NPs that are apparently different from pristine *S. oneidensis* MR-1 cells. Compared with the dark area in pristine *S. oneidensis* MR-1 cells, the brightness of the cells coated with biosynthesized Pd nanocatalysts reflects higher conductivity, since the brightness in CSAFM images indicates the current flow between the AFM tip and the substrate (*34*).

EDS in SEM and TEM can show the elemental composition but does not give other information, such as chemical and electronic state of the elements. X-ray photoelectron spectroscopy (XPS) is helpful in achieving this additional information. XPS is a qualitative and semiquantitative analysis method for chemical surface elements including the chemical and electronic states. For example, a peak belonging to Se_{3d} emerged in Se nanospheres synthesized by MR-1 (*35*). In another study, the ratio of Pd, Au, P, N, and S elements (32.28:10.83:4.32:47.09:5.48) was obtained using XPS (*17*).

The qualitative and semiquantitative information from XPS analysis is highly valuable, but XPS is limited only to the surface of the sample (e.g., 20 nm into samples). This is not comprehensive enough for bulk materials (i.e. nanomaterial-bacterial hybrids), which are hundreds of nanometers in diameter. To get the crystal structure on or within nanomaterial-bacterial hybrids, X-ray diffraction (XRD) analysis, in which the penetration of the X-ray can be several millimeters, is practical. For example, both peaks belonging to Au and Pd were observed in the PdAu alloy (*17, 26*). Nonetheless, XRD analysis alone is not sufficient to confirm the phase analysis since the XRD pattern of different crystals can be very similar and it is difficult to analyze when different crystals co-exist. Moreover, signals are weak if the nanocatalysts are amorphous or the metals are highly dispersed in the nanomaterials. If possible, XRD, EDS, TEM, and XPS analysis should all be considered (*17, 23*).

The biosynthesized nanocatalysts are different from their chemically synthesized analogues. Bacterial cells are normally covered with EPS, and organic substances can participate in the formation of the nanocatalysts. To quantify these substances, thermogravimetric analyses can be employed. For example, the weight loss from 100 to 400 °C is caused by thermal decomposition of adsorbed organic substances in nanomaterials synthesized by bacteria, while the weight loss from 400 to 800 °C is due to decomposition of intracrystalline organic substances (*26, 77*).

Classes of Biogenic Nanocatalysts

Metals and Alloys

Much attention has been given to the biosynthesis of metallic nanomaterials, (especially noble metal nanomaterials) due to the considerable economic benefits of metal recovery (*54*). A variety of noble metal ions, such as Au(III), Pt(IV), Pd(III), and Ag(I), can be recovered by bacteria (*17, 46, 78, 79*). The local size of the nanomaterials depends on experimental conditions, but typically the microstructures of noble metal nanomaterials are small (i.e., in ranges from a few to tens of nanometers) and uniform. Biogenic Pt NPs synthesized by *Shewanella algae* (*79*) are about 5 nm. Au and Pd NPs synthesized by *S. oneidensis* MR-1 are 5–30 nm (*25*) and 3–10 nm (*34*), respectively. The size of Ag NPs is similar to that of Pd NPs (i.e., about 2–10 nm in diameter) (*46*).

The recovery of noble metal ions is very high within a relatively short treatment time. For example, a recovery of 99.6% of Pd(II) was achieved by *S. oneidensis* MR-1 ($OD_{600} = 2.0$) overnight when 50 mg/L of Pd(II) was used as an electron acceptor and dihydrogen as an electron donor (*18*). The recovery remains high at high concentrations of Pd(II). For instance, only 0.1 mg/L remained in the solution when 1000 mg/L of Pd(II) was introduced (*18*). Au (III) removal can be visually observed within 30 min indicated by the color change from pale yellow to purple when *S. oneidensis* MR-1 is exposed to 100 mg/L of $[AuCl_4]^-$ (*25*). Similarly, about 90% of $[PtCl_6]^{2-}$ was removed by *S. algae* in 60 min (*79*).

Other metal ions can also be recovered by bacteria, forming corresponding metal NPs. The Cu NPs synthesized by *S. oneidensis* MR-1 are relatively larger than the noble metal NPs just noted, with a typical size range of 20 to 40 nm (*19*). The recovery of Cu(II) is also high, with 70% after 3 h, 91% after 24 h, and 100% after 96 h. The core of the NPs is Cu(0), while the surface is Cu_2O due to oxidation caused by exposure to oxygen in the air (*19*). Co NPs of 4-8 nm in length were formed in the outer membrane of *S. putrefaciens* CN32 after the cells were exposed to 195 mg/L of Co^{2+} in pH 3 for 24 h. Moderate recovery of 21% was achieved with an initial concentration of 210.745 mg/L. The limited recovery may be due to the short lifespan of the bacteria under these experimental conditions (*80*).

The ability of EAB to reduce various metal ions offers an approach for the synthesis of alloys. A highly dispersed PdAu alloy was synthesized on the cell surface of *S. oneidensis* MR-1 after successive addition of $[AuCl_4]^-$ and $PdCl_4]^{2-}$. As shown in Figure 6, NPs of this alloy are quite small, with an average diameter of 5 nm (*17*). An extracellular PdAu alloy is also formed when a higher concentration of $[PdCl_4]^{2-}$ and $[AuCl_4]^-$ are used as electron acceptors; for example, a nanocomposite of PdAu alloy and Fe_3O_4 with an NP size of 3–15 nm was obtained from *S. oneidensis* MR-1 after 1 mM of $[PdCl_4]^{2-}$ and 1 mM of $[AuCl_4]^-$ were introduced into the medium containing akaganeite for 48 h (*26*).

Metal and Metalloid Sulfides

EAB can also reduce sulfur and thiosulfate to sulfide (*81*), producing metal sulfides when specific metal ions, sulfur, and thiosulfate as electron acceptors are present simultaneously. For example, brown-colored CuS NPs with a uniform size of about 5 nm are formed extracellularly, when *S. oneidensis* MR-1 is incubated in a HEPES-buffered mineral medium containing 1 mM of $Na_2S_2O_3$, $CuCl_2$, and 20 mM of lactate. The Cu:S ratio was 0.94:1 analyzed from XPS results (23). In another study, CuS nanorods with 17.4 nm and 80.8 nm in diameter and length were embedded in the *S. oneidensis* MR-1 cell membrane, forming a complex hollow shell structure (*24*).

S. oneidensis MR-1 can synthesize Ag_2S NPs of 53.4 ± 12.4 nm. The size of the NPs decreased to 27.6± 6.4 nm when MtrC and OmcA were knocked out (*20*). In contrast, another study showed that the presence of *S. oneidensis* MR-1 cells is not required to form Ag_2S NPs (*82*). These authors inferred that the complex $Na_3[Ag(S_2O_3)_2]$ caused precipitation of Ag_2S. However, MtrC, OmcA, and MtrB from *S. oneidensis* MR-1 can stabilize Ag_2S NPs. The native cell structure stabilizes rather than forms Ag_2S NPs, which means larger Ag_2S NPs are formed in the absence of *S. oneidensis* MR-1 cells (*82*). Smaller Ag_2S NPs can also be obtained from the same bacterium. For example, Ag_2S NPs smaller than 8 nm attached to TiO_2 nanotubes have been produced by *S. oneidensis* MR-1 (*83*).

FeS can also be formed abiotically. For example, poorly crystalline FeS can be obtained by mixing 0.57 M of $FeCl_2$ and 1.1 M of Na_2S (*84*). Different from abiotic FeS with bulk and irregular forms, the biogenic FeS synthesized by *S. putrefaciens* CN32 is mainly comprised of 100-nm NPs. Moreover, the biogenic Fe:S ratio was 2.3, which is different from that of abiotic FeS with a ratio of 1.3 (*85*).

CdS NPs ("quantum dots") with an average diameter of 2 nm can be synthesized by *S. cerevisiae* via yeast cells cultured in 0.1 mM of $CdCl_2$ and 0.05 mM of Na_2S for one day. Notably, the size of the CdS quantum dots increases with longer culture times (*38*). Much larger CdS NPs (about 15 nm in diameter) synthesized by *S. oneidensis* MR-1 have also been reported (*86*). Moreover, the addition of ionic liquid can modify the CdS NPs from agglomerated and irregular shapes to highly ordered spherical structures (*86*).

Biosynthesis also provides morphologies that are not available from chemical synthesis. AsS nanotubes that are 20–100 nm in diameter and about 30 µm in length were produced extracellularly by *Shewanella* sp. strain HN-41 in a medium containing As(V) and $S_2O_3^{2-}$. The nanotube composition nine days after inoculation was As_2S_3, which transformed to AsS after two to three weeks (*87*).

Notably, the addition of metal and sulfur or thiosulfate is sometimes not needed. Some bacteria are able to synthesize metal sulfide-based nanomaterials even in groundwater. Natural biofilms of *Desulfobacteraceae* can, for example, synthesize 2–5-nm ZnS NPs by accumulating a Zn concentration that is 10^6 times over ground water level (*88*).

The biosynthesis of other metal sulfides is also reported, such as MnS (*89*), but the application of these metal sulfides as catalysts is rarely reported. We therefore do not include detailed discussion of these other biogenic metal sulfides.

Metal Oxides and Metal Hydroxides

Fe_3O_4 NPs were obtained under anaerobic conditions by transforming akaganeite to magnetite by culturing *S. oneidensis* MR-1 with lactate as an electron donor for 48 h. The Fe_3O_4 NP size was 3–15 nm in diameter (*26*). Larger Fe_3O_4 NPs with a diameter of 26–38 nm can be acquired from *Shewanella* sp. HN-41 using a similar method (*90*).

In contrast, another study reported that the transformation of akageneite by *Shewanella* sp. HN-41 depends on the amount of akaganeite precursors and Fe(II) in the solution. Akaganeite nanorods that are about 5 nm in width and 20 nm in length were used as precursors and electron acceptors. When 30-mM akaganeite was introduced after 10 days, magnetite NPs up to 100 nm diameter appeared. However, goethite nanowires 15 nm in width and 500 nm in length appeared instead (*91*).

UO_2 NPs formed when *S. oneidensis* MR-1 cells were inoculated in 250 µm of uranyl acetate and 10 mM of sodium lactate under anaerobic conditions. The UO_2 NPs are quite small, only 1–5 nm in diameter. The NPs appeared in three forms, in which some, densely packed with EPS, were complex structures, similar to glycocalyx. This indicates that EPS (possibly with a redox substance inside) plays an important role in the formation of UO_2 NPs (*71*).

Metalloids

Apart from metals and metal compounds, the biosynthesis of metalloid nanomaterials by EAB have been extensively studied (*92*). For example, the size distribution of Se NPs can be adjusted by controlling the biomass concentration of *Shewanella* sp. HN-41 and the initial selenite concentration. Within 2 h, 1–20 nm amorphous Se NPs were produced with low initial biomass under anaerobic conditions. Much larger NPs (about 150 nm) were also observed. Se NPs around 123 nm were produced when a larger initial biomass was exposed for 24 h (*93*). Different bacteria can synthesize Se NPs at different sizes using different mechanisms of Se(VI) reduction. Large, 100–250 nm Se NPs were synthesized by *S. oneidensis* MR-1, and smaller 50–100 nm Se NPs were acquired by *Geobacter sulfurreducens* (ATCC 51573), while the smallest Se NPs (around 50 nm) were synthesized by *Veillonella atypica* (ATCC 14894) (*94*). Another study showed that the bigger Se NPs (around 100 nm) formed inside the cells, while smaller Se NPs (around 20 nm) formed extracellularly; the biosynthesis of Se NPs is potentially controlled by the EPS (*21*).

Another metalloid NP biosynthesis by EAB was also reported (*63*). Tellurite (Te) nanomaterials were formed inside *S. oneidensis* MR-1, and a pathway different from Se reduction was proposed (*74*). According to another detailed study, more than 90% of Te(IV) was recovered after incubating *S. oneidensis* MR-1 in 100 μM of sodium Te and 10 mM sodium lactate for 120 h. The products are single-crystalline Te nanorods with a length of 100–200 nm (*73*). Notably, the presence of Fe(III) will change the synthesis location and the size of the Te nanorods. When Te(IV) and Fe(III) co-exist in the medium, more extracellular Te nanorods accumulated with 240 nm and 25 nm for the length and diameter, respectively. However, when Te(IV) was introduced after Fe(III) was reduced to Fe(II), exclusively extracellular crystalline Te nanorods were formed at a smaller size (i.e., 89 nm and 7.5 nm in the length and diameter, respectively were found) (*95*). In addition, *Shewanella baltica* was reported as having the ability to reduce Te(IV) and form 8–75 nm Te nanorods (*96*).

Applications of Biogenic Nanocatalysts

As noted, different bacteria can synthesize various nanomaterials . The resulting nanomaterials offer numerous applications. The applications of catalytic NPs synthesized by *S. oneidensis* MR-1 and other representative bacteria are summarized in Table 2. Three kinds of applications (i.e., electrocatalysts, photocatalysts, and biocatalysts) are discussed below.

Electrocatalysis

The biosynthesized nanocatalysts not only exhibit some exclusive morphology, but also unique catalytic properties. Pd NPs synthesized by *S. oneidensis* MR-1 show unique selective catalysis to formate electrooxidation, but no electrocatalysis to oxidize other biofuels, such as ethanol, methanol, and acetate, in neutral solution. The selectivity is caused by preferential binding of formate over the other fuels. Moreover, compared to Pd electrodeposited on an electrode, the anodic peak for formate oxidation is more negative by 220 mV (0.10 V vs. a saturated calomel electrode), and exhibit less activation energy (*34*). The poor conductivity of the cell substrate itself is compensated by the PdNP coating, as inferred clearly from the current-sensing AFM images (Figure 7).

EAB can also facilitate nanocatalyst formation not only on the cell surface, but also on electrochemical electrode surfaces. For example, 10–100 nm Pd NPs were coated on a cathode (a piece of carbon cloth) by *S. oneidensis* MR-1 poised at 0.8 V. The size increased to 200–250 nm for abiotic Pd NPs produced using electrochemical method. The smaller size of the NPs and the presence of cells as biocatalysts led to a larger surface area, resulting 90.0 ± 1.4 Coulombs of electron transfer in dihydrogen production with a Pd loading of 40.5 m^2 g^{-1}. This is much higher than the 75.0 ± 1.2 Coulombs in the absence of the bacteria. As a result, the dihydrogen production and recovery (61.8 ± 2.0 L-H_2 m^{-3} day^{-1}, hydrogen volume per reactor volume per day, and 65.5 ± 3.1%, respectively) in the presence of the bacteria were significantly higher than in the absence of bacteria (38.5 ± 2.0 L-H_2 m^{-3} day^{-1} and 47.3 ± 3.9%, respectively). However, the stability of the biodeposited Pd NPs was not satisfactory and the dihydrogen production decreased by 37% after five cycles. Addition of Nafion as binding agents therefore was needed to improve consistent catalytic performance (*98*).

Table 2. Nanocatalysts Synthesized by Bacteria and Their Applications

Nanocatalysts	*Bacteria*	*Applications*	*References*
Au	*S. oneidensis* MR-1	Biocatalysts	(*26*)
Cu	*S. oneidensis* MR-1	Biocatalysts	(*19*)
Pt	*S. oneidensis* MR-1	Biocatalysts	(*97*)
Pd	*S. oneidensis* MR-1	Electrocatalysts	(*34, 98*)
		Biocatalysts	(*18, 26, 97*)
Se	*Lysinibacillus* sp. ZYM- 1	Photocatalysts	(*29*)
Te	*S. baltica*	Photocatalysts	(*96*)
Ag_2S	*S. oneidensis* MR-1	Photocatalysts	(*82*)
		Biocatalysts	(*20*)
FeS	*S. putrefaciens* CN32	Biocatalysts	(*85*)
CdS	*Moorella thermoacetica*	Biocatalysts	(*36*)
	S. oneidensis MR-1	Antibacterial agent	(*86*)
ZnS	*S. oneidensis* MR-1	Photocatalysts	(*99*)
PdAu	*S. oneidensis* MR-1	Electrocatalysts	(*17*)
		Biocatalysts	(*26*)
PdPt	*S. oneidensis* MR-1	Biocatalysts	(*97*)

The nanocatalysts synthesized by EAB are small and uniform but have poor crystallinity and conductivity, which may prevent electrocatalysis. Other subsequent treatments can be employed to improve the performance. A highly efficient electrocatalyst with a hybrid PdAu alloy covered with graphene oxide has been designed. The hybrid was initially synthesized by *S. oneidensis* MR-1 and underwent subsequent hydrothermal treatment. As shown in Figure 8, the hybrid showed a 6.15-fold higher mass electrocatalytic activity for ethanol oxidation in alkaline condition and a 6.58-fold higher activity for formate oxidation in acid condition compared to a commercial Pd/C catalyst with the same Pd loading of 4 μg cm^{-2} (*17*). The hybrid also showed better stability and continued outperforming the commercial Pd/C catalyst after 2000 s. The high catalytic activity was attributed to the three-dimensional porous structure and the carbon support as well as doping elements from *S. oneidensis* MR-1 cells, the nature of PdAu alloy, and the enhanced conductivity of reduced graphene oxide (*17*).

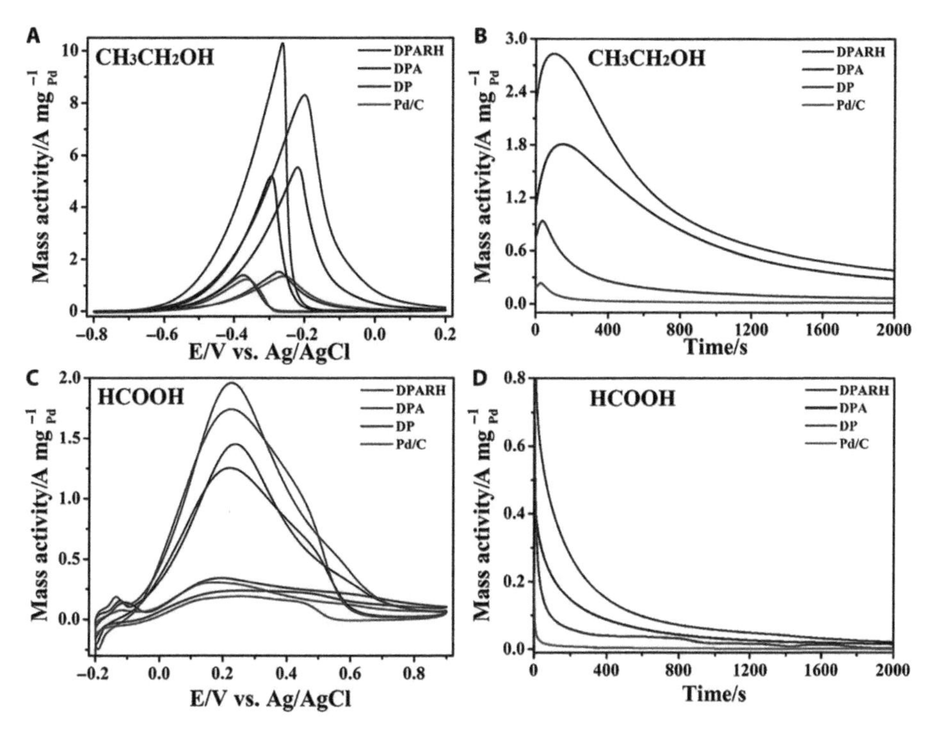

Figure 8. Electrocatalytic performances of as-prepared biogenic catalysts. (A) Cyclic voltammetry (CV) of the carbonized hybrids of S. oneidensis MR-1 cells, PdAu alloy with reduced graphene oxide (DPARH), carbonized hybrids of S. oneidensis MR-1 cells and PdAu alloy (DPA), carbonized hybrids of S. oneidensis MR-1 cells and Pd NPs (DP), and commercial Pd/C catalyst–modified electrodes in 1 M KOH + 1 M ethanol. Scan rate is 50 mV s^{-1}. Potentials versus Ag/AgCl (saturated KCl). (B) Chronoamperometric curves of the catalyst-modified electrodes in 1 M KOH + 1 M of ethanol at −0.3 V for 2000 s. (C) CVs of these catalyst-modified electrodes in 0.5 M H_2SO_4 + 0.5 M HCOOH. Scan rate is 50 mV s^{-1}. (D) Chronoamperometric curves of the catalyst-modified electrodes in 0.5 M of H_2SO_4 + 0.5 M HCOOH at 0.1 V for 2000 s. The Pd mass amounts of all the catalysts were about 1 μg in each electrode (17). Reproduced with permission from reference (17). Copyright 2016 by the American Association for the Advancement of Science.

Photocatalysis

The nanocatalysts formed on the EAB surface possess an ability to degrade pollutants that are resistant to biodegradation. As shown in Figure 9A, a hybrid of *S. oneidensis* MR-1 and 50 mg of biosynthesized ZnS NPs with a diameter of 5 nm totally degraded 20 mg L^{-1} of rhodamine B (RhB) in 3 h under UV irradiation. The 554-nm peak belonging to RhB vanished, and a new, blue-shifted peak (from 550 nm to 500 nm) appeared with the degradation, indicative of de-ethylation of the N,N,N′,N′-tetraethylrhodamine structure in RhB (Figure 9B). Further investigation concluded that the photogenerated holes generated by the biosynthesized ZnS, not the hydroxyl radicals, contributed to the photocatalysis (99).

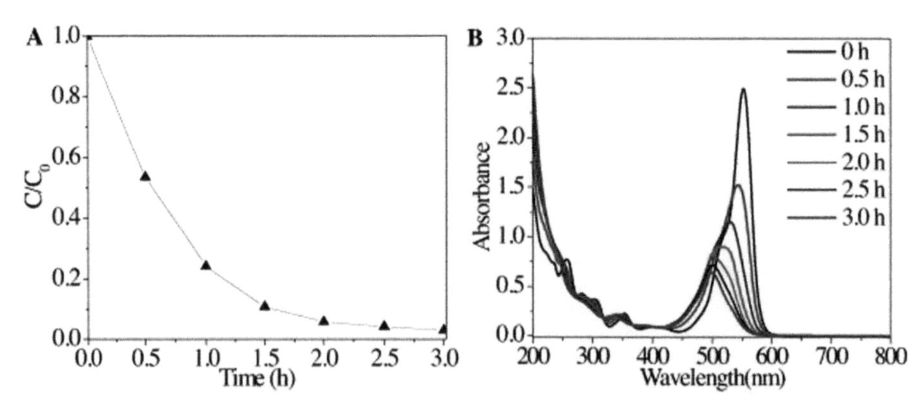

Figure 9. (A) Photocatalytic activity of synthesized ZnS nanocrystals for the photodegradation of rhodamine B (RhB) in aqueous solution (20 mg/L) in air. C_0 and C represent the initial concentration and residual concentrations of RhB, respectively. (B) UV-vis absorption changes of a RhB aqueous solution at room temperature in the presence of ZnS NPs under UV irradiation. Reproduced with permission from reference (99). Copyright 2015 Elsevier.

In a recent study, a mixture of chemically synthesized Ag_3PO_4 NPs and *S. oneidensis* MR-1 can degrade RhB with visible light irradiation under anaerobic conditions. After five days of light irradiation, 15 mg/L of RhB had been completely degraded by the Ag_3PO_4 NPs (0.5g/L) and *S. oneidensis* MR-1 cells. In this case, *S. oneidensis* MR-1 cells significantly enhanced the photocatalytic efficiency. As shown in Figure 10, the main role of *S. oneidensis* MR-1 was to provide electrons to the Ag_3PO_4 photocatalyst after Ag_3PO_4 excitation by light to produce photogenerated electrons in stepwise decomposition of RhB to rhodamine (*100*).

RhB can also be photodegraded by Se nanocatalysts synthesized by *Lysinibacillus* sp. ZYM-1 under visible light irradiation in combination with H_2O_2. *Lysinibacillus* sp. ZYM-1 produces Se nanorods, nanocubes, and nanospheres with different initial concentrations of selenite, but only the nanospheres showed photocatalytic performance. Ten milligrams of Se nanospheres photodecomposed RhB (10 mg L^{-1}, 50 mL) in 5 hours with a reaction rate constant of 0.0048 min^{-1} outperforming chemogenic Se nanomaterials. Both chromophore cleavage and N-de-ethylation contributed to the photodegradation (*29*).

The biogenic Ag_2S nanocatalysts produced by *S. oneidensis* MR-1 were also found to degrade pollutants under visible light irradiation. For example, coated on TiO_2 nanotubes, 20-mg Ag_2S nanocatalysts can photodecompose 4-nitrophenol (0.12 mmol L^{-1}, 50 mL) to 4-aminophenol completely within five hours. The photocatalytic activity increased with increasing molar ratio of Ag/Ti until 1/10, where the excess Ag_2S conglomerates and hinders the activity. One of the key factors is the electron transfer between Ag_2S NPs and TiO_2 nanotubes (*83*).

The reduction of methylene blue dye under sunlight is relatively slow, and only 20% reduction was detected after four hours. However, 90% reduction with 10 μg/mL Te NPs synthesized by *S. baltica* is achieved within the same period. Compared to the Te NPs obtained from chemical synthesis, the option of recycling is a notable advantage (*96*).

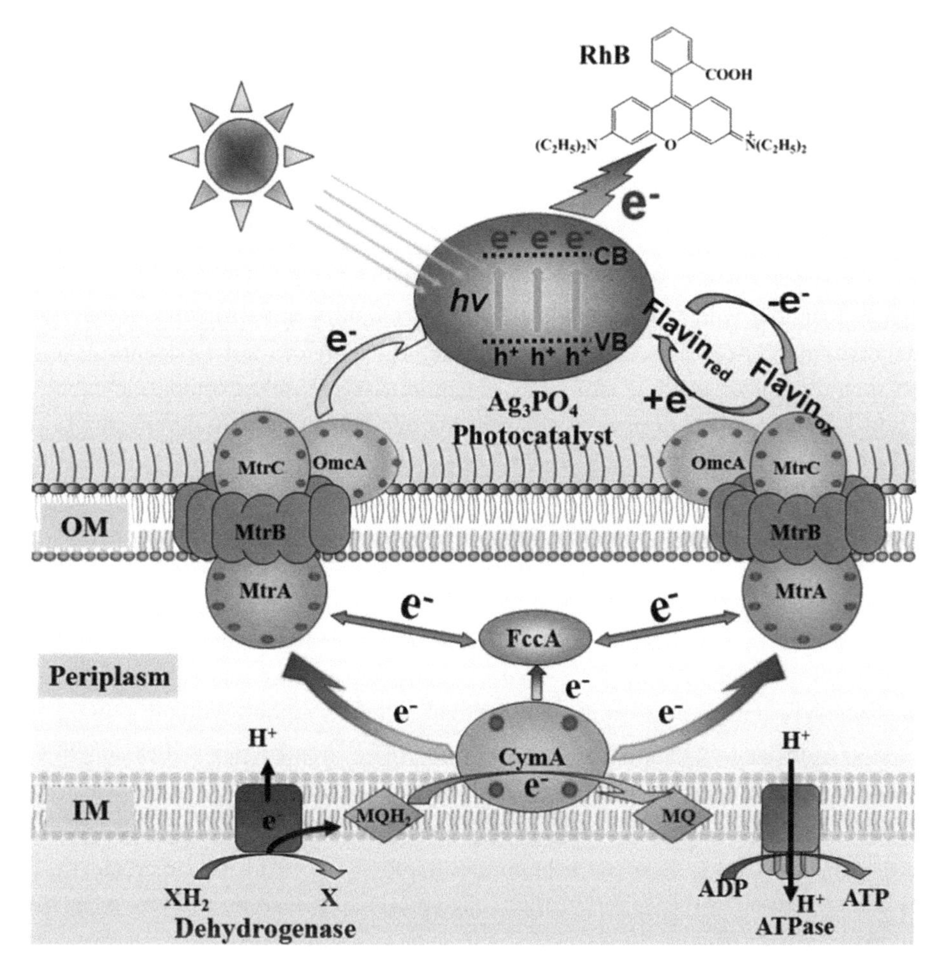

Figure 10. Proposed photocatalytic RhB degradation mechanisms in a biophotoelectric reductive degradation system. Arrows represent the electron flow. OM, outer membrane; IM, inner membrane. Adapted with permission from reference (100). Copyright 2019 Elsevier.

Biocatalysis

Biogenic Pd NPs also act as biocatalysts in other ways, such as in reductive dechlorination of polychlorinated biphenyls both in aqueous solution and in sediment matrices. Using formate as an electron donor, the hybrid of *S. oneidensis* MR-1 cells exposed to 500 mg/L of Pd(II) decomposed the polychlorinated 1 mg L^{-1} of 2,3,4-chlorobiphenyl to undetectable levels in 1 h at room temperature. The biocatalysis is enhanced when incubated in contaminated sediments. The hybrid from 50 mg/L of Pd(II) achieved dechlorination of seven polychlorinated biphenyls in 48 h, which is comparable to 500 mg/L of commercial Pd(0) powder (*18*).

Dechlorination of carbon tetrachloride was achieved using FeS biosynthesized by *S. putrefaciens* CN32. The hybrid of *S. putrefaciens* CN32 and biogenic FeS showed eight- and five-fold increases in dechlorination compared to *S. putrefaciens* CN32 and chemogenic FeS NPs, respectively. The efficient catalysis was attributed to the even distribution of FeS nanocatalysts and the larger amount of Fe(II) and disulfide. The addition of Fe(III) can enhance the catalytic efficiency further. The main role of *S. putrefaciens* CN32 is to produce FeS NPs that are well dispersed on the cell surface. The cell contribution is minimal after FeS formation (*85*).

The conventional method of biosynthesizing nanocatalysts is to retain the nanocatalyst coating on the cell membranes ("metallized" cells such as "palladized" cells) and ignore the non-cell-associated nanocatalysts in the bulk medium. However, one study showed that the extracellular nanocatalysts produced by *S. oneidensis* MR-1 outperformed the palladized cells. The initial rate of reduction of methyl viologen to methyl viologen cations radical of non-cell-associated Ag_2S (0.26 mM/s) is thus three-fold higher than that of cell-associated Ag_2S (0.26 mM/s). Another notable result is that the non-cell-associated Ag_2S from the mutant lacking OmcA and MtrC shows better performance than the wild-type strain (*20*). The morphology and structure of cell-associated nanocatalysts and non-cell-associated nanocatalysts may therefore not be identical, and equal attention should be given to these two kinds of nanocatalysts. Unlike chemogenic nanocatalysts, no additional capping agents or protecting agents are needed during biosynthesis due to the EPS secreted by the bacteria.

Photoautotrophic microorganisms can harvest light and carbon sources to produce food and energy. However, these processes are slow. On the other hand, although solid-state semiconductors can efficiently absorb light, semiconductors still face the challenge of converting photoexcited electrons into chemical bonds. Moreover, chemogenic semiconductors pose a threat to the environment during the synthesis process. A recent study combined the high efficiency of light harvesting by semiconductors with the low cost, self-replication, and self-repairing of the biology process. CdS NPs (<10 nm) were biodeposited on a nonphotosynthetic bacterium (*M. thermoacetica*) by incubating the bacteria in in a solution of Cd^{2+} and cysteine. The CdS NPs act as a photocatalyst and collect photons under light irradiation. The excited CdS NPs then deliver electrons to *M. thermoacetica*, which act as a biocatalyst and produce acetic acid from CO_2. The CdS NPs have three roles in the overall process: to harvest photons, to provide electrons, and to protect the bacteria from the damage caused by the light irradiation. High production of acetic acid was in fact harvested from this hybrid system over several days with light-dark cycles (*36*).

Outlook

In summary, biogenic nanocatalysts are small, well-dispersed, environmentally friendly, biocompatible, narrow-sized, and of low cost. In addition, some morphologies and properties are exclusive to biogenic nanocatalysts. However, there are also limitations on the biosynthesis of nanomaterials including nanocatalysts. Compared to their well-developed chemical synthesis counterparts, current biosynthesis of nanocatalysts are normally trial-and-error efforts, especially regarding morphology and structure control. The challenges of biosynthesis of nanocatalysts come from the complexity of biological processes. Different organisms, for example, bacteria, fungi, yeasts, and even plants, are able to achieve biosynthesis, but even for the same microorganisms, the processes proceed differently in different growth phases. The main contributing parts in biosynthesis are proteins, enzymes, polysaccharides, and specific functional groups in the cell membrane (such as carboxylate groups), most of which evolve during the lifetime of microbes. Another challenge is the scaling-up of the biosynthesis. Most biosynthesis is in the millimole scale, as a high concentration of metal ions can harm the microbes or inhibit the biosynthesis. Moreover, the biosynthesis of nanomaterials under mild temperature is time-consuming and unlikely to be accelerated by higher temperature and pressure as in chemical synthesis. The low concentration and relatively long synthesis time are bottlenecks for large-scale synthesis of biogenic nanocatalysts.

Considerable research efforts are needed to further explore the potential of biogenic nanocatalysts, with a focus on several key directions. First, the complete biosynthesis processes need to be mapped in much greater detail, addressing, for example, whether or not metal ions pass through the outer cell membrane and which specific components of the microorganism (organelles, protein complexes, DNA) are key players in the biosynthesis. The answers to these questions offer clues to the synthesis sites of biogenic nanocatalysts, the constituents of the nanocatalysts, and the genes involved in the synthesis processes. Secondly, efforts should also be spent on exploring and designing applications for nanocatalysts. The main advantages of nanocatalyst synthesis and operations are mild experimental conditions, self-replication, and self-rehabilitation, but they are also fraught with low yields and long synthesis durations. A suitable application should make use of the advantages of biogenic nanocatalysts and avoid the disadvantages. Finally, subsequent treatments should be considered to overcome the shortcomings of biogenic nanocatalysts. Biosynthesis is therefore not the end of the synthesis process. Instead, the biogenic nanocatalysts act as precursors, and further processes should be mobilized to optimize the morphologies and structures of the nanocatalysts, aiming at better performance.

List of Abbreviations

AFM	Atomic force microscopy
CSAFM	Current-sensing atomic force microscopy
DP	Carbonized hybrids of *S. oneidensis* MR-1 cells and Pd NPs
DPA	Carbonized hybrids of *S. oneidensis* MR-1 cells and PdAu alloy
DPARH	Carbonized hybrids of *S. oneidensis* MR-1 cells, PdAu alloy with reduced graphene oxide
EAB	Electrochemically active bacteria
EDS	Energy-dispersive X-ray spectroscopy
EET	Extracellular electron transfer
EPS	Extracellular polymeric substances
HEPES	Sodium 4-(2-hydroxyethyl)-1-piperazineethanesulfonic acid
NP(s)	Nanoparticle(s)
PCB	Polychlorinated biphenyls
RhB	Rhodamine B
SEM	Scanning electron microscopy
STC	Small tetraheme cytochromeTEMTransmission electron microscopy
XPS	X-ray photoelectron spectroscopy
XRD	X-ray diffraction

Acknowledgments

Financial support from the China Scholarship Council (CSC) (No. 201606130019), Carlsberg Foundation (CF15-0164), the National Natural Science Foundation of China (41471260, 51478451), and Otto Mønsted Foundation is greatly appreciated.

References

1. ISO. *International Organization for Standardization. Nanotechnologies—vocabulary—part 1: Core Terms*; ISO/TS 80004-1; 2015.

2. Fu, Y.; Liang, F.; Tian, H.; Hu, J. Nonenzymatic Glucose Sensor Based on ITO Electrode Modified with Gold Nanoparticles by Ion Implantation. *Electrochim. Acta* **2014**, *120*, 314–318.
3. Jukk, K.; Kozlova, J.; Ritslaid, P.; Sammelselg, V.; Alexeyeva, N.; Tammeveski, K. Sputter-Deposited Pt Nanoparticle/Multi-Walled Carbon Nanotube Composite Catalyst for Oxygen Reduction Reaction. *J. Electroanal. Chem.* **2013**, *708*, 31–38.
4. Patel, N.; Fernandes, R.; Guella, G.; Kale, A.; Miotello, A.; Patton, B.; Zanchetta, C. Structured and Nanoparticle Assembled Co–B Thin Films Prepared by Pulsed Laser Deposition: A Very Efficient Catalyst for Hydrogen Production. *J. Phys. Chem. C* **2008**, *112*, 6968–6976.
5. Ariga, K.; Mori, T.; Hill, J. P. Mechanical Control of Nanomaterials and Nanosystems. *Adv. Mater.* **2012**, *24*, 158–176.
6. Xu, J.; Sun, J.; Wang, Y.; Sheng, J.; Wang, F.; Sun, M. Application of Iron Magnetic Nanoparticles in Protein Immobilization. *Molecules* **2014**, *19*, 11465–11486.
7. Gupta, S.; Kershaw, S. V.; Rogach, A. L. 25th Anniversary Article: Ion Exchange in Colloidal Nanocrystals. *Adv. Mater.* **2013**, *25*, 6923–6944.
8. Schwamborn, S.; Etienne, M.; Schuhmann, W. Local Electrocatalytic Induction of Sol–Gel Deposition at Pt Nanoparticles. *Electrochem. Commun.* **2011**, *13*, 759–762.
9. Zheng, Z.; Zheng, Y.; Tian, X.; Yang, Z.; Jiang, Y.; Zhao, F. Interactions Between Iron Mineral-Humic Complexes and Hexavalent Chromium and the Corresponding Bio-Effects. *Environ. Pollut.* **2018**, *241*, 265–271.
10. Cao, H.; Wang, X.; Gu, H.; Liu, J.; Luan, L.; Liu, W.; Wang, Y.; Guo, Z. Carbon Coated Manganese Monoxide Octahedron Negative-Electrode for Lithium-Ion Batteries with Enhanced Performance. *RSC Adv.* **2015**, *5*, 34566–34571.
11. Cao, H.; Wu, N.; Liu, Y.; Wang, S.; Du, W.; Liu, J. Facile Synthesis of Rod-Like Manganese Molybdate Crystallines with Two-Dimensional Nanoflakes for Supercapacitor Application. *Electrochim. Acta.* **2017**, *225*, 605–613.
12. Engelbrekt, C.; Sørensen, K. H.; Lübcke, T.; Zhang, J.; Li, Q.; Pan, C.; Bjerrum, N. J.; Ulstrup, J. 1.7 nm Platinum Nanoparticles: Synthesis with Glucose Starch, Characterization and Catalysis. *ChemPhysChem* **2010**, *11*, 2844–2853.
13. Engelbrekt, C.; Sorensen, K. H.; Zhang, J.; Welinder, A. C.; Jensen, P. S.; Ulstrup, J. Green Synthesis of Gold Nanoparticles with Starch-Glucose and Application in Bioelectrochemistry. *J. Mater. Chem.* **2009**, *19*, 7839–7847.
14. Hulkoti, N. I.; Taranath, T. C. Biosynthesis of Nanoparticles Using Microbes—A Review. *Colloids Surf., B* **2014**, *121*, 474–483.
15. Mohanpuria, P.; Rana, N. K.; Yadav, S. K. Biosynthesis of Nanoparticles: Technological Concepts and Future Applications. *J. Nanopart. Res.* **2008**, *10*, 507–517.
16. Roy, K.; Mao, H. Q.; Huang, S. K.; Leong, K. W. Oral Gene Delivery with Chitosan–DNA Nanoparticles Generates Immunologic Protection in a Murine Model of Peanut Allergy. *Nat. Med.* **1999**, *5*, 387–391.
17. Liu, J.; Zheng, Y.; Hong, Z.; Cai, K.; Zhao, F.; Han, H. Microbial Synthesis of Highly Dispersed Pdau Alloy for Enhanced Electrocatalysis. *Sci. Adv.* **2016**, *2*, e1600858.

18. Windt, W. D.; Aelterman, P.; Verstraete, W. Bioreductive Deposition of Palladium (0) Nanoparticles on *Shewanella oneidensis* with Catalytic Activity Towards Reductive Dechlorination of Polychlorinated Biphenyls. *Environ. Microbiol.* **2005**, 7, 314–325.
19. Kimber, R. L.; Lewis, E. A.; Parmeggiani, F.; Smith, K.; Bagshaw, H.; Starborg, T.; Joshi, N.; Figueroa, A. I.; van der Laan, G.; Cibin, G.; Gianolio, D.; Haigh, S. J.; Pattrick, R. A. D.; Turner, N. J.; Lloyd, J. R. Biosynthesis and Characterization of Copper Nanoparticles Using *Shewanella oneidensis:* Application for Click Chemistry. *Small* **2018**, *14*, 1703145.
20. Ng, C. K.; Sivakumar, K.; Liu, X.; Madhaiyan, M.; Ji, L.; Yang, L.; Tang, C.; Song, H.; Kjelleberg, S.; Cao, B. Influence of Outer Membrane *c*-Type Cytochromes on Particle Size and Activity of Extracellular Nanoparticles Produced by *Shewanella oneidensis*. *Biotechnol. Bioeng.* **2013**, *110*, 1831–1837.
21. Li, D. B.; Cheng, Y. Y.; Wu, C.; Li, W. W.; Li, N.; Yang, Z. C.; Tong, Z. H.; Yu, H. Q. Selenite Reduction by *Shewanella oneidensis* MR-1 is Mediated by Fumarate Reductase in Periplasm. *Sci. Rep.* **2014**, *4*, 3735.
22. Ng, C. K.; Cai Tan, T. K.; Song, H.; Cao, B. Reductive Formation of Palladium Nanoparticles by *Shewanella oneidensis*: Role of Outer Membrane Cytochromes and Hydrogenases. *RSC Adv.* **2013**, *3*, 22498–22503.
23. Zhou, N. Q.; Tian, L. J.; Wang, Y. C.; Li, D. B.; Li, P. P.; Zhang, X.; Yu, H. Q. Extracellular Biosynthesis of Copper Sulfide Nanoparticles by *Shewanella oneidensis* MR-1 as a Photothermal Agent. *Enzyme Microb. Technol.* **2016**, *95*, 230–235.
24. Xiao, X.; Liu, Q. Y.; Lu, X. R.; Li, T. T.; Feng, X. L.; Li, Q.; Liu, Z. Y.; Feng, Y. J. Self-Assembly of Complex Hollow CuS Nano/Micro Shell by an Electrochemically Active Bacterium *Shewanella oneidensis* MR-1. *Int. Biodeterior. Biodegrad.* **2017**, *116*, 10–16.
25. Wu, R.; Cui, L.; Chen, L.; Wang, C.; Cao, C.; Sheng, G.; Yu, H.; Zhao, F. Effects of Bio-Au Nanoparticles on Electrochemical Activity of *Shewanella oneidensis* Wild Type and *ΔomcA/MtrC* Mutant. *Sci. Rep.* **2013**, *3*, 3307.
26. Tuo, Y.; Liu, G.; Dong, B.; Zhou, J.; Wang, A.; Wang, J.; Jin, R.; Lv, H.; Dou, Z.; Huang, W. Microbial Synthesis of Pd/Fe_3O_4, Au/Fe_3O_4 and PdAu/Fe_3O_4 Nanocomposites for Catalytic Reduction of Nitroaromatic Compounds. *Sci. Rep.* **2015**, *5*, 13515.
27. Miller, A.; Singh, L.; Wang, L.; Liu, H. Linking Internal Resistance with Design and Operation Decisions in Microbial Electrolysis Cells. *Environ. Int.* **2019**, *126*, 611–618.
28. Huang, B. C.; Yi, Y. C.; Chang, J. S.; Ng, I. S. Mechanism Study of Photo-Induced Gold Nanoparticles Formation by *Shewanella oneidensis* MR-1. *Sci. Rep.* **2019**, *9*, 7589.
29. Che, L.; Dong, Y.; Wu, M.; Zhao, Y.; Liu, L.; Zhou, H. Characterization of Selenite Reduction by *Lysinibacillus* sp. ZYM-1 and Photocatalytic Performance of Biogenic Selenium Nanospheres. *ACS Sustainable Chem. Eng.* **2017**, *5*, 2535–2543.
30. Søbjerg, L. S.; Gauthier, D.; Lindhardt, A. T.; Bunge, M.; Finster, K.; Meyer, R. L.; Skrydstrup, T. Bio-Supported Palladium Nanoparticles as a Catalyst for Suzuki–Miyaura and Mizoroki–Heck Reactions. *Green Chem.* **2009**, *11*, 2041–2046.
31. Gauthier, D.; Søbjerg, L. S.; Jensen, K. M.; Lindhardt, A. T.; Bunge, M.; Finster, K.; Meyer, R. L.; Skrydstrup, T. Environmentally Benign Recovery and Reactivation of Palladium from Industrial Waste by Using Gram-Negative Bacteria. *ChemSusChem* **2010**, *3*, 1036–1039.

32. Bunge, M.; Søbjerg, L. S.; Rotaru, A. E.; Gauthier, D.; Lindhardt, A. T.; Hause, G.; Finster, K.; Kingshott, P.; Skrydstrup, T.; Meyer, R. L. Formation of Palladium(0) Nanoparticles at Microbial Surfaces. *Biotechnol. Bioeng.* **2010**, *107*, 206–215.
33. Kowshik, M.; Vogel, W.; Urban, J.; Kulkarni, S. K.; Paknikar, K. M. Microbial Synthesis of Semiconductor PbS Nanocrystallites. *Adv. Mater.* **2002**, *14*, 815–818.
34. Wu, R.; Tian, X.; Xiao, Y.; Ulstrup, J.; Molager Christensen, H. E.; Zhao, F.; Zhang, J. Selective Electrocatalysis of Biofuel Molecular Oxidation Using Palladium Nanoparticles Generated on *Shewanella oneidensis* MR-1. *J. Mater. Chem. A* **2018**, *6*, 10655–10662.
35. Wang, T.; Yang, L.; Zhang, B.; Liu, J. Extracellular Biosynthesis and Transformation of Selenium Nanoparticles and Application in H_2O_2 Biosensor. *Colloids Surf., B* **2010**, *80*, 94–102.
36. Sakimoto, K. K.; Wong, A. B.; Yang, P. Self-Photosensitization of Nonphotosynthetic Bacteria for Solar-to-Chemical Production. *Science* **2016**, *351*, 74.
37. Zhang, H.; Liu, H.; Tian, Z.; Lu, D.; Yu, Y.; Cestellos-Blanco, S.; Sakimoto, K. K.; Yang, P. Bacteria Photosensitized by Intracellular Gold Nanoclusters for Solar Fuel Production. *Nat. Nanotechnol.* **2018**, *13*, 900–905.
38. Wu, R.; Wang, C.; Shen, J.; Zhao, F. A Role for Biosynthetic CdS Quantum Dots in Extracellular Electron Transfer of *Saccharomyces cerevisiae. Process Biochem.* **2015**, *50*, 2061–2065.
39. Beveridge, T. J.; Murray, R. G. Sites of Metal Deposition in the Cell Wall of *Bacillus subtilis. J. Bacteriol.* **1980**, *141*, 876–887.
40. Sweeney, R. Y.; Mao, C.; Gao, X.; Burt, J. L.; Belcher, A. M.; Georgiou, G.; Iverson, B. L. Bacterial Biosynthesis of Cadmium Sulfide Nanocrystals. *Chem. Biol.* **2004**, *11*, 1553–1559.
41. Derby, H.; Hammer, B. Bacteriology of Butter Iv. Bacteriological Studies on Surface Taint Butter. *Research Bulletin (Iowa Agriculture and Home Economics Experiment Station)* **1931**, *11*, 1.
42. MacDonell, M. T.; Colwell, R. R. Phylogeny of the Vibrionaceae, and Recommendation for Two New Genera, *Listonella* and *Shewanella*. *Syst. Appl. Microbiol.* **1985**, *6*, 171–182.
43. Myers, C. R.; Nealson, K. H. Bacterial Manganese Reduction and Growth with Manganese Oxide as the Sole Electron Acceptor. *Science* **1988**, *240*, 1319–1321.
44. Venkateswaran, K.; Moser, D. P.; Dollhopf, M. E.; Lies, D. P.; Saffarini, D. A.; MacGregor, B. J.; Ringelberg, D. B.; White, D. C.; Nishijima, M.; Sano, H.; Burghardt, J.; Stackebrandt, E.; Nealson, K. H. Polyphasic Taxonomy of the Genus *Shewanella* and Description of *Shewanella* oneidensis sp. Nov. *Int. J. Syst. Evol. Microbiol.* **1999**, *49*, 705–724.
45. Heidelberg, J. F.; Paulsen, I. T.; Nelson, K. E.; Gaidos, E. J.; Nelson, W. C.; Read, T. D.; Eisen, J. A.; Seshadri, R.; Ward, N.; Methe, B.; Clayton, R. A.; Meyer, T.; Tsapin, A.; Scott, J.; Beanan, M.; Brinkac, L.; Daugherty, S.; DeBoy, R. T.; Dodson, R. J.; Durkin, A. S.; Haft, D. H.; Kolonay, J. F.; Madupu, R.; Peterson, J. D.; Umayam, L. A.; White, O.; Wolf, A. M.; Vamathevan, J.; Weidman, J.; Impraim, M.; Lee, K.; Berry, K.; Lee, C.; Mueller, J.; Khouri, H.; Gill, J.; Utterback, T. R.; McDonald, L. A.; Feldblyum, T. V.; Smith, H. O.; Venter, J. C.; Nealson, K. H.; Fraser, C. M. Genome Sequence of the Dissimilatory Metal Ion–Reducing Bacterium *Shewanella oneidensis*. *Nat. Biotechnol.* **2002**, *20*, 1118–1123.
46. Suresh, A. K.; Pelletier, D. A.; Wang, W.; Moon, J. W.; Gu, B.; Mortensen, N. P.; Allison, D. P.; Joy, D. C.; Phelps, T. J.; Doktycz, M. J. Silver Nanocrystallites: Biofabrication Using

Shewanella oneidensis, and an Evaluation of Their Comparative Toxicity on Gram-Negative and Gram-Positive Bacteria. *Environ. Sci. Technol.* **2010**, *44*, 5210–5215.
47. Kumar, R.; Singh, L.; Wahid, Z. A.; Din, M. F. M. Exoelectrogens in Microbial Fuel Cells toward Bioelectricity Generation: A Review. *Int. J. Energy Res.* **2015**, *39*, 1048–1067.
48. Zheng, Z.; Xiao, Y.; Wu, R.; Mølager Christensen, H. E.; Zhao, F.; Zhang, J. Electrons Selective Uptake of a Metal-Reducing Bacterium *Shewanella oneidensis* MR-1 from Ferrocyanide. *Biosens. Bioelectron.* **2019**, *142*, 111571.
49. Xiao, Y.; Zhang, E.; Zhang, J.; Dai, Y.; Yang, Z.; Christensen, H. E. M.; Ulstrup, J.; Zhao, F. Extracellular Polymeric Substances Are Transient Media for Microbial Extracellular Electron Transfer. *Sci. Adv.* **2017**, *3*, e1700623.
50. Tian, X.; Zhao, F.; You, L.; Wu, X.; Zheng, Z.; Wu, R.; Jiang, Y.; Sun, S. Interaction between *in vivo* Bioluminescence and Extracellular Electron Transfer in *Shewanella woodyi* via Charge and Discharge. *Phys. Chem. Chem. Phys.* **2017**, *19*, 1746–1750.
51. El-Naggar, M. Y.; Wanger, G.; Leung, K. M.; Yuzvinsky, T. D.; Southam, G.; Yang, J.; Lau, W. M.; Nealson, K. H.; Gorby, Y. A. Electrical Transport Along Bacterial Nanowires from *Shewanella oneidensis* MR-1. *Proc. Natl. Acad. Sci. U. S. A.* **2010**, *107*, 18127–18131.
52. Marsili, E.; Baron, D. B.; Shikhare, I. D.; Coursolle, D.; Gralnick, J. A.; Bond, D. R. *Shewanella* Secretes Flavins that Mediate Extracellular Electron Transfer. *Proc. Natl. Acad. Sci. U. S. A.* **2008**, *105*, 3968–3973.
53. Mal, J.; Nancharaiah, Y. V.; van Hullebusch, E. D.; Lens, P. N. L. Metal Chalcogenide Quantum Dots: Biotechnological Synthesis and Applications. *RSC Adv.* **2016**, *6*, 41477–41495.
54. Park, T. J.; Lee, K. G.; Lee, S. Y. Advances in Microbial Biosynthesis of Metal Nanoparticles. *Appl. Microbiol. Biotechnol.* **2016**, *100*, 521–534.
55. Ghosh, S. Copper and Palladium Nanostructures: A Bacteriogenic Approach. *Appl. Microbiol. Biotechnol.* **2018**, *102*, 7693–7701.
56. Khatami, M.; Alijani, H. Q.; Sharifi, I. Biosynthesis of Bimetallic and Core–Shell Nanoparticles: Their Biomedical Applications – A Review. *IET Nanobiotechnol.* **2018**, *12*, 879–887.
57. Kim, T. Y.; Kim, M. G.; Lee, J. H.; Hur, H. G. Biosynthesis of Nanomaterials by *Shewanella* Species for Application in Lithium Ion Batteries. *Front. Microbiol.* **2018**, *9*, 2817.
58. Vaseghi, Z.; Nematollahzadeh, A.; Tavakoli, O. Green Methods for the Synthesis of Metal Nanoparticles Using Biogenic Reducing Agents: A Review. *Rev. Chem. Eng.* **2018**, *34*, 529.
59. Gahlawat, G.; Choudhury, A. R. A Review on the Biosynthesis of Metal and Metal Salt Nanoparticles by Microbes. *RSC Adv.* **2019**, *9*, 12944–12967.
60. Salunke, B. K.; Sawant, S. S.; Lee, S. I.; Kim, B. S. Microorganisms as Efficient Biosystem for the Synthesis of Metal Nanoparticles: Current Scenario and Future Possibilities. *World J. Microbiol. Biotechnol.* **2016**, *32*, 88.
61. van der Wal, A.; Minor, M.; Norde, W.; Zehnder, A. J. B.; Lyklema, J. Electrokinetic Potential of Bacterial Cells. *Langmuir* **1997**, *13*, 165–171.
62. Li, X.; Xu, H.; Chen, Z. S.; Chen, G. Biosynthesis of Nanoparticles by Microorganisms and Their Applications. *J. Nanomater.* **2011**, *2011*, 16.

63. Narayanan, K. B.; Sakthivel, N. Biological Synthesis of Metal Nanoparticles by Microbes. *Adv. Colloid Interface Sci.* **2010**, *156*, 1–13.
64. Beveridge, T. J. Role of Cellular Design in Bacterial Metal Accumulation and Mineralization. *Annu. Rev. Microbiol.* **1989**, *43*, 147–171.
65. Gorby, Y. A.; Yanina, S.; McLean, J. S.; Rosso, K. M.; Moyles, D.; Dohnalkova, A.; Beveridge, T. J.; Chang, I. S.; Kim, B. H.; Kim, K. S.; Culley, D. E.; Reed, S. B.; Romine, M. F.; Saffarini, D. A.; Hill, E. A.; Shi, L.; Elias, D. A.; Kennedy, D. W.; Pinchuk, G.; Watanabe, K.; Ishii, S. I.; Logan, B.; Nealson, K. H.; Fredrickson, J. K. Electrically Conductive Bacterial Nanowires Produced by *Shewanella oneidensis* Strain MR-1 and Other Microorganisms. *Proc. Natl. Acad. Sci. U. S. A.* **2006**, *103*, 11358–11363.
66. Kumar, R.; Singh, L.; Zularisam, A. W. Exoelectrogens: Recent Advances in Molecular Drivers Involved in Extracellular Electron Transfer and Strategies Used to Improve It for Microbial Fuel Cell Applications. *Renewable Sustainable Energy Rev.* **2016**, *56*, 1322–1336.
67. Shi, L.; Dong, H.; Reguera, G.; Beyenal, H.; Lu, A.; Liu, J.; Yu, H. Q.; Fredrickson, J. K. Extracellular Electron Transfer Mechanisms Between Microorganisms and Minerals. *Nat. Rev. Microbiol.* **2016**, *14*, 651–662.
68. Shi, L.; Squier, T. C.; Zachara, J. M.; Fredrickson, J. K. Respiration of Metal (Hydr)Oxides by *Shewanella* and *Geobacter*: A Key Role for Multihaem C-Type Cytochromes. *Mol. Microbiol.* **2007**, *65*, 12–20.
69. Kumar, A.; Hsu, L. H. H.; Kavanagh, P.; Barrière, F.; Lens, P. N. L.; Lapinsonnière, L.; Lienhard V, J. H.; Schröder, U.; Jiang, X.; Leech, D. The Ins and Outs of Microorganism–Electrode Electron Transfer Reactions. *Nat. Rev. Chem.* **2017**, *1*, 0024.
70. Belchik, S. M.; Kennedy, D. W.; Dohnalkova, A. C.; Wang, Y.; Sevinc, P. C.; Wu, H.; Lin, Y.; Lu, H. P.; Fredrickson, J. K.; Shi, L. Extracellular Reduction of Hexavalent Chromium by Cytochromes MtrC and OmcA of *Shewanella oneidensis* MR-1. *Appl. Environ. Microbiol.* **2011**, *77*, 4035.
71. Marshall, M. J.; Beliaev, A. S.; Dohnalkova, A. C.; Kennedy, D. W.; Shi, L.; Wang, Z.; Boyanov, M. I.; Lai, B.; Kemner, K. M.; McLean, J. S.; Reed, S. B.; Culley, D. E.; Bailey, V. L.; Simonson, C. J.; Saffarini, D. A.; Romine, M. F.; Zachara, J. M.; Fredrickson, J. K. *c*-Type Cytochrome-Dependent Formation of U(IV) Nanoparticles by *Shewanella oneidensis*. *PLoS Biol.* **2006**, *4*, 1324–1333.
72. Meshulam-Simon, G.; Behrens, S.; Choo, A. D.; Spormann, A. M. Hydrogen Metabolism in *Shewanella oneidensis* MR-1. *Appl. Environ. Microbiol.* **2007**, *73*, 1153–1165.
73. Kim, D. H.; Kanaly, R. A.; Hur, H. G. Biological Accumulation of Tellurium Nanorod Structures via Reduction of Tellurite by *Shewanella oneidensis* MR-1. *Bioresour. Technol.* **2012**, *125*, 127–131.
74. Klonowska, A.; Heulin, T.; Vermeglio, A. Selenite and Tellurite Reduction by *Shewanella oneidensis*. *Appl. Environ. Microbiol.* **2005**, *71*, 5607.
75. Xiao, Y.; Wu, S.; Zhang, F.; Wu, Y.; Yang, Z.; Zhao, F. Promoting Electrogenic Ability of Microbes with Negative Pressure. *J. Power Sources* **2013**, *229*, 79–83.
76. Wang, Z.; Zheng, Z.; Zheng, S.; Chen, S.; Zhao, F. Carbonized Textile with Free-Standing Threads as an Efficient Anode Material for Bioelectrochemical Systems. *J. Power Sources* **2015**, *287*, 269–275.

77. Perez-Gonzalez, T.; Jimenez-Lopez, C.; Neal, A. L.; Rull-Perez, F.; Rodriguez-Navarro, A.; Fernandez-Vivas, A.; Iañez-Pareja, E. Magnetite Biomineralization Induced by *Shewanella oneidensis*. *Geochim. Cosmochim. Acta* **2010**, *74*, 967–979.
78. Wu, X.; Zhao, F.; Rahunen, N.; Varcoe, J. R.; Avignone-Rossa, C.; Thumser, A. E.; Slade, R. C. T. A Role for Microbial Palladium Nanoparticles in Extracellular Electron Transfer. *Angew. Chem. Int. Ed.* **2010**, *50*, 427–430.
79. Konishi, Y.; Ohno, K.; Saitoh, N.; Nomura, T.; Nagamine, S.; Hishida, H.; Takahashi, Y.; Uruga, T. Bioreductive Deposition of Platinum Nanoparticles on the Bacterium *Shewanella algae*. *J. Biotechnol.* **2007**, *128*, 648–653.
80. Varia, J.; Zegeye, A.; Roy, S.; Yahaya, S.; Bull, S. *Shewanella putrefaciens* for the Remediation of Au^{3+}, Co^{2+} and Fe^{3+} Metal Ions from Aqueous Systems. *Biochem. Eng. J.* **2014**, *85*, 101–109.
81. Burns, J. L.; DiChristina, T. J. Anaerobic Respiration of Elemental Sulfur and Thiosulfate by *Shewanella oneidensis* MR-1 Requires *psrA*, a Homolog of the *phsA* Gene of *Salmonella enterica* Serovar Typhimurium LT2. *Appl. Environ. Microbiol.* **2009**, *75*, 5209–5217.
82. Voeikova, T. A.; Shebanova, A. S.; Ivanov, Y. D.; Kaysheva, A. L.; Novikova, L. M.; Zhuravliova, O. A.; Shumyantseva, V. V.; Shaitan, K. V.; Kirpichnikov, M. P.; Debabov, V. G. The Role of Proteins of the Outer Membrane of *Shewanella oneidensis* MR-1 in the Formation and Stabilization of Silver Sulfide Nanoparticles. *Appl. Biochem. Microbiol.* **2016**, *52*, 769–775.
83. Yang, M.; Shi, X. Biosynthesis of Ag_2s/Tio_2 Nanotubes Nanocomposites by *Shewanella oneidensis* MR-1 for the Catalytic Degradation of 4-Nitrophenol. *Environ. Sci. Pollut. Res.* **2019**, *26*, 12237–12246.
84. Butler, E. C.; Hayes, K. F. Effects of Solution Composition and Ph on the Reductive Dechlorination of Hexachloroethane by Iron Sulfide. *Environ. Sci. Technol.* **1998**, *32*, 1276–1284.
85. Huo, Y. C.; Li, W. W.; Chen, C. B.; Li, C. X.; Zeng, R.; Lau, T. C.; Huang, T. Y. Biogenic Fes Accelerates Reductive Dechlorination of Carbon Tetrachloride by *Shewanella putrefaciens* CN32. *Enzyme Microb. Technol.* **2016**, *95*, 236–241.
86. Wang, L.; Chen, S.; Ding, Y.; Zhu, Q.; Zhang, N.; Yu, S. Biofabrication of Morphology Improved Cadmium Sulfide Nanoparticles Using *Shewanella oneidensis* Bacterial Cells and Ionic Liquid: For Toxicity Against Brain Cancer Cell Lines. *J. Photochem. Photobiol., B* **2018**, *178*, 424–427.
87. Lee, J. H.; Kim, M. G.; Yoo, B.; Myung, N. V.; Maeng, J.; Lee, T.; Dohnalkova, A. C.; Fredrickson, J. K.; Sadowsky, M. J.; Hur, H. G. Biogenic Formation of Photoactive Arsenic-Sulfide Nanotubes by *Shewanella* sp. Strain HN-41. *Proc. Natl. Acad. Sci. U. S. A.* **2007**, *104*, 20410–20415.
88. Labrenz, M.; Druschel, G. K.; Thomsen-Ebert, T.; Gilbert, B.; Welch, S. A.; Kemner, K. M.; Logan, G. A.; Summons, R. E.; Stasio, G. D.; Bond, P. L.; Lai, B.; Kelly, S. D.; Banfield, J. F. Formation of Sphalerite (ZnS) Deposits in Natural Biofilms of Sulfate-Reducing Bacteria. *Science* **2000**, *290*, 1744–1747.
89. Lee, J.-H.; Kennedy, D. W.; Dohnalkova, A.; Moore, D. A.; Nachimuthu, P.; Reed, S. B.; Fredrickson, J. K. Manganese Sulfide Formation via Concomitant Microbial Manganese Oxide and Thiosulfate Reduction. *Environ. Microbiol.* **2011**, *13*, 3275–3288.

90. Lee, J. H.; Roh, Y.; Hur, H. G. Microbial Production and Characterization of Superparamagnetic Magnetite Nanoparticles by *Shewanella* sp. HN-41. *J. Microbiol. Biotechnol.* **2008**, *18*, 1572–1577.
91. Jiang, S.; Kim, M. G.; Kim, I. Y.; Hwang, S. J.; Hur, H. G. Biological Synthesis of Free-Standing Uniformed Goethite Nanowires by *Shewanella* sp. HN-41. *J. Mater. Chem. A* **2013**, *1*, 1646–1650.
92. Zannoni, D.; Borsetti, F.; Harrison, J. J.; Turner, R. J. The Bacterial Response to the Chalcogen Metalloids Se and Te. *Adv. Microb. Physiol.* **2007**, *53*, 1–312.
93. Tam, K.; Ho, C. T.; Lee, J. H.; Lai, M.; Chang, C. H.; Rheem, Y.; Chen, W.; Hur, H. G.; Myung, N. V. Growth Mechanism of Amorphous Selenium Nanoparticles Synthesized by *Shewanella* sp. HN-41. *Biosci. Biotechnol. Biochem.* **2010**, *74*, 696–700.
94. Pearce, C. I.; Pattrick, R. A. D.; Law, N.; Charnock, J. M.; Coker, V. S.; Fellowes, J. W.; Oremland, R. S.; Lloyd, J. R. Investigating Different Mechanisms for Biogenic Selenite Transformations: *Geobacter sulfurreducens*, *Shewanella oneidensis* and *Veillonella atypica*. *Environ. Technol.* **2009**, *30*, 1313–1326.
95. Kim, D. H.; Kim, M. G.; Jiang, S.; Lee, J. H.; Hur, H. G. Promoted Reduction of Tellurite and Formation of Extracellular Tellurium Nanorods by Concerted Reaction between Iron and *Shewanella oneidensis* MR-1. *Environ. Sci. Technol.* **2013**, *47*, 8709–8715.
96. Vaigankar, D. C.; Dubey, S. K.; Mujawar, S. Y.; D'Costa, A.; S. K, S. Tellurite Biotransformation and Detoxification by *Shewanella baltica* with Simultaneous Synthesis of Tellurium Nanorods Exhibiting Photo-Catalytic and Anti-Biofilm Activity. *Ecotoxicol. Environ. Saf.* **2018**, *165*, 516–526.
97. Tuo, Y.; Liu, G.; Dong, B.; Yu, H.; Zhou, J.; Wang, J.; Jin, R. Microbial Synthesis of Bimetallic PdPt Nanoparticles for Catalytic Reduction of 4-Nitrophenol. *Environ. Sci. Pollut. Res.* **2017**, *24*, 5249–5258.
98. Wang, W.; Zhang, B.; He, Z. Bioelectrochemical Deposition of Palladium Nanoparticles as Catalysts by *Shewanella oneidensis* MR-1 Towards Enhanced Hydrogen Production in Microbial Electrolysis Cells. *Electrochim. Acta* **2019**, *318*, 794–800.
99. Xiao, X.; Ma, X. B.; Yuan, H.; Liu, P. C.; Lei, Y. B.; Xu, H.; Du, D. L.; Sun, J. F.; Feng, Y. J. Photocatalytic Properties of Zinc Sulfide Nanocrystals Biofabricated by Metal-Reducing Bacterium *Shewanella oneidensis* MR-1. *J. Hazard. Mater.* **2015**, *288*, 134–139.
100. Xiao, X.; Ma, X. L.; Liu, Z. Y.; Li, W. W.; Yuan, H.; Ma, X. B.; Li, L. X.; Yu, H. Q. Degradation of Rhodamine B in a Novel Bio-Photoelectric Reductive System Composed of *Shewanella oneidensis* MR-1 and Ag_3PO_4. *Environ. Int.* **2019**, *126*, 560–567.

Chapter 6

Current Trends in Development of Photosynthetic Bioelectrochemical Systems for Light Energy Conversion

Dmitry Pankratov,[1] Galina Pankratova,[2] and Lo Gorton[*,1]

[1]Department of Biochemistry and Structural Biology, Lund University, PO Box 124, SE-22100 Lund, Sweden

[2]National Centre of Nano Fabrication and Characterization, DTU Nanolab, Technical University of Denmark, DK-2800 Kongens Lyngby, Denmark

[*]E-mail: lo.gorton@biochemistry.lu.se.

During the past few decades, intensive research has been focused on the application of biological photosynthetic processes for "green" electrical power generation. In this contribution, we consider different possible photobiocatalysts, such as isolated photosystems, thylakoid membranes (TMs), and whole photosynthetic organisms suitable for building bioelectrochemical systems for light energy conversion. We discuss numerous approaches to immobilization and wiring of the photobioagents on conductive surfaces and various electrode materials and mediators employed in state-of-the-art photobioelectrochemical systems, and we highlight the latest achievements in photocurrent intensity and power output of the solar-driven biodevices.

Introduction

Today's society faces needs for an ecologically clean and cheap energy alternative to fossil and oil fuels. Solar energy arriving at our planet is the most abundant and sustainable source of power and potentially can cover all annual energy needs of humanity (*1*). Though it is considered unlimited, natural sunlight is diluted and distributed irregularly over the surface of the earth (*2*). The development and employment of efficient ways to capture and store solar energy have gained great scientific interest over the last few decades.

Currently, there are a number of photovoltaic technologies for conversion of energy available from the Sun into electrical power (*3*). Traditional photovoltaics use molecular photosensitizers or semiconductor materials as photoactive agents and produce relatively high electrical currents on an annual basis; however, such devices suffer from not including storage of the electrical energy produced. Additionally, they are still limited by the lack of transportation and storage, scarcity, high cost of material production, and need for waste disposal (*4–6*).

© 2020 American Chemical Society

Green plants and cyanobacteria are organisms equipped with an evolutionary highly developed and highly efficient photosynthetic apparatus and are able to harvest and convert light energy and store it in organic molecules. Natural photosynthesis operates at an extreme quantum efficiency, nearly 100% under optimum conditions (*7*), and is considered as a more efficient process regarding short-term energy conversion yields (*2*). The application of the biological photosynthetic machinery in energy-converting bioelectrochemical systems (BESs) is a low-cost, environmentally friendly, and renewable alternative, which holds promising potential and clear advantages for solar energy production.

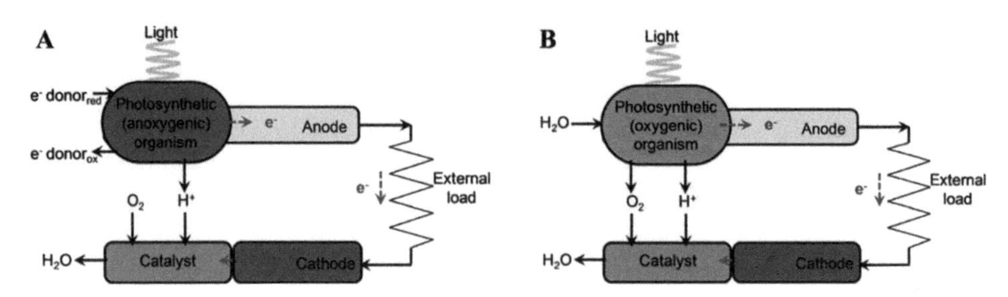

Figure 1. Schematic representations of (A) photosynthetic microbial fuel cells and (B) cellular biophotovoltaic cells. Adapted with permission from reference (8). Copyright 2015 Royal Society of Chemistry.

BESs employing various photosynthetic biocatalysts constitute a relatively new technology, which is still poorly defined in terms of nomenclature. Formally they can be divided into photosynthetic microbial fuel cells (PMFCs; Figure 1A) and biophotovoltaic cells (BPVs; Figure 1B) (*9*). Both types of devices exploit an electron-producing phototroph at the anode; however, the ultimate source of those electrons is different. PMFCs are colonized by anaerobic chemoautotrophic microbes, which use chemical fuel as the primary electron source and generate electric currents in a light-dependent manner (*8*). True BPVs, which we mainly focus on in this chapter, can be defined as systems employing various oxygenic photosynthetic microorganisms (*8*) or functional subfractions of the photosynthetic apparatus (*8*, *10*), which transfer electrons to an anode by light-driven water splitting. The produced electrons can be delivered to the anode directly (through a direct physical conductive contact between a biocatalyst and electrode surface (*8*, *11*)) or via various mediators (excreted by the microorganisms themselves, but through a largely unknown mechanism or externally provided mediators, freely diffusing or electron-conducting redox polymers (*8*, *11*, *12*)).

This chapter gives an overview of different types of light-driven BESs and current achievements reported up to now in BPV research. We summarize the state of the art of biophotoelectrochemical systems and recent progress in optimization and improvement of light-to-current efficiencies and operating stabilities by employing various photobiocatalysts, electrode materials and modifications, redox mediators, and setup designs and configurations. The field of *solar fuels* research, that is, employment of photosynthetic machinery to produce H_2 or reduce CO_2 to CO or hydrocarbons, is beyond the scope of this chapter, and the interested reader is referred to the recent review (*13*) covering state-of-the-art advances in that rapidly developing field. Another important concept not addressed here is related to coupling bioelements with semiconductive materials to create semiartificial photosynthetic systems aimed at overcoming the limitations of natural photosynthesis, which has been thoroughly discussed by Kornienko et al. (*14*)

Principles of Photosynthesis

Photosynthesis is a unique series of complex reactions, by which light energy is directly converted into the energy of chemical bonds in organic compounds. This process makes a connection between light and life and is the primary origin of energy for all living organisms. The whole process is divided into two stages: light reactions, where light is trapped by photosynthetic pigments and adenosine triphosphate (ATP) and nicotinamide adenine dinucleotide phosphate (NADPH) are formed. The produced ATP and NADPH are further used in the dark reactions for synthesis of various carbohydrates from carbon dioxide and water.

Photosynthetic eukaryotes and cyanobacteria perform oxygenic photosynthesis and produce molecular oxygen as a byproduct from the water splitting in the light reactions. The light stage is a complex of enzymatic reactions occurring in membrane structures, thylakoids, which are located in the stroma of the chloroplasts in eukaryotic cells and simply in the cytoplasm in the prokaryotic cyanobacteria (*15*). The cascade of the enzymatic reactions is mainly catalyzed by four protein complexes (Figure 2): photosystem II (PSII), photosystem I (PSI), cytochrome *b/f* complex, and ATP synthase.

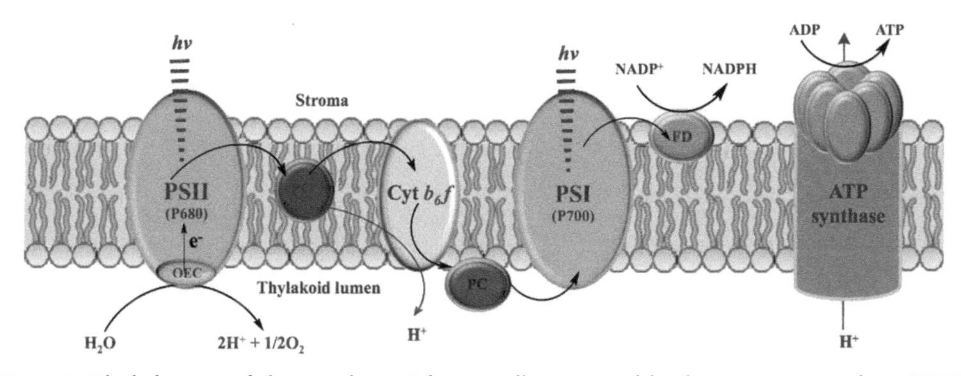

Figure 2. The light stage of photosynthesis. Schematic illustration of the electron transport chain (ETC) located in a thylakoid membrane (TM). Water is the source of electrons, produced in the oxygen-evolving complex (OEC). The electrons further flow through the ETC consisting of three main complexes, PSI, PSII, and cytochrome b_6f, which are connected by plastoquinone (PQ) and plastocyanin (PC) mobile electron carriers. The electrons continuously travel from water to $NADP^+$ to conserve reducing power in the form of NADPH. This process is coupled to a unidirectional proton pumping across the TM from the stroma to the lumen. The generated proton motive force is used by ATP synthase to make ATP.

The central process is light absorption accomplished by a number of pigments; the primary energy-transduction pigment that traps solar energy is chlorophyll. There are several types of chlorophyll in oxygenic phototrophs: *a*, *b*, *c*, *d*, and *f*. The most important and widely spread are chlorophyll *a* (absorption maxima at 420 and 660 nm) and chlorophyll *b* (absorption maxima at 450 and 640 nm) and have broad absorption spectra (*16*). Most photosynthetic organisms additionally contain accessory pigments, carotenoids and phycobilins, which capture light in the range not absorbed by chlorophylls, to allow a broader light-trapping spectrum and further transmit this light energy to chlorophylls. The pigments are assembled in highly organized antennas of large area to efficiently capture light quanta and transfer them to a photosynthetic reaction center by a hopping mechanism (*17*). In oxygenic phototrophs, the antenna complexes are associated within PSI (absorbs long and short wavelengths of red light with an absorption maximum of 700 nm) and PSII (absorbs only short wavelengths of red light with an absorption maximum of 680 nm) with reaction centers

P700 and P680, respectively (Figure 3). Absorption of light by PSI and PSII brings electrons up to the highest level of the photosynthetic electron transport chain.

Water is an abundant electron donor in oxygenic photosynthesis. The light-driven oxidation of water with production of protons and molecular oxygen is widely known as water photolysis and appeared about 2 to 3 billion years ago in cyanobacteria. When the oxygen-evolving complex, associated with PSII and assembled of manganese ions and proteins, catalyzes water photooxidation, each water molecule donates two electrons to oxidized P680. The PSII antennas absorb and transfer light energy to the reaction center P680, which turns into an excited state P^*_{680} with a negative potential of about -0.80 V. The excited high-energy electrons are further passed to the reaction center P700 through specific electron carriers (Figure 3). When P700 turns into an excited state P^*_{700} with a very negative reduction potential of -1.30 V, this allows donation of the photoexcited electrons from the reaction center to ferredoxin through an iron–sulfur cluster or a third chlorophyll *a* molecule (Figure 3).

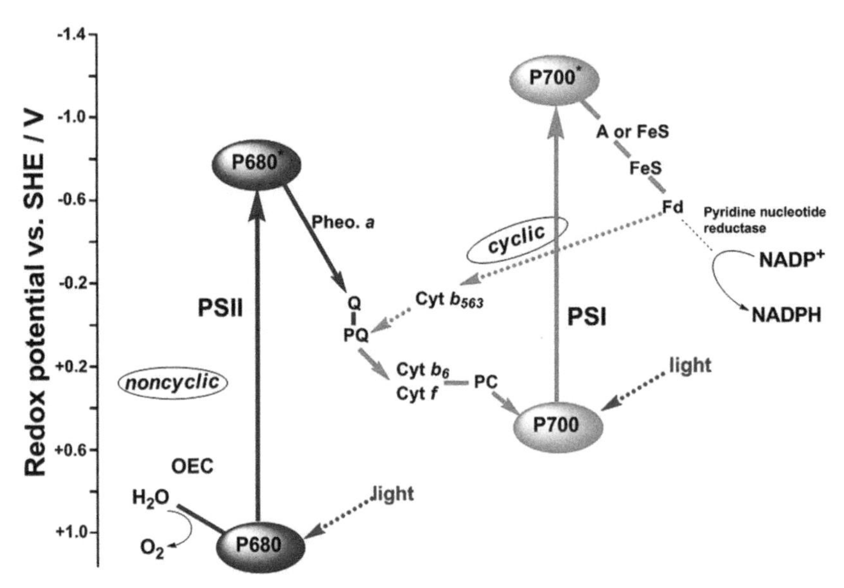

Figure 3. Scheme of electron flow in oxygenic photosynthesis. Both PSI and PSII participate in the noncyclic electron transfer pathway; the cyclic electron flow is realized only with aid of PSI (green lines). The electron carriers involved are pheophytin a (Pheo. a), quinones (Q), plastoquinone (PQ), cytochrome b_6 (Cyt b_6), cytochrome f (Cyt f), iron–sulfur (FeS) cluster, chlorophyll a (A), and copper-containing plastocyanin (PC).

At this point, two optional electron transport routes can be accomplished by the living cell in order to balance NADPH and ATP synthesis depending on the precise energy needs. When NADPH consumption is low in the cell, the cyclic pathways can dominate. In this case, the flow of electrons bypasses PSII and circulates from the P700 reaction center through a number of electron carriers in the ETC and back to the oxidized P700 (Figure 3). This allows the formation of a proton-motive force for ATP synthesis; no water oxidation and molecular oxygen release are observed because only PSI participates. The noncyclic electron flow is active in the cell when NADPH and ATP are consumed by a variety of metabolic pathways and involves both photosystems. The transfer of electrons from ferredoxin to $NADP^+$ is catalyzed by ferredoxin-$NADP^+$ reductase. The electrons continuously travel from water to $NADP^+$ down the ETC, and ATP synthesis, so-called noncyclic photophosphorylation, is driven by the electrochemical proton gradient. However, the amount of ATP generated is slightly less.

Research in photosynthesis made dramatic progress during the last century. Several great discoveries awarded the Nobel prize (to P. Mitchel in 1978; to J. Deisenhofer, H. Michel, and R. Huber in 1988; to R. Marcus in 1992; and to P. Boyer and J. Walker in 1997) gave us a deep understanding of the photosynthetic processes in biological systems. Today, photosynthetic energy conversion is one of the most studied biological processes in both molecular and atomic terms (*18*). The accumulated knowledge allows for a wide range of photosynthetic biocatalysts to be employed in various BESs for energy generation.

Photosynthetic Bioelements

The central component in PBV systems is a photosynthetic biomaterial ranging from purified protein complexes to whole microbiological organisms. The main advantage in using isolated photosystems is their extreme quantum efficiency. They have less influence from the surrounding redox active compounds compared to whole TMs, so the current output is expected to be higher. Furthermore, the electron transfer (ET) sites in the reaction centers are much closer to the electrode surfaces, which may facilitate the ET processes (*19, 20*).

The P^*_{700} reaction center is the strongest biological reductant (-1.3 V) and functions as an electron pump with a charge separation of about 1 V. In biological systems, PSI requires an electron donor and is dependent on the electron flow from PSII and the PSI/PSII ratio (*21*), but its stability is much higher than for PSII, so the turnover number is also higher (*22*). PSI is a robust complex with promising properties, which provide a great potential for its application in BESs and to furthermore outperform semiconductor-based devices (*23–25*).

The P^*_{680} reaction center is the strongest oxidant (+1.2 V) reported in biological systems with a solar conversion efficiency of about 34% (*26*). The advantage of PSII is that it is autonomous in ET, has an oxygen-evolving complex, and gains electrons from water oxidation. In such a way, PSII requires only water and light, and both are abundant in the environment (*27*). The main disadvantage is the short life of isolated PSII, which is a challenge in any application of the protein complex. The integration of PSII into electrochemical systems still lies within fundamental research; the findings give a better understanding of the associated ET processes and an insight into functional principles of the water-splitting anodes as such.

Though photosynthetic reaction centers exhibit a general structure and functions involved in most organisms, they still may differ in terms of the molecular compositions and mechanisms. It should be mentioned that bacterial photosynthetic reaction centers, such as bacterial reaction center–light harvesting 1 complex (RC-LH1) in purple *Rhodobacter sphaeroides*, implicating the reaction center encircled by light-harvesting pigments, are distinct in their structure and biochemistry from the ones in algae and plants. They have a simple molecular structure, display a broad spectral sensitivity, and are able to drive a special photoinduced ET chain, which no doubt make them an attractive object of study for employment in bioelectrochemical solar energy conversion devices (*28, 29*).

TMs are cellular suborganelles, which may be easily isolated from green plants (typically from spinach leaves). The photosynthetic protein complexes remain in their natural environment, which potentially may increase the operational stability and power output in comparison to isolated photosystems. Carpentier et al. (*30*) for the first time demonstrated an application of TMs for current production. Recent works have reported (*31*) that both noncyclic and cyclic photosynthetic pathways contribute to photocurrent output, which is one of the benefits of using TMs since various ET pathways for communication with the electrode are available. When thylakoids are immobilized directly on the electrode surface, PSI, PSII, plastoquinone, cytochrome b_6f, and plastocyanin are involved in the extracellular ET to electrodes (*31*). However, under mediated ET conditions, thylakoid-based bioanodes show no involvement of PSI in outward ET processes, since the mediator interacts only with one photosystem (*32*). During the last few decades, dramatic progress has been made, and today TMs are one of the best studied and characterized photobiocatalysts. However, due to the relative instability and short life of the photosynthetic membranes, large-scale practical application of the BPV systems based on TMs remains a challenge.

In whole-cell BPVs, oxygenic photosynthetic microorganisms, cyanobacteria, and algae are able to produce electrons via photolysis of water in light and pass electrons to electrodes. It should be noted that living microorganisms generate currents also in the dark via the respiratory chain (*33, 34*). In contrast to individual subcomponents of the photosynthetic apparatus, photosynthetic microbes are robust, less susceptible to dehydration, self-regulating and self-repairing organisms (*35*). Cyanobacteria are the most preferable and well-characterized microbes in cellular BPVs due to the simplicity of their metabolism and physiology, such as in contrast to eukaryotic green algae (*36*). A number of species have been demonstrated in association with anodes for light-dependent electricity production: unicellular *Synechococcus* ssp., *Synechocystis* spp., filamentous *Arthrospira maxima*, *Spirulina platens*, *Phormidium* spp., *Nostoc* sp., *Anabaena* spp., and others (*37*). The main disadvantage and limitation of such systems are the low capacity of ET in the microorganisms in comparison with traditional microbial fuel cells or PMFCs. A fundamental understanding of the physiology and the intracellular ET mechanisms of bacteria in BPVs is urgently needed to improve the exoelectrogenic activity and optimize the electrochemical connection between the bacteria and the conductive surfaces.

Electrode Materials

Carbon-based materials were widely used to provide high loading of the bioelement and achieve efficient nondiffusional ET between the photosynthetic centers and electrode surfaces. Numerous advantages of carbon-based materials, such as robustness, low fabrication cost, great variety of surface morphologies, broad possibilities for chemical functionalization to facilitate interaction between biomolecules and electrode surfaces, and biocompatibility, make them the most widely used electrode materials in the field of photobioelectrochemistry (*38, 39*). In spite of early studies conducted on planar and low-porosity carbon and mainly dedicated to the fundamental investigation of ET processes between the bioelement and electrode and proof-of-principle demonstration of possibilities of light-to-energy conversion using the photosynthetic machinery (*31, 40, 41*), further development of photobioelectrochemical systems (PBESs) requires higher performance, and therefore materials with expanded surface area without loss of physiochemical advantages are of special interest.

Carbon nanotubes (CNTs) fulfill all of the requirements, including an impressive specific surface area of 1000 m^2 g^{-1} (*42*), and were used in PBESs for light energy conversion. Calkins et al. (*32*) achieved a maximum photobiocurrent density of 68 μA cm^{-2} (light intensity [E_e] of 80 mW cm^{-2}, 0.2 V vs. Ag|AgCl) for TMs immobilized on multiwalled CNTs using dissolved ferrocyanide as mediator. The current output achieved for the TM/CNT composite was significantly higher than that reported for analogous systems at that moment.

Fundamental investigations of the possibility of direct electron transfer communication between the TMs and the CNT surface were conducted by Dewi et al. (*43*) using various spectroscopic techniques and further converted into practice by fabrication of a mediatorless TM/CNT photobioanode displaying a photobiocurrent density of ~2 μA cm^{-2} (0.6 V vs. standard hydrogen electrode [SHE], E_e of 40 mW cm^{-2}; see Figure 4A) (*44*).

Incorporation of single-walled CNTs modified with cerium oxide nanoparticles within the lipid envelope of chloroplasts enables higher ET rates with depressed generation of damaging reactive oxygen species (*45*). CNT-modified hybrids exhibit over threefold higher photosynthetic activity than that of controls and also facilitate leaf electron transport in vivo. A similar effect has recently been demonstrated (*46*) for *Chlorella* cells immobilized with multiwall CNTs using a polyethylene glycol diacrylate hydrogel film in a microfluidic chamber. A successful conjugation of single-walled CNTs with PSI has also been demonstrated recently (*47*).

Buckypaper is a self-supporting plexiform film with a typical thickness of 5–200 μm consisting of densely packed CNTs maintaining close contact due to π–π stacking and interweaving interactions between the nanotubes (*48*). Due to a favorable surface morphology, buckypaper may become a promising support for microbial or TM-based PBESs, where the size of the photosynthetic bioelement is comparable with the dimensions of the CNTs (*49*). Currently, Toray carbon paper, carbon cloth, and carbon brush are the most common carbon supports for these kinds of systems, due to their low price, availability, scalability, and attractiveness for the formation of bacterial and cyanobacterial biofilms (*8*). Kirchhofer et al. (*50*) demonstrated an up to 1.7-fold increase in photobiocurrent when conjugated oligoelectrolytes were used to improve the contact between the TMs and the carbon surface.

Graphene (Gr)-based surfaces became a popular support to host various bioelements due to the unique physiochemical and structural features of Gr, namely, good electrical conductivity and high mechanical strength and specific surface area (*51–53*).

Pankratova et al. (*54*) utilized amidated three-dimensional (3D) Gr prepared by simultaneous electrodeposition and electroreduction of graphene oxide to create an optimized TM-based photobioanode. A maximum current density of 5.24 μA cm^{-2} achieved at 0.6 V vs. SHE under an E_e of 40 mW cm^{-2} is the highest direct electron transfer photobiocurrent output for a TM-based system reported so far (Figure 4B).

Feifel et al. (*55*) demonstrated high-performance Gr/PSI biohybrid light-harvesting electrodes employing a strong π–π interaction between various pyrene and anthracene derivatives and the Gr surface. The reported systems display a unidirectional photocurrent generation up to 23 and 135 μA cm^{-2} (E_e of 60 mW cm^{-2}) in the absence and in the presence of an electron acceptor (methyl viologen), respectively.

In contrast to the many opportunities provided by opaque carbon, the conductive indium tin oxide (ITO) surface has advantages for manufacturing transparent bioelectrodes, opening new opportunities in development of closed PBESs, in which light can reach the immobilized photosynthetic components in the shortest way.

Dewi et al. (*56*) used ITO as the support for direct immobilization of TMs and demonstrated a bidirectional photocurrent arising from the reduction of hydrogen peroxide or oxygen (cathodic current) or from ET between the TMs and ITO (anodic current) at potentials lower and higher than ~0.37 V vs. SHE, respectively.

ITO has also been demonstrated as a possible support for growing photosynthetic biofilms of algae and cyanobacteria that enable the production of a notable photocurrent output for extended periods (*34, 57, 58*).

Utilization of carbon quantum dots in PBESs results in a significant enhancement of the photocurrent output (*59, 60*). A small particle size (>10 nm) allows their penetration close to the photosynthetic reaction center and a high electrochemical stability, and the opportunity for structural and surface functionalization allows targeted modification. Small fullerene nanoparticles can also serve as electrical wiring agents for PSI to improve the communication of the protein with the electrode surface (*61*). A notable shift in the onset potential and an increase in the photobiocurrent output up to 15 μA cm^{-2} (−300 mV vs. Ag|AgCl, E_e of 50 mW cm^{-2}) have been achieved for a monolayer coverage, confirming a highly efficient communication in the PSI-electrode system.

Incorporation of the photosynthetic machinery into a conducting polymer matrix has also been considered as a promising tool to improve the conductivity and the orientational control in the photoactive layer. Employment of polyaniline (*62, 63*), poly(3,4-ethylenedioxythiophene) (*64*), and cationic poly(fluorene-*co*-phenylene) (*65*) resulted in a notable increase in photobiocurrent output and operational stability of the developed bioelectrodes.

The relatively high cost and limited availability of transparent conductive oxides have stimulated screening for possible alternatives to these supports (*66*). Fabrication of semitransparent carbon-on-quartz chips through the pyrolysis of a photoresist allowed a thorough and detailed investigation of the optimal relationship between the transparent nonconductive area and the opaque carbon support and the influence of the size of the quartz "windows" on the photobiocurrent generation, and it allowed the development of a theoretical model describing this relationship, which might be an important step toward development of scalable PBESs with a controlled surface area (*67*).

The morphology of the electrode surface plays an important role in the optimization of the flux of electrons to and from the bioelement. Kanso et al. (*68*) investigated the influence of gold microparticles and gold nanoparticles electrodeposited onto gold and carbon surfaces on the activity of TMs, "wired" via an osmium-complex-containing redox polymer. A relatively high photobiocurrent density of 62.5 μA cm^{-2} was registered for a gold microparticle-modified carbon electrode at 0.4 V vs. Ag pseudoreference under an E_e of 400 W m^{-2}, which confirmed the importance of the material–structure relationship in the development of PBESs.

In spite of the promising examples of electrochemical communication between the bioelements and different planar and porous conductive surfaces, their practical application is limited by the low loading of the photosynthetic components. Switching from an active surface area to an active volume by the development of hierarchical transparent 3D electrodes, combining the advantages of interconnected macroporous and mesoporous structures, can be considered as a possible solution for this problem. Utilization of hierarchical mesoporous (*70*) and inverse opal (IO) ITO electrodes (*69*) allowed an increase in the PSII loading up to 19 and 1020 pmol cm^{-2}, respectively, compared to 0.74 pmol cm^{-2} estimated for an ideal monolayer (Figure 4C) (*71*). A similar loading has been demonstrated by Stieger et al. (*72*) for PSI-based multilayered mesoporous ITO electrodes prepared by spin coating. A surface concentration of 4.5 pmol cm^{-2} per layer was achieved for PSI, which is close to an entire monolayer coverage of 0.2 pmol cm^{-2} related to the electrochemically active surface area.

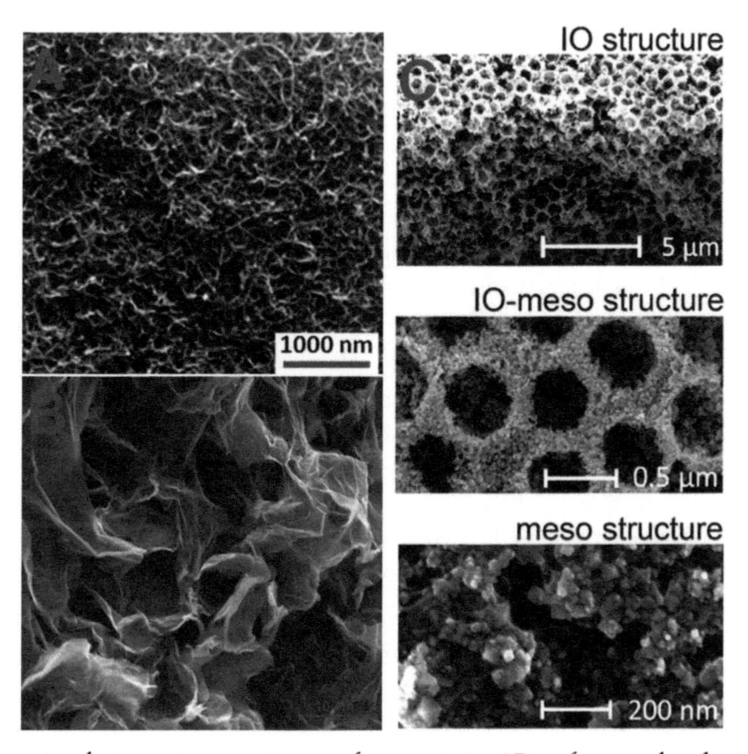

Figure 4. Scanning electron microscopy images of representative 3D surfaces employed as a support in photobioelectrochemical systems. (A) Amidated multiwalled CNTs. Adapted with permission from reference (44). Copyright 2017 American Chemical Society. (B) Electrodeposited reduced graphene oxide. Adapted with permission from reference (54). Copyright 2018 American Chemical Society. (C) Hierarchical mesoporous inverse opal–ITO. Adapted with permission from reference (69). Copyright 2015 American Chemical Society.

A peak photobiocurrent density of -166 µA cm^{-2} (-50 mV vs. Ag/AgCl, E_e of 100 mW cm^{-2}) was achieved for photocathodes based on the *R. sphaeroides* bacterial reaction center RC-LH1 self-assembled on a nanostructured silver substrate decaying to a steady-state level of -80 µA cm^{-2} after 90 s of operation (73). Remarkably, a 6.1-fold decrease of the photocurrent output was registered on a silver surface with a lower roughness factor, allowing 2.4-fold less protein loading, which indicates that the nanostructuring of the substrate not only results in an enhanced surface but also provides opportunities for more efficient light harvesting.

A two order of magnitude increase in photobiocurrent generation compared to their planar surface counterparts (~1 and 0.004 µA cm^{-2}, respectively) has been demonstrated for the cyanobacteria *Nostoc punctiforme* and *Synechocystis* sp. PCC6803 on nanostructured ITO electrodes with a porosity suitable for hosting entire living cells (74).

Fang et al. (75) conducted a systematic study of the structure–activity relationship for IO-ITO and IO-graphene electrodes with a variable morphology biomodified with PSII. It has been demonstrated that mesopores larger than the size of PSII combined with small macropores are preferable for a higher loading of the bioelement, whereas the activity of adsorbed PSII is mainly determined by the light intensity and the electronic communication at the interface of bioelement and electrode. Various strategies for surface modification to achieve a better communication with the immobilized bioelement are discussed in recent reviews (*76*, *77*) and can potentially be adapted for the development of PBSs.

Mediators

To overcome the challenges of achieving a sufficient ET rate between the bioelement and the electrode surface, artificial mediators can be used as charge carriers to shuttle the electrons to and from the active center of the immobilized or suspended photobioelement and the electrode surface. A wide range of soluble mediators can be employed in PBESs, either to achieve a better photocurrent output or to perform fundamental studies of the ET processes, such as metal complexes (*32*, *67*), quinones (*78–80*), or metallocenes (Figure 5A). However, in spite of the clear effect on the ET rate, a real employment of soluble mediators in BESs has a number of limitations, including the introduction of a membrane between the cathodic and anodic compartments, a necessity in a closed system to avoid mediator leakage and poisoning of the bioelement and short circuiting of the system.

Redox polymer hydrogels (RPs) have received considerable attention toward "wiring" photosynthetic bioelements, due to their good solubility, availability of hydrophobic, charged, and hydrogen-binding sites, flexible 3D network, broad range of redox potentials and hydrophilicity depending on the structure of the RP (Figure 5B), and the stability and relatively fast ET through the RP matrix (*81*). They consist of a polymer backbone (poly(vinyl imidazole)s and poly(vinyl pyridine)s for the first-generation RPs (*82*)) and redox centers incorporated into the polymer through flexible tethers. The ET ability of the RP is related to the self-exchange of electrons or holes in the water-swollen hydrogels from the reduced to the oxidized mobile redox centers, or vice versa.

Osmium-complex-containing redox polymers (OsRPs) were initially used for the development of enzymatic fuel cells (*83*) and biosensors (*84*), and afterward they were successfully introduced into the field of PBESs. A significant contribution to electrochemical wiring of PSI and PSII to the various electrode materials using OsRPs was made by Schuhman's group. From a proof-of-principle demonstration of the possibility of ET between PSII (*85*) and PSI (*86*) with the electrode surface through the OsRP matrix, to the recent examples of rational wiring of photosynthetic proteins (*87*, *88*), the photobiocurrent density was increased by an order of magnitude up to several hundred μA cm^{-2}, displaying a rapid and simultaneous development of materials science techniques, polymer chemistry methodology, and fundamental investigation of the factors limiting photobioelectrochemical performance (*89*, *90*).

Gorton's group introduced OsRPs for fabrication of photobioanodes employing TMs (*91*), cyanobacteria (*92*), purple bacteria *Rhodobacter capsulatus* (*93*), and eukaryotic algae *Paulschulzia pseudovolvox* (*92*, *94*). The maximum photobiocurrent output achieved for the bioelement/OsRP communication varies in these works from several μA cm^{-2} for whole living cells up to ~100 μA cm^{-2} for the TM/OsRP/CNT composite (*95*).

Utilization of OsRP in a hybrid PSII electrode mimicking the photosynthetic Z-scheme by incorporation of light-sensitive semiconductive PbS quantum dots was demonstrated by Riedel et al. (*96*) (Figure 5C). Coupling of both entities into one photobioanode results in the formation of a multistep signal chain and oxidation of water starting at −0.55 V vs. Ag|AgCl under illumination. Coupling of this hybrid photobioanode with an oxygen-reducing enzymatic biocathode based on the biocatalytic activity of immobilized bilirubin oxidase (BOx) gives an exceptionally high open circuit voltage (OCV) close to 1 V.

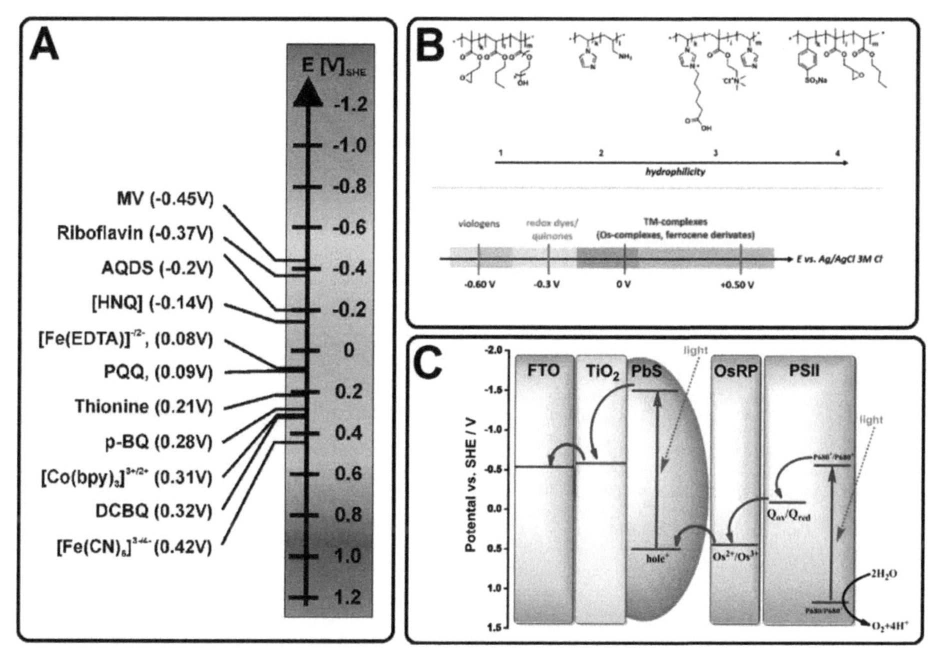

Figure 5. (A) Representative soluble mediators suitable for utilization in PBESs: AQDS, 9,10-anthraquinone-2,6-disulfonate; $[Co(bpy)_3]^{3+/2+}$, cobalt(II/III) tris(2,2′-bipyridine); DCBQ, 2,6-dichloro-1,4-benzoquinone; HNQ, 2-hydroxy-1,4-naphthoquinone; MV, methyl-viologen; p-BQ, p-benzoquinone; PQQ, pyrroloquinoline quinone. Reproduced with permission from reference (37). Copyright 2019 Frontiers Media S.A. (B) Relative hydrophilicity of functional polymer backbones of different structures and the scale of redox potentials of RPs containing covalently anchored redox groups. Adapted with permission from reference (82). Copyright 2017 Elsevier. (C) Schematic representation of a hybrid photobioanode developed by Riedel et al. (96) depicting energetic levels of the components.

Minteer and coauthors introduced naphthoquinone-based redox polymer hydrogels (NQ-RP) into BESs (*97*) and PBESs in particular (*98*). Because the ability of the NQ-RP to penetrate through the cell wall of a Gram-positive bacterium (*Enterococcus faecalis*) and directly incorporate into the respiratory chain of the bacterium resulted in a system that recently was shown to outperform in ET features a comparable system based on a monomeric mediator (*99*), further employment of RPs based on the principles of the natural needs of the particular microorganism is also a promising strategy for further development of microbial PBESs.

Natural redox proteins (cytochrome *c* in particular) were also successfully employed to facilitate ET between the electrode surface and PSI or PSII. Employment of small redox proteins as mediators leads to improvement in photocurrent density through orientation of photoactive proteins on the electrodes and formation of a biprotein or multilayered signal chains (*72*, *100*, *101*).

Apart from immobilization of the photobioelements on the electrode surface, various concepts of PBESs employing suspensions of photosynthetic cells or cellular components electrochemically connected to a current collector via a dissolved redox mediator have also been introduced. Longatte and coworkers (*102*, *103*) and Sayegh et al. (*104*) developed an electrochemical configuration, where a *Chlamydomonas reinhardtii* algae suspension was mixed with exogenous quinones with thorough evaluation of the parameters determining the long-term photobioelectrochemical performance. Photocurrent densities up to 60 μA cm^{-2} at 0.38 V vs. Ag|AgCl were registered under an E_e of 60 mW cm^{-2}.

Pinhassi et al. (*105*) constructed a PBES for water splitting, using a TM suspension as an anodic photobiocatalyst and ferri/ferrocyanide redox couple as mediator, yielding a photocurrent density equal to several hundred μA cm^{-2}, sufficient to create an external bias of voltage for autonomous hydrogen evolution.

Rational Design of Photosynthetic Proteins

The performance of photosynthetic bioelements in terms of increasing the concentration of photosynthetic protein complexes, widening of the spectral range of photon absorption and the opportunity for multiple electron transfer pathways, and the stability or ease of immobilization on the electrode material can be significantly enhanced by using methods of protein engineering (Figure 6) (*106*). Targeted incorporation of additional functional groups may significantly enhance an oriented immobilization of the protein monolayer and accomplish high electron transfer rates between the bioelement and the electrode surfaces (*107–110*).

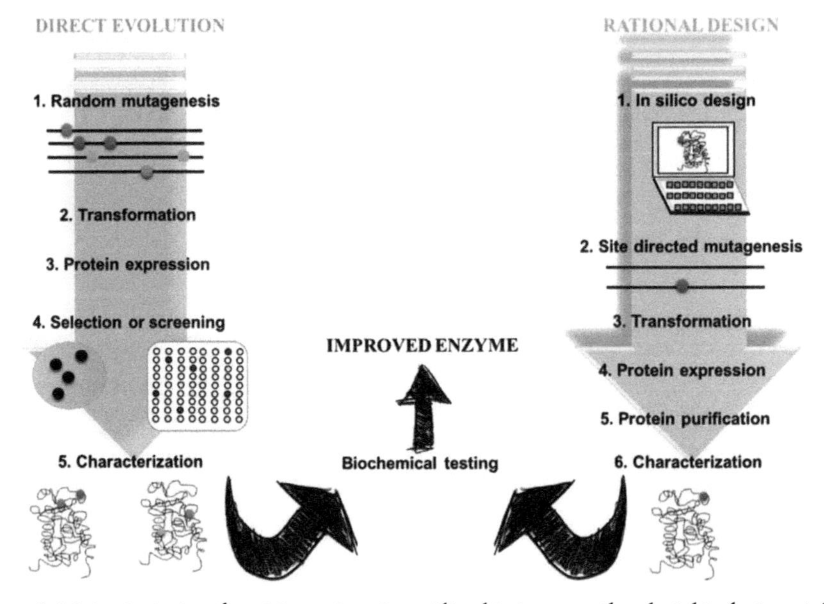

Figure 6. Main strategies of protein engineering utilized to improve the photobioelectrocatalytic performance. Reproduced with permission from reference (111). Copyright 2019 Elsevier.

Sekar et al. (*112*) engineered the cyanobacterium *Synechococcus elongatus* PCC 7942 to enhance its extracellular ET by the ability to express a nonnative outer membrane cytochrome. The mutant cyanobacterium exhibited a ninefold increase in the photocurrent output compared to the corresponding wild-type cells.

Since both photosystems compete for the same range of wavelengths in the light spectrum, replacement of PS1 by a reaction center containing bacteriochlorophylls may double the efficiency of photosynthesis by improvement in the solar photon capture by broadening absorption maxima from ~730 to ~1100 nm (2).

Hybrid Photobioelectrochemical Systems

The performance of separate bioelectrodes and complete electrochemical cells can be significantly improved by separation of the processes of conversion of the light energy into electric charges and extraction of the accumulated power.

The concept of supercapacitive bioelectrodes previously developed for microbial (*113–119*) and enzymatic BESs (*120–123*) was adapted for the PBESs by Pankratova et al. (*124*) (Figure 7A). In this work, additional inert electrodes with a variable capacitance were connected to the TM-based photobioanodes employing OsRP as mediator. The supercapacitive bioanodes displayed up to a fivefold higher photocurrent density and a significantly lower charge transfer resistance compared to the noncapacitive electrodes with the same photobioelectrochemically active surface area. The supercapacitive PBESs fabricated by interconnection of the photobioanode with a BOx-based biocathode have an OCV of 0.4 V and produce a maximum power density of 2.5 μW cm^{-2} at 0.15 V (E_e of 40 mW cm^{-2}) in conventional continuous mode, which can be increased to 56 μW cm^{-2} if tested in the self-charge/discharge regime by applying constant current pulses.

The concept of double-featured electrodes for simultaneous conversion and storage of electric power was also employed in a recent example of a light-driven Nernstian biosupercapacitor (*125*). The principle of a so-called Nernstian BES has been previously disclosed for enzymatic devices (*126*), where the anodic and cathodic bioelements were incorporated within a redox polymer matrix and act simultaneously as a mediator and a charge-storing component. The self-charging process occurring in the presence of fuel and oxidant results in an increase in the relative amounts of oxidized and reduced forms of the RP on the anode and cathode, respectively, and consequently to the growth of an OCV and the accumulation of electric charge (*127–129*). A fully light-driven device developed by Zhao et al. (*125*) comprised transparent porous ITO electrodes modified with isolated PSII and PS1 for photobioanode and photobiocathode, respectively, embedded in the same OsRP for both electrodes. An assembled BES exhibited an OCV of ~0.18 V and a maximum power output of 1.0 μW cm^{-2} (red light, an E_e of 6.5 mW cm^{-2}).

The first DET supercapacitive PBES was built by González-Arribas et al. (*130*) An anodic photobioelement (TM) and a cathodic oxygen-reducing enzyme (BOx) were directly physisorbed on a nanoparticle-coated transparent ITO surface, and the inherent capacitance of the bioelectrodes was utilized to accumulate the generated charge. The PBES was tested under ambient daylight (~23 kLux on average), displaying an OCV of 0.2 V in the fully charged state, and a peak power of 0.6 μW cm^{-2} was extracted by applying a constant load of 1 MΩ, whereas the power output achieved in the continuous mode was 0.005 μW cm^{-2}, which indicated a poor electrochemical "wiring" of TMs to the ITO surface.

Another example of a supercapacitive PBES employed amidated CNTs as the support for immobilization of TMs and as a charge-storing element (*44*). A favorable charge of the electrode surface promotes a proper orientation of the photobioelement, which results in a low open circuit potential of the photobioanode, a relatively low charge transfer resistance, and a remarkable current output. The PBES built by coupling of the photobioanode with the BOx-based biocathode had an OCV of 0.45 V and a maximum power output of 0.66 μW cm^{-2} at 0.21 V (E_e of 40 mW cm^{-2}) as a biosolar cell, which can be increased up to 155 μW cm^{-2} under self-charge/discharge cycling.

An advanced approach for decoupling the energy conversion part and the power delivery part has been proposed by Saar et al. (*131*) (Figure 7B). In this work, *Synechocystis* cyanobacterial cells were utilized as the photobiocatalyst in the charging unit of a two-chamber flow-controlled system, in which potassium ferricyanide was used as the charge carrier transferring exoelectrogenic electrons from the cyanobacterial cells to the anode in the power delivery unit. Oxidation of ferrocyanide in the power delivery chamber was coupled to the oxygen-reduction part, delivering a power density of over 50 μW cm^{-2}, which is remarkable for PBESs employing cyanobacteria.

Ravi et al. (*132*) employed RC-LH1 from *R. sphaeroides* to create a charge-storing biophotonic power cell. Light-induced self-charging of a constructed device was attributed to oxidation of RC-LH1 at the fluorine-doped tin oxide glass electrode (positively charged protein, Figure 7C) and reduction of RC-LH1 by the photoactive n-type silicon (negatively charged protein, Figure 7C). Charge separation was established by the insulating function of the central protein layers. The maximum photovoltage achieved was as high as 0.45 V with a capacitance of 10–20 μF cm^{-2}.

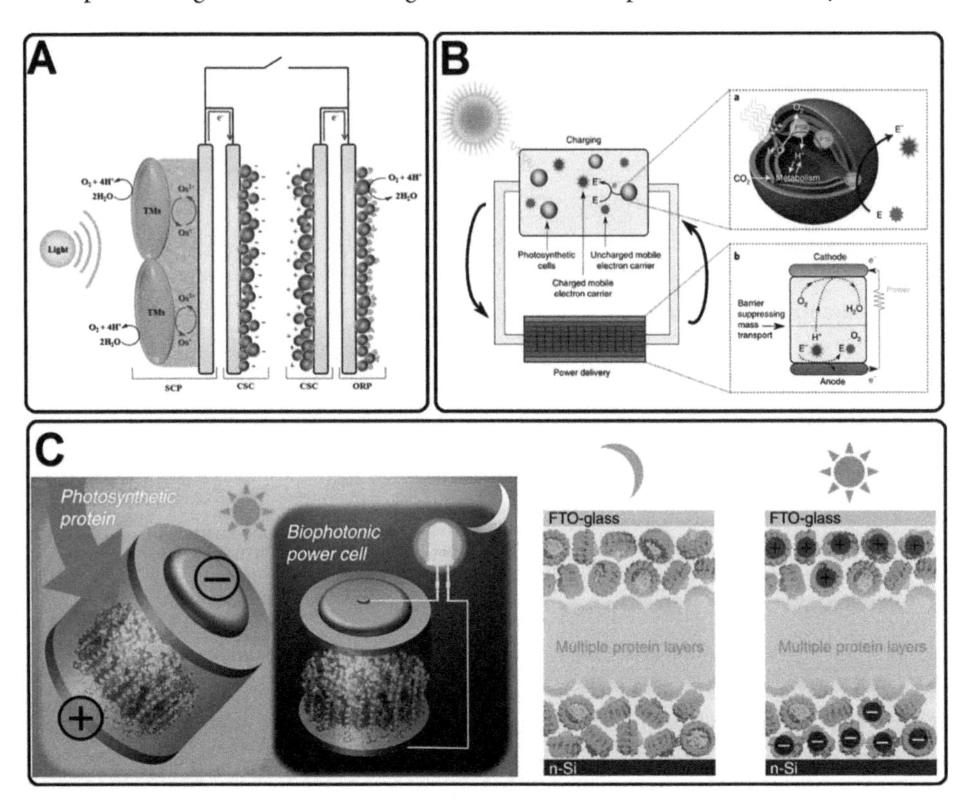

Figure 7. (A) Schematic representation of a supercapacitive PBES combining a solar energy converting part (SCP, TMs/OsRP-based photobioanode), an oxygen-reducing part (ORP), and anodic and cathodic charge storing components (CSCs). Adapted with permission from reference (124). Copyright 2017 John Wiley and Sons.(B) A two-chamber flow-controlled system consisting of the charging unit with cyanobacterial cells and a soluble mediator and the power delivery unit with an oxygen-reducing cathode and an anode, where the oxidation of the charge carrier occurs. Reproduced with permission from reference (131). Copyright 2018 Springer Nature. (C) Charge-storing biophotonic power cell based on the bacterial RC-LH1. Adapted with permission from reference (132). Copyright 2019 Springer Nature.

Conclusions and Future Perspectives

Fabrication of BESs employing an oxygenic photosynthetic biocatalyst is a rapidly growing research field. Though a significant step forward has been made during the last few decades, the technological progress of those systems is still largely limited. Low current outputs and often a short operating stability are obvious objects for further systematic optimization and improvements. A fundamental understanding of the physiological needs and requirements for the biocatalyst might be a key to overcoming the poor characteristics of the currently existing PBESs. Methods of genetic manipulation and synthetic biology are viable ways to improve the ET efficiency and to adapt the photobiocatalyst to the electrode environment.

Abbreviations

3D	three-dimensional
ATP	adenosine triphosphate
BES	bioelectrochemical system
BOx	bilirubin oxidase
BPV	biophotovoltaic cells
CNT	carbon nanotube
E_e	light intensity
ET	electron transfer
ETC	electron transport chain
Gr	graphene
IO	inverse opal
ITO	indium tin oxide
NADPH	nicotinamide adenine dinucleotide phosphate
OCV	open circuit voltage
OsRP	osmium-complex-containing redox polymer
PBES	photobioelectrochemical system
PMFC	photosynthetic microbial fuel cells
PSI	photosystem I
PSII	photosystem II
RC-LH1	RC-LH1, reaction center-light harvesting 1 complex
RP	redox polymer
SHE	standard hydrogen electrode
TM	thylakoid membrane

Acknowledgments

The authors thank the following agencies for financial support: the Swedish Research Council (grant 2014-5908) and the European Research Council, EU project 772370 PHOENEEX.

References

1. Lewis, N. S.; Nocera, D. G. Powering the planet: Chemical challenges in solar energy utilization. *Proc. Natl. Acad. Sci. U. S. A.* **2006**, *103*, 15729–15735.
2. Blankenship, R. E.; Tiede, D. M.; Barber, J.; Brudvig, G. W.; Fleming, G.; Ghirardi, M.; Gunner, M. R.; Junge, W.; Kramer, D. M.; Melis, A.; Moore, T. A.; Moser, C. C.; Nocera, D. G.; Nozik, A. J.; Ort, D. R.; Parson, W. W.; Prince, R. C.; Sayre, R. T. Comparing

photosynthetic and photovoltaic efficiencies and recognizing the potential for improvement. *Science* **2011**, *332*, 805–809.

3. Green, M. A. Commercial progress and challenges for photovoltaics. *Nat. Energy* **2016**, *1*, 15015.
4. Tao, C. S.; Jiang, J. C.; Tao, M. Natural resource limitations to terawatt-scale solar cells. *Sol. Energy Mater. Sol. Cells* **2011**, *95*, 3176–3180.
5. Peter, L. M. Towards sustainable photovoltaics: The search for new materials. *Philos. Trans. A Math. Phys. Eng. Sci.* **2011**, *369*, 1840–1856.
6. Mazzio, K. A.; Luscombe, C. K. The future of organic photovoltaics. *Chem. Soc. Rev.* **2015**, *44*, 78–90.
7. Wraight, C. A.; Clayton, R. K. The absolute quantum efficiency of bacteriochlorophyll photooxidation in reaction centres of *Rhodopseudomonas spheroides*. *Biochim. Biophys. Acta* **1974**, *333*, 246–260.
8. McCormick, A. J.; Bombelli, P.; Bradley, R. W.; Thorne, R.; Wenzele, T.; Howe, C. J. Biophotovoltaics: Oxygenic photosynthetic organisms in the world of bioelectrochemical systems. *Energy Environ. Sci.* **2015**, *8*, 1092–1109.
9. Laureanti, J. A.; Jones, A. K. Photosynthetic microbial fuel cells. In *Biophotoelectrochemistry: From Bioelectrochemistry to Biophotovoltaics*; Jeuken, L. J. C., Ed.; Springer International Publishing: Cham, Switzerland, 2016; pp 159–175.
10. Plumeré, N.; Nowaczyk, M. M. Biophotoelectrochemistry of photosynthetic proteins. In *Biophotoelectrochemistry: From Bioelectrochemistry to Biophotovoltaics*; Jeuken, L. J. C., Ed.; Springer International Publishing: Cham, Switzerland, 2016; pp 111–136.
11. Pankratova, G.; Gorton, L. Electrochemical communication between living cells and conductive surfaces. *Curr. Opin. Electrochem.* **2017**, *5*, 193–202.
12. Rosenbaum, M. A.; Henrich, A. W. Engineering microbial electrocatalysis for chemical and fuel production. *Curr. Opin. Biotechnol.* **2014**, *29*, 93–98.
13. Evans, R. M.; Siritanaratkul, B.; Megarity, C. F.; Pandey, K.; Esterle, T. F.; Badiani, S.; Armstrong, F. A. The value of enzymes in solar fuels research: Efficient electrocatalysts through evolution. *Chem. Soc. Rev.* **2019**, *48*, 2039–2052.
14. Kornienko, N.; Zhang, J. Z.; Sakimoto, K. K.; Yang, P. D.; Reisner, E. Interfacing nature's catalytic machinery with synthetic materials for semi-artificial photosynthesis. *Nat. Nanotechnol.* **2018**, *13*, 890–899.
15. Willey, J.; Sherwood, L.; Woolverton, C. J. *Prescott's Microbiology*, 9th ed.; McGraw-Hill: New York, NY, 2014.
16. Hardin, J.; Bertoni, G. P.; Kleinsmith, L. J. *Becker's World of the Cell*, 8th ed.; Benjamin Cummings: Boston, MA, 2012.
17. Qin, X. C.; Suga, M.; Kuang, T. Y.; Shen, J. R. Structural basis for energy transfer pathways in the plant PSI-LHCI supercomplex. *Science* **2015**, *348*, 989–995.
18. Govindjee. Advances in photosynthesis and respiration. In *Regulation of Photosynthesis*; Aro, E.-M., Andersson, B., Eds.; Kluwer Academic Publishers: Dordrecht, 2001; pp v–vi.
19. Voloshin, R. A.; Kreslavski, V. D.; Zharmukhamedov, S. K.; Bedbenov, V. S.; Ramakrishna, S.; Allakhverdiev, S. I. Photoelectrochemical cells based on photosynthetic systems: A review. *Biofuel Res. J.* **2015**, *2*, 227–235.

20. Friebe, V. M.; Frese, R. N. Photosynthetic reaction center-based biophotovoltaics. *Curr. Opin. Electrochem.* **2017**, *5*, 126–134.
21. Lubner, C. E.; Applegate, A. M.; Knorzer, P.; Ganago, A.; Bryant, D. A.; Happe, T.; Golbeck, J. H. Solar hydrogen-producing bionanodevice outperforms natural photosynthesis. *Proc. Natl. Acad. Sci. U. S. A.* **2011**, *108*, 20988–20991.
22. Yao, D. C. I.; Brune, D. C.; Vermaas, W. F. J. Lifetimes of photosystem I and II proteins in the cyanobacterium *Synechocystis sp.* PCC 6803. *FEBS Lett.* **2012**, *586*, 169–173.
23. Gerster, D.; Reichert, J.; Bi, H.; Barth, J. V.; Kaniber, S. M.; Holleitner, A. W.; Visoly-Fisher, I.; Sergani, S.; Carmeli, I. Photocurrent of a single photosynthetic protein. *Nat. Nanotechnol.* **2012**, *7*, 673–676.
24. Nguyen, K.; Bruce, B. D. Growing green electricity: Progress and strategies for use of Photosystem I for sustainable photovoltaic energy conversion. *Biochim. Biophys. Acta* **2014**, *1837*, 1553–1566.
25. Robinson, M. T.; Gizzie, E. A.; Mwambutsa, F.; Cliffel, D. E.; Jennings, G. K. Mediated approaches to Photosystem I-based biophotovoltaics. *Curr. Opin. Electrochem.* **2017**, *5*, 211–217.
26. Dau, H.; Zaharieva, I. Principles, efficiency, and blueprint character of solar-energy conversion in photosynthetic water oxidation. *Accounts Chem. Res.* **2009**, *42*, 1861–1870.
27. Sekar, N.; Ramasamy, R. P. Recent advances in photosynthetic energy conversion. *J. Photochem. Photobiol. C* **2015**, *22*, 19–33.
28. Yehezkeli, O.; Tel-Vered, R.; Michaeli, D.; Willner, I.; Nechushtai, R. Photosynthetic reaction center-functionalized electrodes for photo-bioelectrochemical cells. *Photosynth. Res.* **2014**, *120*, 71–85.
29. Ravi, S. K.; Tan, S. C. Progress and perspectives in exploiting photosynthetic biomolecules for solar energy harnessing. *Energy Environ. Sci.* **2015**, *8*, 2551–2573.
30. Carpentier, R.; Lemieux, S.; Mimeault, M.; Purcell, M.; Goetze, D. C. A photoelectrochemical cell using immobilized photosynthetic membranes. *Bioelectrochem Bioenerg.* **1989**, *22*, 391–401.
31. Rasmussen, M.; Minteer, S. D. Investigating the mechanism of thylakoid direct electron transfer for photocurrent generation. *Electrochim. Acta* **2014**, *126*, 68–73.
32. Calkins, J. O.; Umasankar, Y.; O'Neill, H.; Ramasamy, R. P. High photo-electrochemical activity of thylakoid-carbon nanotube composites for photosynthetic energy conversion. *Energy Environ. Sci.* **2013**, *6*, 1891–1900.
33. Bombelli, P.; Bradley, R. W.; Scott, A. M.; Philips, A. J.; McCormick, A. J.; Cruz, S. M.; Anderson, A.; Yunus, K.; Bendall, D. S.; Cameron, P. J.; Davies, J. M.; Smith, A. G.; Howe, C. J.; Fishera, A. C. Quantitative analysis of the factors limiting solar power transduction by *Synechocystis sp.* PCC 6803 in biological photovoltaic devices. *Energy Environ. Sci.* **2011**, *4*, 4690–4698.
34. McCormick, A. J.; Bombelli, P.; Scott, A. M.; Philips, A. J.; Smith, A. G.; Fisher, A. C.; Howe, C. J. Photosynthetic biofilms in pure culture harness solar energy in a mediatorless bio-photovoltaic cell (BPV) system. *Energy Environ. Sci.* **2011**, *4*, 4699–4709.

35. Lea-Smith, D. J.; Bombelli, P.; Vasudevan, R.; Howe, C. J. Photosynthetic, respiratory and extracellular electron transport pathways in cyanobacteria. *Biochim. Biophys. Acta* **2016**, *1857*, 247–255.
36. Schultze, M.; Forberich, B.; Rexroth, S.; Dyczmons, N. G.; Roegner, M.; Appel, J. Localization of cytochrome b6f complexes implies an incomplete respiratory chain in cytoplasmic membranes of the cyanobacterium *Synechocystis sp*. PCC 6803. *Biochim. Biophys. Acta* **2009**, *1787*, 1479–1485.
37. Tschörtner, J.; Lai, B.; Krömer, J. O. Biophotovoltaics: Green power generation from sunlight and water. *Front. Microbiol.* **2019**, *10*, 866–866.
38. Kannan, M. V.; Kumar, G. G. Current status, key challenges and its solutions in the design and development of graphene based ORR catalysts for the microbial fuel cell applications. *Biosens. Bioelectron.* **2016**, *77*, 1208–1220.
39. Liang, P.; Wang, H. Y.; Xia, X.; Huang, X.; Mo, Y. H.; Cao, X. X.; Fan, M. Z. Carbon nanotube powders as electrode modifier to enhance the activity of anodic biofilm in microbial fuel cells. *Biosens. Bioelectron.* **2011**, *26*, 3000–3004.
40. Maly, J.; Krejci, J.; Ilie, M.; Jakubka, L.; Masojídek, J.; Pilloton, R.; Sameh, K.; Steffan, P.; Stryhal, Z.; Sugiura, M. Monolayers of photosystem II on gold electrodes with enhanced sensor response: Effect of porosity and protein layer arrangement. *Anal. Bioanal. Chem.* **2005**, *381*, 1558–1567.
41. Lee, J.; Im, J.; Kim, S. Mediatorless solar energy conversion by covalently bonded thylakoid monolayer on the glassy carbon electrode. *Bioelectrochemistry* **2016**, *108*, 21–27.
42. Cosnier, S.; Holzinger, M.; Le Goff, A. Recent advances in carbon nanotube-based enzymatic fuel cells. *Front. Bioeng. Biotech.* **2014**, *2*, 1–6.
43. Dewi, H. A.; Sun, G. Z.; Zheng, L. X.; Lim, S. Interaction and charge transfer between isolated thylakoids and multi-walled carbon nanotubes. *Phys. Chem. Chem. Phys.* **2015**, *17*, 3435–3440.
44. Pankratov, D.; Pankratova, G.; Dyachkova, T. P.; Falkman, P.; Åkerlund, H.-E.; Toscano, M. D.; Chi, Q. J.; Gorton, L. Supercapacitive biosolar cell driven by direct electron transfer between photosynthetic membranes and CNT networks with enhanced performance. *ACS Energy Lett.* **2017**, *2*, 2635–2639.
45. Giraldo, J. P.; Landry, M. P.; Faltermeier, S. M.; McNicholas, T. P.; Iverson, N. M.; Boghossian, A. A.; Reuel, N. F.; Hilmer, A. J.; Sen, F.; Brew, J. A.; Strano, M. S. Plant nanobionics approach to augment photosynthesis and biochemical sensing. *Nat. Mater.* **2014**, *13*, 400–408.
46. You, S.; Song, Y. S.; Bai, S. J. Characterization of a photosynthesis-based bioelectrochemical film fabricated with a carbon nanotube hydrogel. *Biotechnol. Bioprocess Eng.* **2019**, *24*, 337–342.
47. Nii, D.; Miyachi, M.; Shimada, Y.; Nozawa, Y.; Ito, M.; Homma, Y.; Ikehira, S.; Yamanoi, Y.; Nishihara, H.; Tomo, T. Conjugates between photosystem I and a carbon nanotube for a photoresponse device. *Photosynth. Res.* **2017**, *133*, 155–162.
48. Gross, A. J.; Holzinger, M.; Cosnier, S. Buckypaper bioelectrodes: Emerging materials for implantable and wearable biofuel cells. *Energy Environ. Sci.* **2018**, *11*, 1670–1687.

49. Lambreva, M. D.; Lavecchia, T.; Tyystjarvi, E.; Antal, T. K.; Orlanducci, S.; Margonelli, A.; Rea, G. Potential of carbon nanotubes in algal biotechnology. *Photosynth. Res.* **2015**, *125*, 451–471.
50. Kirchhofer, N. D.; Rasmussen, M. A.; Dahlquist, F. W.; Minteer, S. D.; Bazan, G. C. The photobioelectrochemical activity of thylakoid bioanodes is increased via photocurrent generation and improved contacts by membrane-intercalating conjugated oligoelectrolytes. *Energy Environ. Sci.* **2015**, *8*, 2698–2706.
51. Gadipelli, S.; Guo, Z. X. Graphene-based materials: Synthesis and gas sorption, storage and separation. *Prog. Mater. Sci.* **2015**, *69*, 1–60.
52. Geim, A. K.; Novoselov, K. S. The rise of graphene. *Nat. Mater.* **2007**, *6*, 183–191.
53. Hashemi, R.; Weng, G. J. A theoretical treatment of graphene nanocomposites with percolation threshold, tunneling-assisted conductivity and microcapacitor effect in AC and DC electrical settings. *Carbon* **2016**, *96*, 474–490.
54. Pankratova, G.; Pankratov, D.; Di Bar, C.; Goni-Urtiaga, A.; Toscano, M. D.; Chi, Q.; Pita, M.; Gorton, L.; De Lacey, A. L. Three-dimensional graphene matrix-supported and thylakoid membrane-based high-performance bioelectrochemical solar cell. *ACS Appl. Energy Mater.* **2018**, *1*, 319–323.
55. Feifel, S. C.; Stieger, K. R.; Lokstein, H.; Lux, H.; Lisdat, F. High photocurrent generation by photosystem I on artificial interfaces composed of π-system-modified graphene. *J. Mater. Chem. A* **2015**, 3, 12188–12196.
56. Dewi, H. A.; Meng, F. B.; Sana, B.; Guo, C. X.; Norling, B.; Chen, X. D.; Lim, S. R. Investigation of electron transfer from isolated spinach thylakoids to indium tin oxide. *RSC Adv.* **2014**, *4*, 48815–48820.
57. Zhang, J. Z.; Bombelli, P.; Sokol, K. P.; Fantuzzi, A.; Rutherford, A. W.; Howe, C. J.; Reisner, E. Photoelectrochemistry of photosystem II *in vitro vs in vivo*. *J. Am. Chem. Soc.* **2018**, *140*, 6–9.
58. Ng, F. L.; Phang, S. M.; Periasamy, V.; Yunus, K.; Fisher, A. C. Evaluation of algal biofilms on indium tin oxide (ITO) for use in biophotovoltaic platforms based on photosynthetic performance. *PLOS One* **2014**, *9*, e97643.
59. Gong, Y.; Zhao, J. Small carbon quantum dots, large photosynthesis enhancement. *J. Agr. Food Chem.* **2018**, *66*, 9159–9161.
60. Li, W.; Wu, S. S.; Zhang, H. R.; Zhang, X. J.; Zhuang, J. L.; Hu, C. F.; Liu, Y. L.; Lei, B. F.; Ma, L.; Wang, X. J. Enhanced biological photosynthetic efficiency using light-harvesting engineering with dual-emissive carbon dots. *Adv. Funct. Mater.* **2018**, *28*, 1804004.
61. Ciornii, D.; Kolsch, A.; Zouni, A.; Lisdat, F. Exploiting new ways for a more efficient orientation and wiring of PSI to electrodes: A fullerene C70 approach. *Electrochim. Acta* **2019**, *299*, 531–539.
62. Gizzie, E. A.; LeBlanc, G.; Jennings, G. K.; Cliffel, D. E. Electrochemical preparation of photosystem I-polyaniline composite films for biohybrid solar energy conversion. *ACS Appl. Mater. Interfaces* **2015**, 7, 9328–9335.
63. Gizzie, E. A.; Niezgoda, J. S.; Robinson, M. T.; Harris, A. G.; Jennings, G. K.; Rosenthal, S. J.; Cliffel, D. E. Photosystem I-polyaniline/TiO_2 solid-state solar cells: Simple devices for biohybrid solar energy conversion. *Energy Environ. Sci.* **2015**, *8*, 3572–3576.

64. Robinson, M. T.; Simons, C. E.; Cliffel, D. E.; Jennings, G. K. Photocatalytic photosystem I/PEDOT composite films prepared by vapor-phase polymerization. *Nanoscale* **2017**, *9*, 6158–6166.
65. Zhou, X.; Zhou, L.; Zhang, P.; Lv, F.; Liu, L.; Qi, R.; Wang, Y.; Shen, M.-Y.; Yu, H.-H.; Bazan, G.; Wang, S. Conducting polymers–thylakoid hybrid materials for water oxidation and photoelectric conversion. *Adv. Electron. Mater.* **2019**, *5*, 1800789.
66. Kumar, A.; Zhou, C. W. The race to replace tin-doped indium oxide: Which material will win? *ACS Nano* **2010**, *4*, 11–14.
67. Bunea, A. I.; Heiskanen, A.; Pankratova, G.; Tesei, G.; Lund, M.; Akerlund, H. E.; Leech, D.; Larsen, N. B.; Keller, S. S.; Gorton, L.; Emnéus, J. Micropatterned carbon-on-quartz electrode chips for photocurrent generation from thylakoid membranes. *ACS Appl. Energy Mater.* **2018**, *1*, 3313–3322.
68. Kanso, H.; Pankratova, G.; Bollella, P.; Leech, D.; Hernandez, D.; Gorton, L. Sunlight photocurrent generation from thylakoid membranes on gold nanoparticle modified screen-printed electrodes. *J. Electroanal. Chem.* **2018**, *816*, 259–264.
69. Mersch, D.; Lee, C. Y.; Zhang, J. Z.; Brinkert, K.; Fontecilla-Camps, J. C.; Rutherford, A. W.; Reisner, E. Wiring of photosystem II to hydrogenase for photoelectrochemical water splitting. *J. Am. Chem. Soc.* **2015**, *137*, 8541–8549.
70. Kato, M.; Cardona, T.; Rutherford, A. W.; Reisner, E. Photoelectrochemical water oxidation with photosystem II integrated in a mesoporous indium tin oxide electrode. *J. Am. Chem. Soc.* **2012**, *134*, 8332–8335.
71. Vittadello, M.; Gorbunov, M. Y.; Mastrogiovanni, D. T.; Wielunski, L. S.; Garfunkel, E. L.; Guerrero, F.; Kirilovsky, D.; Sugiura, M.; Rutherford, A. W.; Safari, A.; Falkowski, P. G. Photoelectron generation by photosystem II core complexes tethered to gold surfaces. *ChemSusChem* **2010**, *3*, 471–475.
72. Stieger, K. R.; Feifel, S. C.; Lokstein, H.; Hejazi, M.; Zouni, A.; Lisdat, F. Biohybrid architectures for efficient light-to-current conversion based on photosystem I within scalable 3D mesoporous electrodes. *J. Mater. Chem. A* **2016**, *4*, 17009–17017.
73. Friebe, V. M.; Delgado, J. D.; Swainsbury, D. J. K.; Gruber, J. M.; Chanaewa, A.; van Grondelle, R.; von Hauff, E.; Millo, D.; Jones, M. R.; Frese, R. N. Plasmon-enhanced photocurrent of photosynthetic pigment proteins on nanoporous silver. *Adv. Funct. Mater.* **2016**, *26*, 285–292.
74. Wenzel, T.; Hartter, D.; Bombelli, P.; Howe, C. J.; Steiner, U. Porous translucent electrodes enhance current generation from photosynthetic biofilms. *Nat. Commun.* **2018**, *9*, 1299.
75. Fang, X.; Sokol, K. P.; Heidary, N.; Kandiel, T. A.; Zhang, J. Z.; Reisner, E. Structure-activity relationships of hierarchical three-dimensional electrodes with photosystem II for semiartificial photosynthesis. *Nano Lett.* **2019**, *19*, 1844–1850.
76. Saboe, P. O.; Conte, E.; Farell, M.; Bazan, G. C.; Kumar, M. Biomimetic and bioinspired approaches for wiring enzymes to electrode interfaces. *Energy Environ. Sci.* **2017**, *10*, 14–42.
77. Yates, N. D. J.; Fascione, M. A.; Parkin, A. Methodologies for "wiring" redox proteins/enzymes to electrode surfaces. *Chem. - Eur. J.* **2018**, *24*, 12164–12182.
78. Grattieri, M.; Rhodes, Z.; Hickey, D. P.; Beaver, K.; Minteer, S. D. Understanding biophotocurrent generation in photosynthetic purple bacteria. *ACS Catal.* **2019**, *9*, 867–873.

79. Hasan, K.; Dilgin, Y.; Emek, S. C.; Tavahodi, M.; Åkerlund, H.-E.; Ålbertsson, P.-Å.; Gorton, L. Photoelectrochemical communication between thylakoid membranes and gold electrodes through different quinone derivatives. *ChemElectroChem* **2014**, *1*, 131–139.
80. Longatte, G.; Rappaport, F.; Wollman, F. A.; Guille-Collignon, M.; Lemaitre, F. Mechanism and analyses for extracting photosynthetic electrons using exogenous quinones: What makes a good extraction pathway? *Photochem. Photobiol. Sci.* **2016**, *15*, 969–979.
81. Yuan, M.; Minteer, S. D. Redox polymers in electrochemical systems: From methods of mediation to energy storage. *Curr. Opin. Electrochem.* **2019**, *15*, 1–6.
82. Ruff, A. Redox polymers in bioelectrochemistry: Common playgrounds and novel concepts. *Curr. Opin. Electrochem.* **2017**, *5*, 66–73.
83. Heller, A. Miniature biofuel cells. *Phys. Chem. Chem. Phys.* **2004**, *6*, 209–216.
84. Heller, A.; Feldman, B. Electrochemistry in diabetes management. *Acc. Chem. Res.* **2010**, *43*, 963–973.
85. Badura, A.; Guschin, D.; Esper, B.; Kothe, T.; Neugebauer, S.; Schuhmann, W.; Rögner, M. Photo-induced electron transfer between photosystem 2 via cross-linked redox hydrogels. *Electroanalysis* **2008**, *20*, 1043–1047.
86. Badura, A.; Guschin, D.; Kothe, T.; Kopczak, M. J.; Schuhmann, W.; Rögner, M. Photocurrent generation by photosystem 1 integrated in crosslinked redox hydrogels. *Energy Environ. Sci.* **2011**, *4*, 2435–2440.
87. Sokol, K. P.; Mersch, D.; Hartmann, V.; Zhang, J. Z.; Nowaczyk, M. M.; Rögner, M.; Ruff, A.; Schuhmann, W.; Plumeré, N.; Reisner, E. Rational wiring of photosystem II to hierarchical indium tin oxide electrodes using redox polymers. *Energy Environ. Sci.* **2016**, *9*, 3698–3709.
88. Kothe, T.; Pöller, S.; Zhao, F.; Fortgang, P.; Rögner, M.; Schuhmann, W.; Plumeré, N. Engineered electron-transfer chain in Photosystem 1 based photocathodes outperforms electron-transfer rates in natural photosynthesis. *Chem. - Eur. J.* **2014**, *20*, 11029–11034.
89. Zhao, F.; Hardt, S.; Hartmann, V.; Zhang, H.; Nowaczyk, M. M.; Rögner, M.; Plumeré, N.; Schuhmann, W.; Conzuelo, F. Light-induced formation of partially reduced oxygen species limits the lifetime of photosystem 1-based biocathodes. *Nat. Commun.* **2018**, *9*, 1973.
90. Zhao, F.; Ruff, A.; Rogner, M.; Schuhmann, W.; Conzuelo, F. Extended operational lifetime of a photosystem-based bioelectrode. *J. Am. Chem. Soc.* **2019**, *141*, 5102–5106.
91. Hamidi, H.; Hasan, K.; Emek, S. C.; Dilgin, Y.; Åkerlund, H.-E.; Albertsson, P.-Å.; Leech, D.; Gorton, L. Photocurrent generation from thylakoid membranes on osmium-redox-polymer-modified electrodes. *ChemSusChem* **2015**, *8*, 990–993.
92. Hasan, K.; Grippo, V.; Sperling, E.; Packer, M. A.; Leech, D.; Gorton, L. Evaluation of photocurrent generation from different photosynthetic organisms. *ChemElectroChem* **2017**, *4*, 412–417.
93. Hasan, K.; Reddy, K. V. R.; Eßmann, V.; Górecki, K.; Ó Conghaile, P.; Schuhmann, W.; Leech, D.; Hägerhäll, C.; Gorton, L. Electrochemical communication between electrodes and *Rhodobacter capsulatus* grown in different metabolic modes. *Electroanalysis* **2015**, *27*, 118–127.
94. Hasan, K.; Çevik, E.; Sperling, E.; Packer, M. A.; Leech, D.; Gorton, L. Photoelectrochemical wiring of *Paulschulzia pseudovolvox* (algae) to osmium polymer modified electrodes for harnessing solar energy. *Adv. Energy Mater.* **2015**, *5*, 1501100.

95. Pankratov, D.; Zhao, J.; Nur, M. A.; Shen, F.; Leech, D.; Chi, Q.; Pankratova, G.; Gorton, L. The influence of surface composition of carbon nanotubes on the photobioelectrochemical activity of thylakoid bioanodes mediated by osmium-complex modified redox polymer. *Electrochim. Acta* **2019**, *310*, 20–25.
96. Riedel, M.; Wersig, J.; Ruff, A.; Schuhmann, W.; Zouni, A.; Lisdat, F. A Z-scheme-inspired photobioelectrochemical H_2O/O_2 cell with a 1 V open-circuit voltage combining photosystem II and PbS quantum dots. *Angew. Chem. Int. Ed.* **2019**, *58*, 801–805.
97. Milton, R. D.; Hickey, D. P.; Abdellaoui, S.; Lim, K.; Wu, F.; Tan, B.; Minteer, S. D. Rational design of quinones for high power density biofuel cells. *Chem. Sci.* **2015**, *6*, 4867–4875.
98. Hasan, K.; Milton, R. D.; Grattieri, M.; Wang, T.; Stephanz, M.; Minteer, S. D. Photobioelectrocatalysis of intact chloroplasts for solar energy conversion. *ACS Catal.* **2017**, *7*, 2257–2265.
99. Pankratova, G.; Pankratov, D.; Milton, R. D.; Minteer, S. D.; Gorton, L. Following nature: Bioinspired mediation strategy for Gram-positive bacterial cells. *Adv. Energy Mater.* **2019**, *9*, 1900215.
100. Efrati, A.; Tel-Vered, R.; Michaeli, D.; Nechushtai, R.; Willner, I. Cytochrome c-coupled photosystem I and photosystem II (PSI/PSII) photo-bioelectrochemical cells. *Energy Environ. Sci.* **2013**, *6*, 2950–2956.
101. Kiliszek, M.; Harputlu, E.; Szalkowski, M.; Kowalska, D.; Unlu, C. G.; Haniewicz, P.; Abram, M.; Wiwatowski, K.; Niedziółka-Jönsson, J.; Maćkowski, S.; Ocakoglu, K.; Kargul, J. Orientation of photosystem I on graphene through cytochrome c553 leads to improvement in photocurrent generation. *J. Mater. Chem. A* **2018**, *6*, 18615–18626.
102. Longatte, G.; Guille-Collignon, M.; Lemaitre, F. Electrocatalytic mechanism involving Michaelis-Menten kinetics at the preparative scale: Theory and applicability to photocurrents from a photosynthetic algae suspension with quinones. *ChemPhysChem* **2017**, *18*, 2643–2650.
103. Longatte, G.; Sayegh, A.; Delacotte, J.; Rappaport, F.; Wollman, F. A.; Guille-Collignon, M.; Lemaitre, F. Investigation of photocurrents resulting from a living unicellular algae suspension with quinones over time. *Chem. Sci.* **2018**, *9*, 8271–8281.
104. Sayegh, A.; Longatte, G.; Buriez, O.; Wollman, F. A.; Guille-Collignon, M.; Labbe, E.; Delacotte, J.; Lemaitre, F. Diverting photosynthetic electrons from suspensions of *Chlamydomonas reinhardtii* algae: New insights using an electrochemical well device. *Electrochim. Acta* **2019**, *304*, 465–473.
105. Pinhassi, R. I.; Kallmann, D.; Saper, G.; Dotan, H.; Linkov, A.; Kay, A.; Liveanu, V.; Schuster, G.; Adir, N.; Rothschild, A. Hybrid bio-photo-electro-chemical cells for solar water splitting. *Nat. Commun.* **2016**, *7*, 12552.
106. Schuergers, N.; Werlang, C.; Ajo-Franklin, C. M.; Boghossian, A. A. A synthetic biology approach to engineering living photovoltaics. *Energy Environ. Sci.* **2017**, *10*, 1102–1115.
107. Carmeli, I.; Frolov, L.; Carmeli, C.; Richter, S. Photovoltaic activity of photosystem I-based self-assembled monolayer. *J. Am. Chem. Soc.* **2007**, *129*, 12352–12353.
108. Das, R.; Kiley, P. J.; Segal, M.; Norville, J.; Yu, A. A.; Wang, L. Y.; Trammell, S. A.; Reddick, L. E.; Kumar, R.; Stellacci, F.; Lebedev, N.; Schnur, J.; Bruce, B. D.; Zhang, S.; Baldo, M. Integration of photosynthetic protein molecular complexes in solid-state electronic devices. *Nano Lett.* **2004**, *4*, 1079–1083.

109. Faulkner, C. J.; Lees, S.; Ciesielski, P. N.; Cliffel, D. E.; Jennings, G. K. Rapid assembly of photosystem I monolayers on gold electrodes. *Langmuir* **2008**, *24*, 8409–8412.
110. Frolov, L.; Wilner, O.; Carmeli, C.; Carmeli, I. Fabrication of oriented multilayers of photosystem I proteins on solid surfaces by auto-metallization. *Adv. Mater.* **2008**, *20*, 263–266.
111. Antonacci, A.; Scognamiglio, V. Photosynthesis-based hybrid nanostructures: Electrochemical sensors and photovoltaic cells as case studies. *Trends Anal. Chem.* **2019**, *115*, 100–109.
112. Sekar, N.; Jain, R.; Yan, Y.; Ramasamy, R. P. Enhanced photo-bioelectrochemical energy conversion by genetically engineered cyanobacteria. *Biotechnol. Bioeng.* **2016**, *113*, 675–679.
113. Deeke, A.; Sleutels, T. H. J. A.; Hamelers, H. V. M.; Buisman, C. J. N. Capacitive bioanodes enable renewable energy storage in microbial fuel cells. *Environ. Sci. Technol.* **2012**, *46*, 3554–3560.
114. Deeke, A.; Sleutels, T. H. J. A.; Ter Heijne, A.; Hamelers, H. V. M.; Buisman, C. J. N. Influence of the thickness of the capacitive layer on the performance of bioanodes in microbial fuel cells. *J. Power Sources* **2013**, *243*, 611–616.
115. Malvankar, N. S.; Mester, T.; Tuominen, M. T.; Lovley, D. R. Supercapacitors based on c-type cytochromes using conductive nanostructured networks of living bacteria. *ChemPhysChem* **2012**, *13*, 463–468.
116. Houghton, J.; Santoro, C.; Soavi, F.; Serov, A.; Ieropoulos, I.; Arbizzani, C.; Atanassov, P. Supercapacitive microbial fuel cell: Characterization and analysis for improved charge storage/delivery performance. *Bioresour. Technol.* **2016**, *218*, 552–560.
117. Santoro, C.; Abad, F. B.; Serov, A.; Kodali, M.; Howe, K. J.; Soavi, F.; Atanassov, P. Supercapacitive microbial desalination cells: New class of power generating devices for reduction of salinity content. *Appl. Energy* **2017**, *208*, 25–36.
118. Santoro, C.; Kodali, M.; Shamoon, N.; Serov, A.; Soavi, F.; Merino-Jimenez, I.; Gajda, I.; Greenman, J.; Ieropoulos, I.; Atanassov, P. Increased power generation in supercapacitive microbial fuel cell stack using Fe-N-C cathode catalyst. *J. Power Sources* **2019**, *412*, 416–424.
119. Santoro, C.; Soavi, F.; Serov, A.; Arbizzani, C.; Atanassov, P. Self-powered supercapacitive microbial fuel cell: The ultimate way of boosting and harvesting power. *Biosens. Bioelectron.* **2016**, *78*, 229–235.
120. Agnes, C.; Holzinger, M.; Le Goff, A.; Reuillard, B.; Elouarzaki, K.; Tingry, S.; Cosnier, S. Supercapacitor/biofuel cell hybrids based on wired enzymes on carbon nanotube matrices: Autonomous reloading after high power pulses in neutral buffered glucose solutions. *Energy Environ. Sci.* **2014**, *7*, 1884–1888.
121. Pankratov, D.; Blum, Z.; Suyatin, D. B.; Popov, V. O.; Shleev, S. Self-charging electrochemical biocapacitor. *ChemElectroChem* **2014**, *1*, 343–346.
122. Pankratov, D.; Shen, F.; Ortiz, R.; Toscano, M. D.; Thormann, E.; Zhang, J. D.; Gorton, L.; Chi, Q. J. Fuel-independent and membrane-less self-charging biosupercapacitor. *Chem. Commun.* **2018**, *54*, 11801–11804.
123. Knoche, K. L.; Hickey, D. P.; Milton, R. D.; Curchoe, C. L.; Minteer, S. D. Hybrid glucose/O_2 biobattery and supercapacitor utilizing a pseudocapacitive dimethylferrocene redox polymer at the bioanode. *ACS Energy Lett.* **2016**, *1*, 380–385.

124. Pankratova, G.; Pankratov, D.; Hasan, K.; Akerlund, H. E.; Albertsson, P. A.; Leech, D.; Shleev, S.; Gorton, L. Supercapacitive photo-bioanodes and biosolar cells: A novel approach for solar energy harnessing. *Adv. Energy Mater.* **2017**, *7*, 1602285.
125. Zhao, F. Y.; Bobrowski, T.; Ruff, A.; Hartmann, V.; Nowaczyk, M. M.; Rogner, M.; Conzuelo, F.; Schuhmann, W. A light-driven Nernstian biosupercapacitor. *Electrochim. Acta* **2019**, *306*, 660–666.
126. Pankratov, D.; Conzuelo, F.; Pinyou, P.; Alsaoub, S.; Schuhmann, W.; Shleev, S. A Nernstian biosupercapacitor. *Angew. Chem. Int. Ed.* **2016**, *55*, 15434–15438.
127. Xiao, X.; Ó Conghaile, P.; Leech, D.; Ludwig, R.; Magner, E. A symmetric supercapacitor/biofuel cell hybrid device based on enzyme-modified nanoporous gold: An autonomous pulse generator. *Biosens. Bioelectron.* **2017**, *90*, 96–102.
128. Alsaoub, S.; Ruff, A.; Conzuelo, F.; Ventosa, E.; Ludwig, R.; Shleev, S.; Schuhmann, W. An intrinsic self-charging biosupercapacitor comprised of a high-potential bioanode and a low-potential biocathode. *ChemPlusChem* **2017**, *82*, 576–583.
129. Alsaoub, S.; Conzuelo, F.; Gounel, S.; Mano, N.; Schuhmann, W.; Ruff, A. Introducing pseudocapacitive bioelectrodes into a biofuel cell/biosupercapacitor hybrid device for optimized open circuit voltage. *ChemElectroChem* **2019**, *6*, 2080–2087.
130. González-Arribas, E.; Aleksejeva, O.; Bobrowski, T.; Toscano, M. D.; Gorton, L.; Schuhmann, W.; Shleev, S. Solar biosupercapacitor. *Electrochem. Commun.* **2017**, *74*, 9–13.
131. Saar, K. L.; Bombelli, P.; Lea-Smith, D. J.; Call, T.; Aro, E. M.; Muller, T.; Howe, C. J.; Knowles, T. P. J. Enhancing power density of biophotovoltaics by decoupling storage and power delivery. *Nat. Energy* **2018**, *3*, 75–81.
132. Ravi, S. K.; Rawding, P.; Elshahawy, A. M.; Huang, K.; Sun, W. X.; Zhao, F. F.; Wang, J.; Jones, M. R.; Tan, S. C. Photosynthetic apparatus of *Rhodobacter sphaeroides* exhibits prolonged charge storage. *Nat. Commun.* **2019**, *10*, 902.

Chapter 7

Significance of Nanostructures of an Electrode Surface in Direct Electron Transfer-Type Bioelectrocatalysis of Redox Enzymes

Yuki Kitazumi, Osamu Shirai, and Kenji Kano*

Division of Applied Life Sciences, Graduate School of Agriculture, Kyoto University, Sakyo, Kyoto 606-8502, Japan

***E-mail: kano.kenji.5z@kyoto-u.ac.jp.**

Redox enzymes, which are approximately one-quarter of all known proteins, can be distinguished by their extraordinary activity under mild conditions and wide variety of reaction types. During the last quarter of a century, several attempts have been made to utilize these enzymes as electrocatalysts and develop novel electrochemical systems. The direct electron coupling between redox enzyme reactions and electrode reactions without a redox mediator is known as direct electron transfer (DET)-type bioelectrocatalysis, one of the most intriguing subjects studied in recent decades. In a DET-type system, similar to conventional inorganic catalysts, the enzymes merely function as electrocatalysts. Compared with inorganic catalysts, the distinctive characteristics of redox enzymes include high activity, extremely large size, identity, and uniformity by biological regeneration, versatility, and fragility. Furthermore, the redox site of the enzymes that communicates with electrodes can be rigidly assigned, which clearly defines the redox potential of enzymatic catalysts. These factors lead to characteristic features in DET-type bioelectrocatalytic waves by redox enzymes adsorbed (or immobilized) on a nanostructured electrode surface. This chapter provides an overview, additional insights, and discussion of the recent progress on the control of electrode surfaces for DET-type bioelectrocatalysis while considering the importance of nanostructures of the electrode surface.

General Introduction to DET-Type Bioelectrocatalysis

The simplest direct electron transfer (DET)-type bioelectrocatalytic reaction is given by the following electrode and enzymatic reactions (for the substrate oxidation):

$$\mathrm{ER} \underset{k_\mathrm{b}}{\overset{k_\mathrm{f}}{\rightleftarrows}} \mathrm{EO} + n_\mathrm{E}\mathrm{e}^-, \qquad (1)$$

© 2020 American Chemical Society

$$\mathrm{S} + \mathrm{EO} \xrightarrow{k_c} \mathrm{P} + \mathrm{ER}, \tag{2}$$

where ER, EO, S, P, and n_E are a reduced form of an enzyme, an oxidized form of an enzyme, the substrate, the product, and the enzyme's electron number, respectively. Furthermore, k_f, k_c, and k_b are the kinetic constants for the forward electrode reaction, reverse electrode reaction, and the enzymatic reaction, respectively. Under steady-state conditions (defined by $\partial c_i/\partial t = 0$), the steady-state current (i_s) is given as follows (*1*):

$$i_s = \frac{i_s^{\lim}}{1 + \frac{k_c}{k_f} + \frac{k_b}{k_f}}, \tag{3}$$

where $i_s^{\lim}$ is the maximum steady-state current of the system. This example used the Michaelis–Menten-type equation:

$$\frac{i_s^{\lim}}{n_E FA} = \frac{\nu_E k_c \Gamma_E}{1 + \frac{K_S}{c_S^*}}, \tag{4}$$

and the Butler–Volmer equation:

$$k_f = k^{\circ} \exp\left\{\frac{(1-\alpha)n_E'}{RT}\left(E - E_E^{\circ'}\right)\right\}, \tag{5}$$

and

$$k_b = k^{\circ} \exp\left\{-\frac{\alpha n_E' F}{RT}\left(E - E_E^{\circ'}\right)\right\}, \tag{6}$$

to formulate the kinetics of the enzymatic reaction and electrode reactions, respectively. The parameters n'_E and $E_E^{\circ\prime}$ are the electron number and formal potential of the rate-determining step of the enzyme redox reaction on an electrode, respectively, while ν_E ($\equiv n_S/n_E$) is the ratio of the stoichiometric number of the substrate against that of the enzyme, k_c, is the enzyme turnover number in the DET reaction, Γ_E is the amount of the adsorbed enzyme, and c_S* is the substrate concentration in the bulk phase. Equation 3 holds for a steady-state system in the absence of concentration polarization.

One can consider the enzymatic reaction controlled ($k_c < k_f$) and the limiting current conditions ($k_b < k_f$). Under the limiting conditions, Equation 3 is simplified as follows:

$$i_s = n_S FA \frac{k_c \Gamma_E}{1 + \frac{K_S}{c_S^*}}. \tag{7}$$

This situation seems to be easily realized because sufficient potential can be applied easily due to the acceleration of the electrode reaction. Under this situation, the limiting steady-state catalytic current depends on c_S* in a Michaelis–Menten-type relationship.

Orientation Effect on the Electron Transfer Rate in the DET-Type Bioelectrocatalysis

In the DET reaction, the distance (d) between the electrode and the electrochemically active cofactor in the enzyme is one of the critical physical quantities because the electron transfer reaction rate is dominated by the distance between the electron acceptor and donor. The rate constant of a long-range electron transfer reaction depends on d in the following equation:

$$k_0 = k^0_{\max} \exp(-\beta d)\,, \quad (8)$$

where $k^0{}_{\max}$ is the rate constant at the closest approach ($d = 0$) and β is the decay coefficient. Therefore, to determine the DET-type bioelectrocatalytic reaction rate, the orientation of the adsorbed enzyme at the electrode is crucial because the value of d varies with the orientation.

We further demonstrate the simplest orientation model, which is based on the random orientation. Figure 1 shows the schematic of the model, which assumes a spherical enzyme with a radius (r). In the enzyme, the electrode reaction center is located at a distance (r_{as}) from the center of the enzyme. However, when the orientation of the enzyme at a planar electrode is random, the electrode reaction center in the enzyme is located between $r - r_{\text{as}}$ and $r + r_{\text{as}}$ with a constant probability (f), which is expressed in the following equation (2–5):

$$f = \frac{1}{2r_{\text{as}}}, \quad (9)$$

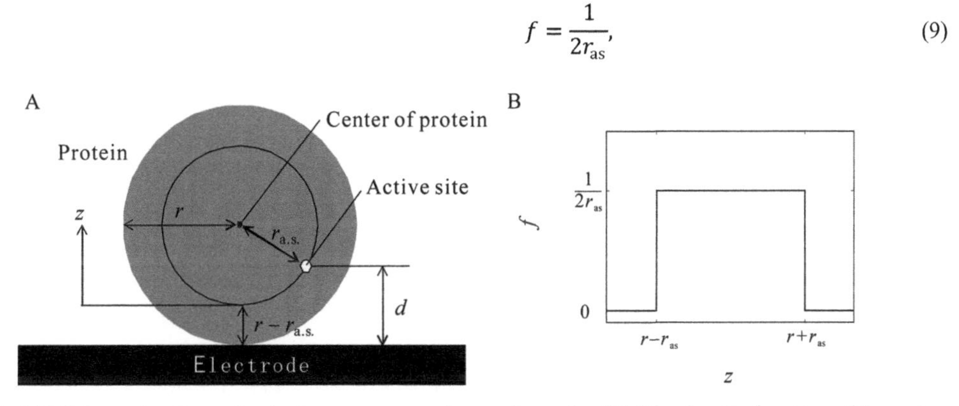

Figure 1. (A) Schematic of an adsorbed enzyme on a planar electrode. (B) The density function of the active site distance in the random orientation. Reprinted with permission from reference (4). Copyright 2016 American Chemical Society.

Substituting Equation 8 into Equation 3 yields the following steady-state DET-type catalytic current by the enzyme located at $z = d - (r - r_{\text{as}})$:

$$i_{\text{s},z=d-(r-r_{\text{as}})} = \frac{i_{\text{S}}^{\lim}}{1 + \eta^{-1} + \dfrac{k_{\text{c}}}{k^0 \eta^{1-\alpha} \exp(-\beta(z + r - r_{\text{as}}))}}, \quad (10)$$

where

$$\eta = \exp\left(\frac{n'_{\text{E}} F(E - E^{o'}_{\text{E}})}{RT}\right). \quad (11)$$

The integration of the current gives the total current for the whole enzyme ($i_{s,random}$) in the following equation:

$$i_{s,random} = \frac{1}{2r_{as}} \int_0^{2r_{as}} i_s \, dz \, . \tag{12}$$

Furthermore, the analytical form of the integration can be expressed by the following equation:

$$i_s = \frac{i_s^{lim}}{2\beta r_{as}(1+\eta^{-1})} \ln \left| \frac{\frac{k_c}{k^0 \eta^{1-\alpha}} + (1+\eta^{-1}) \exp(-\beta(r - r_{as}))}{\frac{k_c}{k^0 \eta^{1-\alpha}} + (1+\eta^{-1}) \exp(-\beta(r + r_{as}))} \right| . \tag{13}$$

The calculated voltammogram is shown by the solid line in Figure 2.

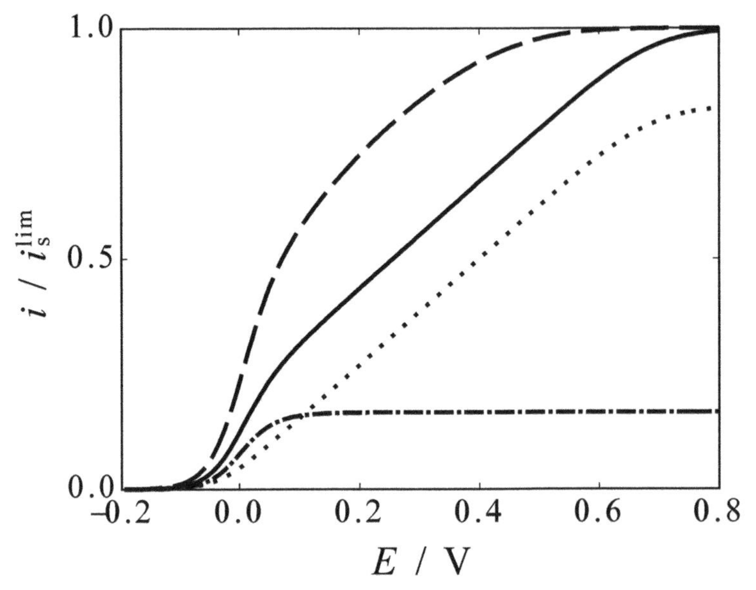

Figure 2. Steady-state bioelectrocatalytic voltammograms calculated by Equation 13 (solid line), by Equation 17 (dashed line), by considering the contribution from only enzymes with poor orientations (dotted line), and by subtracting the dot-dashed line. Parameters are: $r = 2$ nm, $r_{as} = 0.6$ nm, $k_c/k^0_{max} = 10^{-10}$, and $R_p = 2.5$ nm.

Note that the calculated voltammogram has an exponentially increasing region, a linearly increasing region, and a limiting region. In the exponentially increasing region, the electron transfer reaction predominantly occurs between enzymes with suitable orientations and the electrode. The linearly increasing region is caused by the poorly and randomly oriented enzymes, where the exponentially decayed k_0 produced by an increase in d is canceled with an exponential acceleration of k_0 by an increase in E. The slope is called the residual slope (2), which is frequently observed in experimentally recorded voltammograms of the DET-type bioelectrocatalysis.

The random orientation model indicates that the residual slope region in the steady-state catalytic voltammograms is caused by the poor and random orientation of the adsorbed enzymes. Due to this, eliminating the effect of the poorly and randomly oriented enzymes on the voltammogram by subtracting the residual slope seems to be a reasonable approach to identify productively (or suitably) oriented enzymes.

The dot-dashed line in Figure 2 is the calculated voltammogram obtained by subtracting the residual slope part (dotted line) from the entire voltammogram (solid line); it clearly shows reversible characteristics. Moreover, the half-wave potential of the voltammogram agrees with the redox potential of the electrode active redox center in the enzyme. The limiting value is presented in an equation similar to Equation 4; however, Γ_{E} has to be replaced with $\Gamma_{\mathrm{E,effective}}$ (i.e., the surface concentration of productively oriented enzymes). Therefore, subtracting the residual current enables a simple analysis of the thermodynamic and kinetic parameters of the enzyme in DET-type bioelectrocatalysis.

Key Factors for DET-Type Bioelectrocatalysis

In DET-type bioelectrocatalysis, the electron transfer between the redox center in the enzyme and the electrode is the most critical step. To improve electron transfer, several studies have focused on tuning the electrode surface and engineering the enzyme. This section provides an overview of the improvements of the electrode surface while considering the importance of the nanostructures of the electrode surface.

Most studies on DET-type bioelectrocatalysis focused on experiments using porous electrodes with the primary aim of increasing the effective electrode surface area when compared to the projected area. However, several researchers have reported the importance of pore size distribution on the electrode surface (*6–11*). Consequently, researchers have been actively studying nm-sized electrode materials such as carbon black, activated carbon, carbon nanotubes, graphene, and metal nanoparticles (*12–14*). When a redox enzyme is adsorbed in a mesopore with a radius similar to that of an enzyme, the enzyme can contact the inner wall of the mesopore on the electrode surface at several points with a sufficiently small value of d. These multiple contacts seem to have a fundamental effect of realizing DET-type bioelectrocatalysis. The effect of the mesoporous structure on DET-type bioelectrocatalysis is known as the curvature effect. The maximum efficiency of DET-type electrochemical communication should theoretically be realized when the size of the mesopore is approximately the same size as the enzyme (*5*). However, a mesoporous structure with a much larger size (or a macroporous structure) is required for mass transfer in bioelectrocatalysis. Since the quantitative investigation into the effect of the pore on the mass transfer requires a numerical simulation (*15*), the details are not mentioned here. In most cases, the mesoporous structure of the electrode surface is formed by nanomaterials (or primary particles) aggregating on the electrode surface. The next section introduces the simplest model for the curvature effect.

The redox center in enzymes is usually located off-center; moreover, the enzymes are not always spherical. Therefore, in order to determine the value of d and the electrode transfer kinetics, the orientation of enzymes adsorbed on the electrode surface is important. Moreover, to improve DET-type bioelectrocatalysis, it is crucial to control the orientation of the adsorbed enzyme. Furthermore, utilizing the interaction between an enzyme and an electrode is effective to control the orientation of the adsorbed enzyme. Consequently, several researchers have performed chemical modifications on the electrode surfaces (*3*, *16–32*).

Note that the interactions between enzymes and electrodes are classified into three types: electrostatic interactions (*3*, *18*, *20*, *23*, *26*, *32*), hydrophobic interaction (*16*, *22*), and specific chemical interactions (*16*, *21*, *22*, *24*, *27–29*). Electrode modification focusing on the local surface charge near the electrochemical redox center (but not the net charge) of an enzyme can effectively improve the DET-type bioelectrocatalysis (*3*). Hydrophobic interaction focuses on the interaction between the substrate binding pocket in the enzyme and the hydrophobic substrate (*16*, *22*). Furthermore, the interaction was intensified by π–π stacking interactions; however, the specificity may not be high. The specific chemical interaction between an enzyme and an electrode primarily depends on the substrate specificity of the enzyme (i.e., the enzyme surface around the redox center distinguishes its substrate). Therefore, electrode modification with an analogous substrate will effectively control the orientation of an adsorbed enzyme (*24*). The orientation of the adsorbed enzyme at the electrode surface was determined on the basis of the voltammetry (*3*, *32*, *60*) or surface selective spectroscopy (*32*).

We will now discuss the approach to accessing the conductive material in the redox center of an enzyme (Figure 3). This approach uses platinum nanoclusters prepared by reducing $PtCl_6^{2-}$ through flavin adenine dinucleotide (FAD)–dependent glucose oxidase (GOD) as the conductive material near the redox center of the enzyme (*33*). Although the DET from FAD-dependent GOD to the electrode is difficult, enzymatically grown nanoclusters enable electrical contact between the FAD and the electrode. The nanoparticles implanted in the enzyme seem to work as good connectors between the FAD and the electrode. A similar approach has been applied for FAD–dependent glucose dehydrogenase to realize DET-type bioelectrocatalysis using gold nanoclusters that were enzymatically grown (*34*).

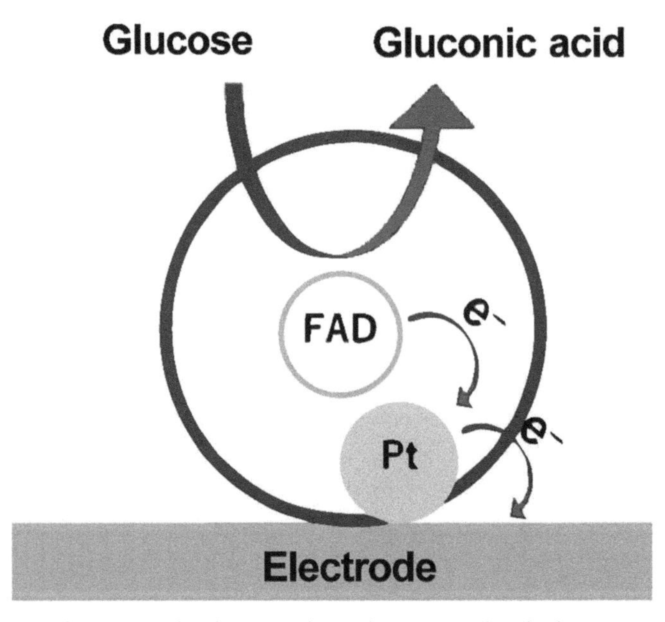

Figure 3. Schematic of DET-type bioelectrocatalysis of GOD wired with platinum nanoclusters at the electrode surface.

The Curvature Effect on DET-Type Bioelectrocatalysis

This section discusses the curvature effect on DET-type bioelectrocatalysis. First, we discuss a simple model developed using a spherical enzyme in the spherical pore. Then, we explain the improvement of DET-type bioelectrocatalysis on a porous electrode prepared from nanoparticles. Finally, we introduce experimentally observed curvature effects.

The Spherical Model for Curvature Effect

Because a porous electrode can effectively enable the occurrence of a DET reaction, we used a mesoporous substrate as the platform for the DET reaction; the consideration of enzyme orientation in a mesopore is crucial. In this model, we considered a spherical enzyme located in a spherical pore. We further assumed that an enzyme with a radius (r) is located in a spherical pore of the radius (R_p) (Figure 4). The distance (d) between the inner surface of the pore and the electrochemically active redox site in the enzyme determines the electron transfer rate constant. The distances in the z, radial, and x directions are defined as:

$$d_1 = \begin{cases} z + r - r_{as} + \sqrt{R_p^2 - 2r_{as}z + z^2} \\ R_p - z - r + r_{as} + \sqrt{R_p^2 - 2r_{as}z + z^2} \end{cases}, \tag{14}$$

$$d_2 = R_p - \sqrt{2r_{as}z - z^2 + \left(R_p - z - r + r_{as}\right)^2}, \tag{15}$$

and

$$d_3 = \sqrt{R_p^2 - \left(R_p - z - r + r_{as}\right)^2} - \sqrt{2r_{as}z - z^2}, \tag{16}$$

respectively. The shortest path seems to be the suitable path for electron transfer in a given orientation. The current is numerically calculated by the following formula:

$$i_{S,\text{spherical}} = \frac{1}{2r_{as}} \sum_i i_{S,z}\, \Delta z\,. \tag{17}$$

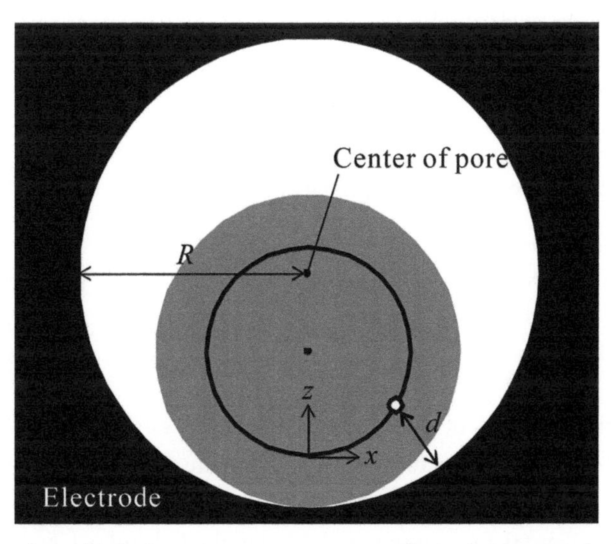

Figure 4. Schematic of an adsorbed enzyme on a mesoporous electrode. Reprinted with permission from reference (4). Copyright 2016 American Chemical Society.

The voltammogram, which was calculated from the spherical model, was observed as a dashed line in Figure 2. The value of the limiting steady-state current on the spherical model is identical with the value calculated for the random orientation model as the calculations were performed for the same amount of adsorbed enzyme in these models. However, the steady-state current calculated using the spherical model is higher than that of the random model outside the limiting region. Moreover, when the increasing part is exponentially enlarged, the residual slope region narrows. Such changes in the steady-state wave indicate an increase in the (apparent) heterogeneous electron transfer kinetics of the enzyme located in a mesopore. Moreover, this is the typical curvature effect of mesoporous electrodes for DET-type bioelectrocatalysis; unfortunately, the precise relationship between the curvature of the mesoporous electrode and the efficiency of the DET-type bioelectrocatalysis seems to be complicated for actual cases because the pore size distribution on the porous electrode is unclear.

Experimental Evidence of Curvature Effects

The importance of the porous structure was experimentally verified to detect or magnify the DET-type catalytic current (*8*). While examining carbon materials for DET-type bioelectrocatalysis of histamine dehydrogenase, Ketjenblack (of which the primary particle has a diameter of ~40 nm) was found to be the most effective material compared to other carbon particles (*6*). Aggregated gold nanoparticles or porous gold electrodes are effective for DET-type bioelectrocatalysis of d-fructose dehydrogenase (FDH) (*10, 17, 35*). Moreover, control of the porous structure of the carbon cryogel has been shown to be effective for improving the catalytic current density of FDH (*7*). For bilirubin oxidase (BOD), the DET-type catalytic wave was shown to be quite small on a planar electrode (*36*) but can easily be detected at the porous electrode (*37*); moreover, the control of the porous structure of the electrode was shown to be effective in improving the catalytic current density (*9*). Similarly, although DET-type bioelectrocatalysis of horseradish peroxidase was shown to be very weak on a planar electrode, it proceeded very clearly on suitably tuned porous electrodes (*37, 38*). Reversible redox reactions of the CO_2 and formate and NAD^+ and NADH were realized by the DET-type bioelectrocatalysis of tungsten-containing formate dehydrogenase (*39*). In order to achieve the DET-type bioelectrocatalytic reactions of the formate dehydrogenase, it was necessary to combine the curvature effect and the chemical interaction by modifying a porous carbon electrode with 4-mercaptopyridine adsorbed gold nanoparticles.

The Curvature Effect on Nanoparticle Aggregates

Some porous electrodes were formed from nanoparticles. Due to the curvature effect, the convex surface was disadvantageous for DET-type bioelectrocatalysis; yet, aggregated gold nanoparticles played a vital role as scaffolds for DET-type reactions (*17, 37, 40, 41*). This subsection introduces a model that explains the structural effect of the aggregated nanoparticles on DET-type bioelectrocatalysis.

Figure 5 shows the structures of the spherical nanoparticles in a close-packed lattice on a planar surface. The packed spheres with a radius (r_s) lead to the formation of several holes between nanoparticles. The tetrahedral and octahedral holes in the close-packed structure are filled with spheres with radii of 0.2 and 0.4 nm, respectively. In tetrahedral and octahedral holes, the invading small spheres will be in contact with packed nanoparticles at four and eight points, respectively. Because the diameters of primary gold particles (*17, 37, 40, 41*) are in the range of 15–50 nm, several enzymes will be embedded in these holes. Furthermore, in these holes, the enzyme makes contact with nanoparticles at various points. Therefore, the probability of achieving an appropriate orientation of the enzyme will increase with an increase in the surface coverage of nanoparticles.

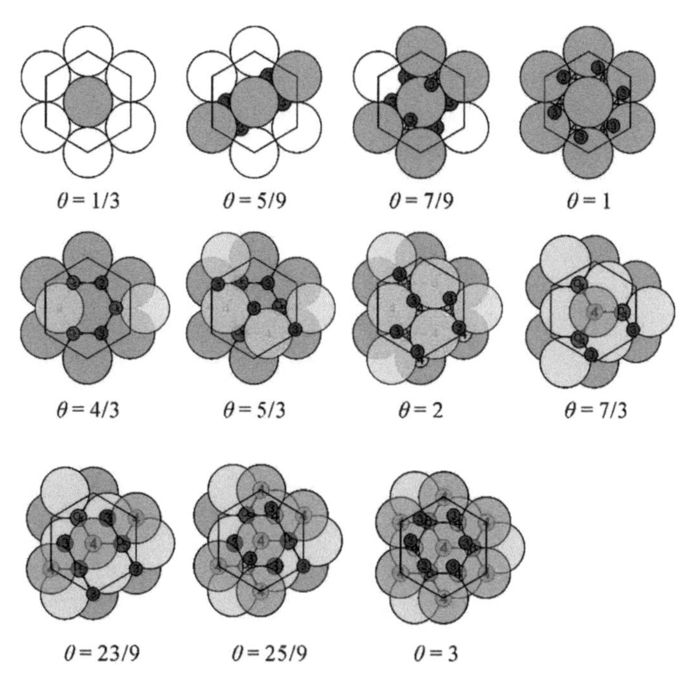

Figure 5. Top views of close-packed structures of spherical nanoparticles at various coverages (θ) on a planar surface. The small circles indicate sites where the number of contacts with the nanoparticles is more than three. The numbers in these small circles indicate the number of contact points with nanoparticles. Circles 4 and 5 indicate four and five points of contact in the octahedral site, respectively. Reprinted with permission from reference (41). Copyright 2019 Elsevier.

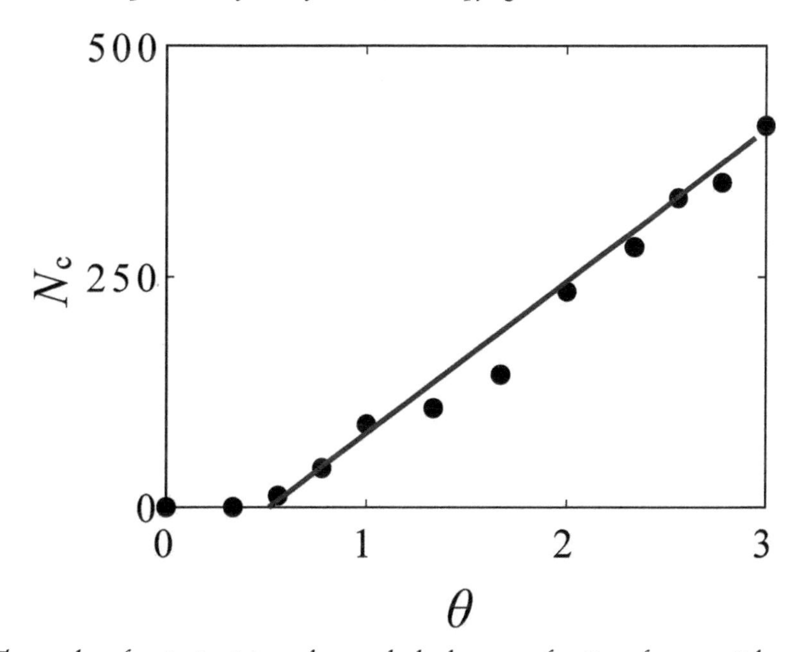

Figure 6. The number of contact points on close-packed spheres as a function of nanoparticles coverage on a planar surface. Reprinted with permission from reference (41). Copyright 2019 Elsevier.

Figure 6 shows that the contact points (for enzymes) increase with the surface coverage (θ) of spherical nanoparticles with negative intercepts. Similar trends were observed in the experiment where the DET-type current is a function of the structure formation degree by gold nanoparticles on the planar surface (*40*) or anodization of the gold surface (*41*). The negative intercept's appearance indicates that the DET-type current does not merely increase with the (electrochemically) effective surface area of the nanoparticle-modified electrodes. The mesoporous structures formed by aggregating nanoparticles form holes or scaffolds that are suitable for DET-type bioelectrocatalysis in a porous structure. Therefore, the curvature effect caused by the aggregated nanoparticles improves the activity of DET-type bioelectrocatalysis.

The Control of the Orientation of Adsorbed Enzymes by Surface Modification of Electrodes

Bilirubin Oxidase (BOD)

Electrode modification using BOD has been utilized to improve the orientation of adsorbed enzymes (*24*). Although the adsorbed amount of BOD is decreased by modification with bilirubin (i.e., the neutral electron donor of BOD), the modification was effective for improving DET-type bioelectrocatalysis of BOD (*24, 42*). This result demonstrates that the control of the orientation of the adsorbed enzyme is very effective for improving the performance of the DET-type reaction.

Studies on the effects of electrode modification on the orientation of BOD showed that an attractive electrostatic interaction between the modifier and BOD was very effective in controlling the orientation of BOD. Interestingly, effective modifiers for the orientation of BOD depend on the origin of BOD. For example, a negatively charged modifier was shown to be effective for orienting BOD from *Myrothecium verrucaria* (*Mv*BOD) (*3, 31, 32*), while a positively charged modifier was effective for orienting BOD from *Bacillus pumilus* (*15, 31*). The entire charge of *Mv*BOD was negative at pH 7; however, the surface of the BOD around the T1 copper site was positively charged at pH 7. The effectiveness of the negatively charged modifier for orienting *Mv*BOD demonstrates that the local charge of the surface of the enzyme is crucial for controlling the orientation of the adsorbed enzyme. Therefore, *Mv*BOD was shown to be effectively orientated on the electrode surface by modifying negatively charged aromatic compounds. The effects of pH and surface charge on the orientation of adsorbed *Mv*BOD were not only electrochemically investigated but also spectroscopically investigated (*32*).

Small molecules and nanomaterials are both effective modifiers of the electrode for DET-type bioelectrocatalysis of BOD. The modified gold nanoparticles (*17, 40, 42–44*) and carbon nanotubes (*31, 45–47*) have been shown to be effective scaffolds for DET-type bioelectrocatalysis of BOD.

Hydrogenase

Hydrogenase is a unique enzyme that catalyzes the bidirectional reaction of hydrogen oxidation and proton reduction. Membrane-bound [NiFe] hydrogenases (MBH_2ase) provide high DET-type bioelectrocatalytic activity (*48, 49*). Moreover, MBH_2ase has several redox centers such as [NiFe] clusters and iron-sulfur clusters (FeS) called the proximal, medial, and distal. The electronic communication between MBH_2ase and the electrode generally occurs at distal FeS (*50, 51*). To improve the performance of DET-type bioelectrocatalysis of MBH, we studied the effects of modifying the electrode surface.

Carbon nanotubes have been shown to be effective scaffolds for DET-type bioelectrocatalysis of MBH_2ases from *Desulfovibrio fructosovorans* and *Aquifex aeolicus* (*15, 52, 53*). Because the shortening process of nanotubes improved DET-type activity, the edge parts or defects on the surface of the carbon nanotubes seem to be conducive for DET-type bioelectrocatalysis. Moreover, chemical interaction between MBH_2ase and modifiers has been investigated. The adsorption of thiol on the gold surface forms a functional self-assembled monolayer (SAM). The SAM formed by a short and positively charged thiol was most effective for DET-type bioelectrocatalysis of MBH_2ase from *A. aeolicus* (*53*).

Furthermore, the positively charged modifier was found to be effective in the DET-type bioelectrocatalysis of MBH_2ase from *Desulfovibrio vulgaris* Miyazaki F. This experimental observation confirms that the distal FeS is located on the negatively charged surface of the MBH_2ase. The increase in the ionic strength of the measurement buffer decreases with the DET-type activity of the MBH_2ase adsorbed at *p*-phenylenediamine-modified electrodes. Moreover, the effect of the ionic strength has been found to support the expectation that the electrostatic interaction between the MBH_2ase and electrode controls the orientation of the adsorbed MBH_2ase (*54, 55*).

D-Fructose Dehydrogenase (FDH)

FDH is a heterotrimeric enzyme for which one subunit containing three heme *c* moieties is the most electrochemically active site (*56*) and exhibits an extremely high DET-type bioelectrocatalytic activity. The mutations on the axial ligands of each heme *c* revealed that one heme *c* moiety (the first one from the N-terminal) does not participate in DET-type bioelectrocatalysis (*57*). Therefore, the variant FDH, which lacks the heme *c* moiety, shows a high DET-type bioelectrocatalytic activity (*58*).

DET-type bioelectrocatalysis of FDH has been studied using a SAM-modified electrode (*17*). SAM formed by 2-mercaptoethanol was found to be more effective for DET-type bioelectrocatalysis of FDH compared to charged and hydrophobic SAMs. The terminal hydroxyl group can change the hydrophobicity of the electrode surface. The coexisting surfactant used to solubilize FDH forms a bilayer on the hydrophilic electrode surface. Therefore, the FDH seems to be firmly embedded in the surfactant bilayer on the electrode surface to communicate with the electrode (*59*).

The natural electron acceptor of FDH is ubiquinone, although FDH was able to donate an electron to various mediators based on a linear free energy relationship (LFER). Ubiquinone derivatives were found to react with FDH at much higher levels than expected in accordance with LFER (*56*). Certain specific interactions of FDH with the structure of ubiquinone (2,3-dimethoxy-1,4-benzoquinone) were expected; thus, we investigated the effects of the surface modifier on DET-type bioelectrocatalysis of FDH at the KB electrode and found that the methoxy group in the modifier improved DET-type bioelectrocatalysis of FDH. Analysis of the catalytic wave based on Equation 13 revealed that modification with 2,4-dimethoxyaniline reduced the $2\beta r_{as}$ values from 2.1 to 1.2 for the orientation of FDH. This indicates that the methoxy-substituent-functionalized surface increases the effectively orientated FDH at the electrode surface (*60*).

The Microstructure and the Electrical Double Layer at the Electrode Surface

Because the interface between the electrode and the electrolyte solution is charged, an electric field (known as the electrical double layer) exists at the interface. Note that a charge transfer reaction occurs between the electrode and the redox species in the electrical double layer of the electrode surface. The Gouy–Chapman theory classically describes the electrical double layer of the planar electrode; however, the theory is unsuitable for the porous electrode surface. Most DET-type bioelectrocatalytic reactions are only observed in mesoporous electrodes; therefore, it is necessary to clarify the effects of the curved surface on the electrical double layer (*61*). This section demonstrates the numerically simulated electrical double layer around the microstructure at the electrode surface based on the Poisson–Boltzmann equation as follows:

$$\nabla^2 \phi = -\frac{F}{\varepsilon\varepsilon_0} \Sigma z_i c_i . \quad (18)$$

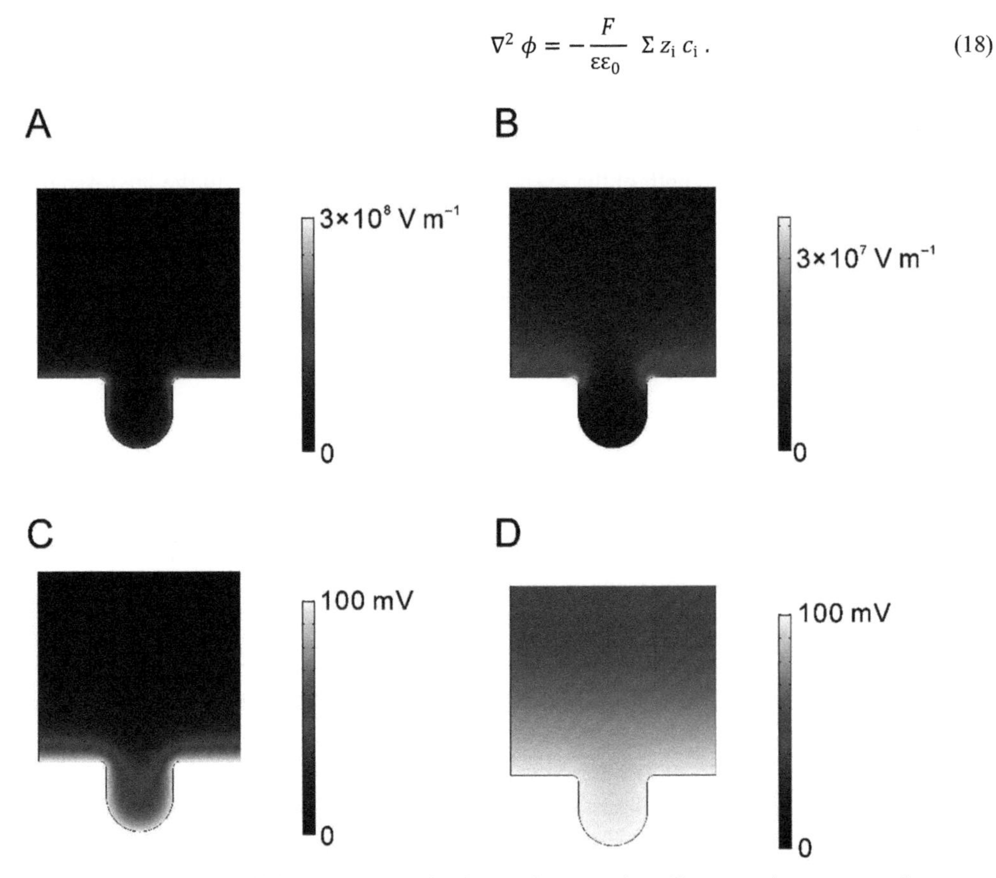

Figure 7. (A, B) The electric fields and (C, D) the electrical potential profiles around a micropore having a diameter of 2 nm with a round bottom in the presence of a 1:1 electrolyte at (A, C) 100 mM and (B, D) 1 mM.

Figure 7 shows the calculated electrical double layer around a micropore with a diameter of 2 nm. The electrical double layer on the curved surface has some interesting features. The thickness of the electrical double layer widens as the electrolyte concentration decreases. Figures 7A and 7B show the electric fields calculated for the electrolyte concentrations of 100 mM and 1 mM, respectively, while Figures 7C and 7D show the electric potential profiles around the micropore calculated for the electrolyte concentrations of 100 mM and 1 mM. At 100 mM, the thickness of the electrical double layer is thinner than the size of the micropore. As a result, the electric field and electric potential profiles seem to be homogeneous at the electrode surface. However, at 10 mM, the thickness of the electrical double layer is thicker than the size of the micropore. Under these conditions, the electrical double layer in the micropore overlaps with itself. Note that the overlapping of the electrical double layer decreases the electric field in the micropore. Moreover, the stretching of the electrical double layer enhances the electric field near the corner at the entrance of the micropore.

Both the overlapping and stretching of the electrical double layer lead to the formation of a homogeneous electric field around the microstructure on the electrode surface. Theoretically, the electric potential of the reaction plane will affect the kinetics of the electrode reaction; these effects are known as the second Frumkin effects.

In conclusion, nanostructures of the electrode surface are very important and likely essential for DET-type bioelectrocatalysis by redox enzymes. Important functions of these nanostructures include the curvature effect of the mesoporous structure, the enhanced electric double layer effect at the edge of the microporous structure, and specific interactions between the electrochemically active site of enzymes and electrode surface as well as the mass transfer effect of the macroporous structure. Several specific interactions are critical for improving the orientation of enzymes. The electrode surface can be tailored to suit each redox enzyme in the near future through a better understanding of the effects described in this chapter. Protein engineering is essential for suitable interaction between enzymes with nanostructured electrode surfaces.

References

1. Tsujimura, S.; Nakagawa, T.; Kano, K.; Ikeda, T. Kinetic Study of Direct Bioelectrocatalysis of Dioxygen Reduction with Bilirubin Oxidase at Carbon Electrodes. *Electrochemistry* **2004**, *72*, 437–439.
2. Léger, C.; Jones, A. K.; Albracht, S. P. J.; Armstrong, F. A. Effect of a Dispersion of Interfacial Electron Transfer Rates on Steady State Catalytic Electron Transport in Hydrogenase and Other Enzymes. *J. Phys. Chem. B* **2002**, *106*, 13058–13063.
3. Xia, H. Q.; Kitazumi, Y.; Shirai, O.; Kano, K. Enhanced Direct Electron Transfer-Type Bioelectrocatalysis of Bilirubin Oxidase on Negatively Charged Aromatic Compound-Modified Carbon Electrode. *J. Electroanal. Chem.* **2016**, *763*, 104109.
4. Sugimoto, Y.; Takeuchi, R.; Kitazumi, Y.; Shirai, O.; Kano, K. Significance of Mesoporous Electrodes for Noncatalytic Faradaic Process of Randomly Oriented Redox Proteins. *J. Phys. Chem. C* **2016**, *120*, 26270–26277.
5. Sugimoto, Y.; Kitazumi, Y.; Shirai, O.; Kano, K. Effects of Mesoporous Structures on Direct Electron Transfer-Type Bioelectrocatalysis: Facts and Simulation on a Three-Dimensional Model of Random Orientation of Enzymes. *Electrochemistry* **2017**, *85*, 82–87.
6. Tsutsumi, M.; Tsujimura, S.; Shirai, O.; Kano, K. Direct Electrochemistry of Histamine Dehydrogenase from *Nocardioides simplex*. *J. Electroanal. Chem.* **2009**, *625*, 144148.
7. Tsujimura, S.; Nishina, A.; Hamano, Y.; Kano, K.; Shiraishi, S. Electrochemical Reaction of Fructose Dehydrogenase on Carbon Cryogel Electrodes with Controlled Pore Sizes. *Electrochem. Commun.* **2010**, *12*, 446–449.
8. de Poulpiquet, A.; Ciaccafava, A.; Lojou, E. New Trends in Enzyme Immobilization at Nanostructured Interfaces for Efficient Electrocatalysis in Biofuel Cells. *Electrochim. Acta* **2014**, *126*, 104114.
9. Funabashi, H.; Takeuchi, S.; Tsujimura, S. Hierarchical Meso/Macro-Porous Carbon Fabricated from Dual MgO Templates for Direct Electron Transfer Enzymatic Electrodes. *Sci. Rep.* **2017**, 7, 45147.

10. Siepenkoetter, T.; Salaj-Kosla, U.; Magner, E. The Immobilization of Fructose Dehydrogenase on Nanoporous Gold Electrodes for the Detection of Fructose. *ChemElectroChem* **2017**, *4*, 905–912.
11. Kizling, M.; Dzwonek, M.; Wieckowska, A.; Bilewicz, R. Size Does Matter — Mediation of Electron Transfer by Gold Clusters in Bioelectrocatalysis. *ChemCatChem* **2018**, *10*, 1988–1992.
12. Walcarius, A.; Minteer, S. D.; Wang, J.; Lin, Y.; Merkoçi, A. Nanomaterials for Bio-Functionalized Electrodes: Recent Trends. *J. Mater. Chem. B* **2013**, *1*, 48784908.
13. Wen, D.; Eychmuller, A. Enzymatic Biofuel Cells on Porous Nanostructures. *Small* **2016**, *12*, 4649–4661.
14. Mazurenko, I.; de Poulpiquet, A.; Lojou, E. Recent Developments in High Surface Area Bioelectrodes for Enzymatic Fuel Cells. *Curr. Opin. Electrochem.* **2017**, *5*, 1–11.
15. Mazurenko, I.; Monsalve, K.; Infossi, P.; Giudici-Orticoni, M. T.; Topin, F.; Mano, N.; Lojou, E. Impact of Substrate Diffusion and Enzyme Distribution in 3D-Porous Electrodes: A Combined Electrochemical and Modelling Study of a Thermostable H_2/O_2 Enzymatic Fuel Cell. *Energy Environ. Sci.* **2017**, *10*, 1966–1982.
16. Blanford, C. F.; Heath, R. S.; Armstrong, F. A. A Stable Electrode for High-Potential, Electrocatalytic O_2 Reduction Based on Rational Attachment of a Blue Copper Oxidase to a Graphite Surface. *Chem. Commun.* **2007**, *17*, 1710–1712.
17. Murata, K.; Kajiya, Y.; Nakamura, N.; Ohno, H. Direct Electrochemistry of Bilirubin Oxidase on Three-Dimensional Gold Nanoparticle Electrodes and Its Application in a Biofuel Cell. *Energy Environ. Sci.* **2009**, *2*, 1280–1285.
18. Tasca, F.; Harreither, W.; Ludwig, R.; Gooding, J. J.; Gorton, L. Cellobiose Dehydrogenase Aryl Diazonium Modified Single Walled Carbon Nanotubes: Enhanced Direct Electron Transfer Through a Positively Charged Surface. *Anal. Chem.* **2011**, *83*, 3042–3049.
19. Cracknell, J. A.; McNamara, T. P.; Lowe, E. D.; Blanford, C. F. Bilirubin Oxidase from *Myrothecium verrucaria*: X-ray Determination of the Complete Crystal Structure and a Rational Surface Modification for Enhanced Electrocatalytic O_2 Reduction. *Dalton Trans.* **2011**, *40*, 6668–6675.
20. Olejnik, P.; Palys, B.; Kowalczyk, A.; Nowicka, A. M. Orientation of Laccase on Charged Surfaces. Mediatorless Oxygen Reduction on Amino- and Carboxyl-Ended Ethylphenyl Groups. *J. Phys. Chem. C* **2012**, *116*, 25911–25918.
21. Karaśkiewicz, M.; Nazaruk, E.; Żelechowska, K.; Biernat, J. F.; Rogalski, J.; Bilewicz, R. Fully Enzymatic Mediatorless Fuel cell with Efficient Naphthylated Carbon Nanotube-laccase Composite Cathodes. *Electrochem. Commun.* **2012**, *20*, 124–127.
22. Giroud, F.; Minteer, S. D. Anthracene-modified Pyrenes Immobilized on Carbon Nanotubes for Direct Electroreduction of O_2 by Laccase. *Electrochem. Commun.* **2013**, *34*, 157–160.
23. Oteri, F.; Ciaccafava, A.; De Poulpiquet, A.; Baaden, M.; Lojou, E.; Sacquin-Mora, S. The Weak, Fluctuating, Dipole Moment of Membrane-bound Hydrogenase from *Aquifex aeolicus* Accounts for Its Adaptability to Charged Electrodes. *Phys. Chem. Chem. Phys.* **2014**, *16*, 1131811322.

24. So, K.; Kawai, S.; Hamano, Y.; Kitazumi, Y.; Shirai, O.; Hibi, M.; Ogawa, J.; Kano, K. Improvement of a Direct Electron Transfer-type Fructose/Dioxygen Biofuel Cell with a Substrate-modified Biocathode. *Phys. Chem. Chem. Phys.* **2014**, *16*, 4823–4829.
25. Tominaga, M.; Sasaki, A.; Togami, M. Laccase Bioelectrocatalyst at a Steroid-Type Biosurfactant-Modified Carbon Nanotube Interface. *Anal. Chem.* **2015**, *87*, 5417–5421.
26. Peng, L.; Utesch, T.; Yarman, A.; Jeoung, J. H.; Steinborn, S.; Dobbek, H.; Scheller, F. W. Surface-Tuned Electron Transfer and Electrocatalysis of Hexameric Tyrosine-Coordinated Heme Protein. *Chem. Eur. J.* **2015**, *21*, 75967602.
27. Lalaoui, N.; Le Goff, A.; Holzinger, M.; Cosnier, S. Fully Oriented Bilirubin Oxidase on Porphyrin-Functionalized Carbon Nanotube Electrodes for Electrocatalytic Oxygen Reduction. *Chem. Eur. J.* **2015**, *21*, 16868–16873.
28. Lalaoui, N.; Rousselot-Pailley, P.; Robert, V.; Mekmouche, Y.; Villalonga, R.; Holzinger, M.; Le Goff, A. Direct Electron Transfer Between a Site-Specific Pyrene-Modified Laccase and Carbon Nanotube/Gold Nanoparticle Supramolecular Assemblies for Bioelectrocatalytic Dioxygen Reduction. *ACS Catal.* **2016**, *6*, 1894–1900.
29. Matanovic, I.; Babanova, S.; Chavez, M. S.; Atanassov, P. Protein-Support Interactions for Rationally Designed Bilirubin Oxidase Based Cathode: A Computational Study. *J. Phys. Chem. B* **2016**, *120*, 3634–3641.
30. So, K.; Kitazumi, Y.; Shirai, O.; Nishikawa, K.; Higuchi, Y.; Kano, K. Direct Electron Transfer-Type Dual Gas Diffusion H_2/O_2 Biofuel Cells. *J. Mater. Chem. A* **2016**, *4*, 8742–8749.
31. Mazurenko, I.; Monsalve, K.; Rouhana, J.; Parent, P.; Laffon, C.; Le Goff, A.; Szunerits, S.; Boukherroub, R.; Giudici-Orticoni, M. T.; Mano, N.; Lojou, E. How the Intricate Interactions Between Carbon Nanotubes and Two Bilirubin Oxidases Control Direct and Mediated O_2 Reduction. *ACS Appl. Mater. Interfaces* **2016**, *8*, 23074–23085.
32. Hitaichi, V. P.; Mazurenko, I.; Harb, M.; Clément, R.; Taris, M.; Castano, S.; Duché, D.; Lecomte, S.; Ilbert, M.; de Poulpiquet, A.; Lojou, E. Electrostatic-Driven Activity, Loading, Dynamics, and Stability of a Redox Enzyme on Functionalized-Gold Electrodes for Bioelectrocatalysis. *ACS Catal.* **2018**, *8*, 12004–12014.
33. Yehezkeli, O.; Raichlin, S.; Tel-Vered, R.; Kesselman, E.; Danino, D.; Willner, I. Biocatalytic Implant of Pt Nanoclusters into Glucose Oxidase: A Method to Electrically Wire the Enzyme and to Transform it from an Oxidase to a Hydrogenase. *J. Phys. Chem. Lett.* **2010**, *1*, 2816–2819.
34. Muguruma, H.; Iwasa, H.; Hidaka, H.; Hirutsuka, A.; Uzawa, H. Mediatorless Direct Electron Transfer Between Flavin Adenine Dinucleotide-Dependent Glucose Dehydrogenase and Single-Walled Carbon Nanotubes. *ACS Catal.* **2016**, *7*, 725–734.
35. Bollella, P.; Hibino, Y.; Kano, K.; Gorton, L.; Antiochia, R. Highly Sensitive Membraneless Fructose Biosensor Based on Fructose Dehydrogenase Immobilized onto Aryl Thiol Modified Highly Porous Gold Electrode: Characterization and Application in Food Samples. *Anal. Chem.* **2018**, *90*, 12131–12136.
36. Kamitaka, Y.; Tsujimura, S.; Ikeda, T.; Kano, K. Electrochemical Quartz Crystal Microbalance Study of Direct Bioelectrocatalytic Reduction of Bilirubin Oxidase. *Electrochemistry* **2006**, *74*, 642–644.

37. Sakai, K.; Kitazumi, Y.; Shirai, O.; Kano, K. Nanostructured Porous Electrodes by the Anodization of Gold for an Application as Scaffolds in Direct-Electron-Transfer-Type Bioelectrocatalysis. *Anal. Sci.* **2018**, *34*, 1317–1322.
38. Xia, H. Q.; Kitazumi, Y.; Shirai, O.; Kano, K. Direct Electron Transfer-Type Bioelectrocatalysis of Peroxidase at Mesoporous Carbon Electrodes and its Application for Glucose Determination Based on Bienzyme System. *Anal. Sci.* **2017**, *33*, 16.
39. Sakai, K.; Kitazumi, Y.; Shirai, O.; Takagi, K.; Kano, K. Direct Electron Transfer-Type Four-Way Bioelectrocatalysis of CO_2/Formate and NAD^+/NADH Redox Couples by Tungsten-Containing Formate Dehydrogenase Adsorbed on Gold Nanoparticle-Embedded Mesoporous Carbon Electrodes Modified with 4-Mercaptopyridine. *Electrochem. Commun.* **2017**, *84*, 75–79.
40. Monsalve, K.; Roger, M.; Gutierrez-Sanchez, C.; Ilbert, M.; Nitsche, S.; Byrne-Kodjabachian, D.; Marchi, V.; Lojou, E. Hydrogen Bioelectrooxidation on Gold Nanoparticle-based Electrodes Modified by *Aquifex aeolicus* Hydrogenase: Application to Hydrogen/Oxygen Enzymatic Biofuel Cells. *Bioelectrochemistry* **2015**, *106*, 47–55.
41. Takahashi, Y.; Wanibuchi, M.; Kitazumi, Y.; Shirai, O.; Kano, K. Improved Direct Electron Transfer-Type Bioelectrocatalysis of Bilirubin Oxidase Using Porous Gold Electrodes. *J. Electroanal. Chem.* **2019**, *843*, 47–53.
42. So, K.; Kitazumi, Y.; Shirai, O.; Kano, K. Analysis of Factors Governing Direct Electron Transfer-Type Bioelectrocatalysis of Bilirubin Oxidase at Modified Electrodes. *J. Electroanal. Chem.* **2016**, *783*, 316–323.
43. Pankratov, D. V.; Zeifman, Y. S.; Dudareva, A. V.; Pankratova, G. K.; Khlupova, M. E; Parunova, Y. M.; Zajtsev, D. N.; Bashirova, N. F.; Popov, V. O.; Shleev, S. V. Impact of Surface Modification with Gold Nanoparticles on the Bioelectrocatalytic Parameters of Immobilized Bilirubin Oxidase. *Acta Naturae* **2014**, *6*, 102–106.
44. Takahashi, Y.; Kitazumi, Y.; Shirai, O.; Kano, K. Improved Direct Electron Transfer-Type Bioelectrocatalysis of Bilirubin Oxidase Using Thiol-Modified Gold Nanoparticles on Mesoporous Cabon Electrode. *J. Electroanal. Chem.* **2019**, *832*, 158–164.
45. Schubert, K.; Goebel, G.; Listat, F. Bilirubin Oxidase Bound to Multi-Walled Carbon Nanotube-Modified Gold. *Electrochim. Acta* **2009**, *54*, 3033–3038.
46. Navaee, A.; Salimi, A.; Jafari, F. Electrochemical Pretreatment of Amino-Carbon Nanotubes on Graphene Support as a Novel Platform for Bilirubin Oxidase with Improved Bioelectrocatalytic Activity Towards Oxygen Reduction. *Chem. Eur. J.* **2015**, *21*, 49494953.
47. Xia, H. Q.; Kitazumi, Y.; Shirai, O.; Ozawa, H.; Onizuka, M.; Komukai, T.; Kano, K. Factors Affecting the Interaction Between Carbon Nanotubes and Redox Enzymes in Direct Electron Transfer-Type Bioelectrocatalysis. *Bioelectrochemistry* **2017**, *118*, 70–74.
48. Léger, C.; Bertrand, P. Direct Electrochemistry of Redox Enzymes as a Tool for Mechanistic Studies. *Chem. Rev.* **2008**, *108*, 2379–2438.
49. Sensi, M.; del Barrio, M.; Baffert, C.; Fourmond, V.; Léger, C. New Perspectives in Hydrogenase Direct Electrochemistry. *Curr. Opin. Electrochem.* **2017**, *5*, 135–145.
50. Ogata, H.; Mizoguchi, Y.; Mizuno, N.; Miki, K.; Adachi, S.; Yasuoka, N.; Yagi, T.; Yamauchi, O.; Hirota, S.; Higuchi, Y. Structural Studies of the Carbon Monoxide Complex of [NiFe]

Hydrogenase from *Desulfovibrio vulgaris* Miyazaki F: Suggestion for the Initial Activation Site for Dihydrogen. *J. Am. Chem. Soc.* **2002**, *124*, 11628–11635.

51. Lojou, E. Hydrogenases as Catalysts for Fuel Cells: Strategies for Efficient Immobilization at Electrode Interfaces. *Electrochim. Acta* **2011**, *58*, 10385–10397.
52. Lojou, E.; Luo, X.; Brugna, M.; Candoni, N.; Dementin, S.; Guidici-Orticoni, M. T. Biocatalysts for Fuel Cells: Efficient Hydrogenase Orientation for H_2 Oxidation at Electrodes Modified with Carbon Nanotubes. *J. Biol. Inorg Chem.* **2008**, *13*, 1157–1167.
53. Luo, X.; Brugna, M.; Tron-Infossi, P.; Giudici-Orticoni, M. T.; Lojou, E. Immobilization of the Hyperthermophilic Hydrogenase from *Aquifex aeolicus* Bacterium onto Gold and Carbon Nanotube Electrodes for Efficient H_2 Oxidation. *J. Biol. Inorg. Chem.* **2009**, *14*, 12751288.
54. Xia, H. Q.; So, K.; Kitazumi, Y.; Shirai, O.; Nishikawa, K.; Higuchi, Y.; Kano, K. Dual Gas-Diffusion Membrane- and Mediatorless Dihydrogen/Air-Breathing Biofuel Cell Operating at Room Temperature. *J. Power Sources* **2016**, *335*, 105112.
55. Sugimoto, Y.; Kitazumi, Y.; Shirai, O.; Nishikawa, K.; Higuchi, Y.; Yamamoto, M.; Kano, K. Electrostatic Roles in Electron Transfer from [NiFe] Hydrogenase to Cytochrome c_3 from *Desulfovibrio vulgaris* Miyazaki F. *Biochim. Biophys. Acta* **2017**, *1865*, 481487.
56. Kawai, S.; Yakushi, T.; Matsushita, K.; Kitazumi, Y.; Shirai, O.; Kano, K. The Electron Transfer Pathway in Direct Electrochemical Communication of Fructose Dehydrogenase with Electrodes. *Electrochem. Commun.* **2014**, *38*, 28–31.
57. Hibino, Y.; Kawai, S.; Kitazumi, Y.; Shirai, O.; Kano, K. Mutation of Heme *c* Axial Ligands in D-Fructose Dehydrogenase for Investigation of Electron Transfer Pathways and Reduction of Overpotential in Direct Electron Transfer-Type Bioelectrocatalysis. *Electrochem. Commun.* **2016**, *67*, 43–46.
58. Hibino, Y.; Kawai, S.; Kitazumi, Y.; Shirai, O.; Kano, K. Construction of a Protein-Engineered Variant of D-Fructose Dehydrogenase for Direct Electron Transfer-Type Bioelectrocatalysis. *Electrochem. Commun.* **2017**, *77*, 112–115.
59. Kawai, S.; Yakushi, T.; Matsushita, K.; Kitazumi, Y.; Shirai, O.; Kano, K. Role of a Non-Ionic Surfactant in Direct Electron Transfer-type Bioelectrocatalysis by Fructose Dehydrogenase. *Electrochim. Acta* **2015**, *152*, 19–24.
60. Xia, H. Q.; Hibino, Y.; Kitazumi, Y.; Shirai, O.; Kano, K. Interaction Between D-Fructose Dehydrogenase and Methoxy-Substituent-Functionalized Carbon Surface to Increase Productive Orientations. *Electrochim. Acta* **2016**, *218*, 41–46.
61. Kitazumi, Y.; Shirai, O.; Yamamoto, M.; Kano, K. Numerical Simulation of Diffuse Double Layer Around Microporous Electrodes Based on the PoissonBoltzmann Equation. *Electrochim. Acta* **2013**, *112*, 171–175.

Chapter 8

Modified Stainless Steel as Anode Materials in Bioelectrochemical Systems

Kai-Bo Pu,[1] Ji-Rui Bai,[1] Qing-Yun Chen,[2] and Yun-Hai Wang[*,1]

[1]Department of Environmental Science and Engineering, Xi'an Jiaotong University, Xi'an 710049, China

[2]State Key Lab of Multiphase Flow in Power Engineering, Xi'an Jiaotong University, Xi'an 710049, China

[*]E-mail: wang.yunhai@mail.xjtu.edu.cn.

As a promising anode material in bioelectrochemical systems (BESs), stainless steel-based bioanodes were reviewed. The anode material must have good biocompatibility, excellent electrical conductivity, large specific surface area, high corrosion resistance, and a low cost. The design of stainless steel-based bioanodes has been the focus of numerous research studies over the past decade. The present chapter summarizes and discusses the recent advances in stainless steel-based bioanodes and their configurations. The stainless steel-based bioanodes will be discussed in terms of optimization of stainless steel structures, carbon modification, metal oxide modification, conductive polymer modification, and others.

Introduction

A microbial fuel cell (MFC) is a typical bioelectrochemical system (BES), which can utilize exoelectrogens as catalysts to oxidize organic matters in wastewater to release electrons (*1*). These electrons are transferred to the cathode via the external circuit and can be utilized by electron acceptors (*2*). MFCs have aroused widespread interest since their invention due to their renewability and lack of pollutant byproducts (*3*). The market uptake of MFCs will rely on the improvements in aspects such as power density, cost effectiveness, and scale-up potential (*4*, *5*). The power generation has significant influence on the popularity and engineering applications of MFCs. There are many factors affecting the power generation performance such as exoelectrogens (*6*), cell configurations (*7*), and electrode materials and structures (*8*). Among these factors, the bioanodes have key impacts on the power generation of MFCs. The bioanodes not only work as support for exoelectrogen attachment but also affect the efficiency of exoelectron transfer (*9*). Additionally, bioanodes account for a large part of the total cost, and they also affect the scaling up of MFCs. Based on previous

© 2020 American Chemical Society

studies, ideal bioanode materials will have high biocompatibility, high electrical conductivity, high specific surface area, high corrosion resistance, and a low cost (*8*). At present, the common bioanode materials are carbon bioanodes (*10*).

Carbon materials, such as carbon brushes (*11*), carbon paper (*12*), carbon cloths (*13*), graphite rods (*14*), graphite paper (*11*), and glassy carbon (*1*), have been widely used as bioanodes in BESs, such as MFCs, microbial electrosynthesis (*15*), and microbial desalination cells (*16*), due to their high biocompatibility, chemical stability, and high specific surface area. For example, traditional carbon-based bioanodes, especially carbon brushes and graphite brushes, have excellent electrical conductivity and biocompatibility, which can significantly decrease the internal resistance and improve the power generation performance of MFCs. Logan et al. used graphite fiber brushes as bioanodes in a single chamber air-cathode MFC and got the maximum power density up to 2400 mW m^{-2} (*17*). For large-scale MFCs, including marine MFCs, carbon-based bioanodes are also the most widely used (*18*). In addition to high biocompatibility and conductivity, the high specific surface area of carbon materials also benefits the attachment and growth of microorganisms. These advantages of carbon-based bioanode have promoted its development in MFCs. However, carbon-based bioanodes like carbon brushes are fragile, which has a negative effect on the scaling up and maintenance of the electrodes (*4*). Carbon-based bioanodes are generally expensive and difficult to be processed, which is not conducive to their large-scale application (*19*). Thus, it is crucial to find cheaper, easy-to-process, and high-performance anode materials.

In addition to carbon materials, some metal bioanodes have been reported in BESs. Metal is relatively cheap with high mechanical strength, flexibility, and good electrical conductivity. It has been employed in structural supports and acts as a current collector in BESs (*20*). For instance, stainless steel has been used as a current collector on the back of carbon electrodes to enhance the electron transfer efficiency (*21*). For many microbial electrolysis cells (MECs) and membrane bioreactors (MBRs), stainless steel can also work as an excellent structural support due to its good mechanical strength (*22–25*). In addition to acting as a structural support, stainless steel has also been widely used as a bioanode, which shows high electrical conductivity and superior mechanical strength in MFCs (*26*). Uwe Schroder et al. analyzed the performance of different kinds of metals including gold, silver, copper, nickel, cobalt, titanium, and stainless steel as anodes in MFCs. The results showed that all metal materials except cobalt and titanium could form high-performing electrochemically active biofilms, which mean that metal-based bioanodes are highly promising anode materials (*27*). Among these metals, silver, gold, and other precious metals are expensive. Among the common metals, copper has been reported to show a better power generation performance compared to others (*28*). However, the metal corrosion doesn't support its engineering application as a bioanode. Similar defects limit the application of other metals with low corrosion resistance like aluminum (*29*). Metal corrosion leads to the electrode's change in electrochemical properties, and the metal ions that dissolved in the electrolytes are usually toxic to microorganisms on bioanodes (*29, 30*). The potential difference between anode and cathode and the biofilm formed on the anode surface, especially in a BES, will accelerate the corrosion of the metal substrate and the dissolution of metal ions. It has been reported that *Geobacter sulfurreducens* can lead to severe metal corrosion (*31–33*). Huang et al. reported a mechanism by which microorganisms affect the corrosion of metal-based bioanodes. The electrons transfer between the microorganisms and the stainless steel, and the secretion of microorganism metabolism can accelerate the metal's corrosion (*34, 35*).

Table 1. BES Performance with Different Anode Materials

Anode Materials	*Reactor Configuration*	*Anode Size*	*Maximum Power or Current Density*	*[a]Cost*	*Ref No.*
Carbon brush	Single chamber, air–cathode	Φ25mm* 142mm	1270 mW m^{-2}	$6.60–$11.30/Piece	(*53*)
Carbon cloth	Single chamber, air–cathode	6 cm^2	822 mW m^{-2}	$12.00–$13.00/m^2	(*54*)
Carbon paper	Single chamber, air–cathode	9 cm^2	30.7 mW m^{-3}	$1.90–$3.90/m^2	(*12*)
Glassy carbon	Two chambers, catholyte exposed to air	6 cm^2	1.33 mW m^{-2}	$50.00–$75.00/m^2	(*55*)
Graphite plate	Single chamber, air–cathode	155 cm^2	1410 mW m^{-2}	$5.00–$50.00/Piece	(*56*)
Graphite filter brush	Single chamber, air–cathode	Φ3cm* 4cm	2400 mW m^{-2}	$1.20–$2.50/Piece	(*17*)
Graphite felt	Two chambers, catholyte $K_3Fe(CN)_6$	Anode chamber 156 mL	386 W m^{-3}	$3.00–$50.00/m^2	(*57*)
Silver	Round-bottom flasks	1.5 cm^2	1.11 mA cm^{-2}	$10.00–$100.00/Kg	(*27*)
Gold	Round-bottom flasks	1.5 cm^2	1.17 mA cm^{-2}	$28.00–$32.00/Kg	(*27*)
Copper mesh	Single chamber, air–cathode		2 mW m^{-2}	$7.50–$8.90/Kg	(*30*)
Titanium	A cube MFC (membrane cathode assembles)	4 cm^2	57 μWm^{-2}	$30.00$50.00/Kg	(*58*)
Stainless steel	Single chamber, air–cathode	4 cm^2	40 mW m^{-2}	$4.50–$9.30/m^2	(*59*)

[a] 2017 values from http://www.alibaba.com and http://china.alibaba.com.

Compared to the previously mentioned metals, stainless steel is much cheaper and has good corrosion resistance. In particular, stainless steel possesses good biocompatibility and is easy to process. Therefore, stainless steel bioanodes are more widely used in BESs (*36–38*). Stainless steel can be used directly as a bioanode in BES. For example, stainless steel worked as the bioanode in a single chamber air-cathode MFC to test its performance (*39–42*). The stainless steel could affect the microbial community structure and the power generation performance of the MFC (*43*). It has also been reported that the amount of microorganisms attached on stainless steel anodes is higher than the amount on carbon anodes especially for some sediment microbial fuel cells (SMFCs) (*44, 45*). Moreover, stainless steel is also widely used as a bioanode in different BESs such as marine MFCs (*46–48*), microbial fuel cell/membrane bioreactor (MFC/MBR) systems (*22*), and MECs (*49*). Table 1 showed the performance of BESs with different types of bioanodes. Stainless steel has been widely used as a bioanode in different kinds of BESs (*50*). Stainless steel is cheap and can be easily customized into various shapes, which may be responsible its large-scale application (*20*, *51*, *52*).

However, there are still a lot of limitations for stainless steel anodes (*10*, *60*). On one hand, the biocompatibility of stainless steel needs further improvements. The components of stainless steel, including chromium (Cr), nickel (Ni), and molybdenum (Mo) among others, may have considerable influence on its biocompatibility (*52*). On the other hand, the corrosion will restrain the growth of microorganisms on the anode surface, which could have certain influences on the performance of the BES (*50*, *61*). Stainless steel generally has excellent corrosion resistance in weak corrosive media such as air, steam, and water. However, when stainless steel operates as an anode in a BES, it is relatively easier to be corroded under the action of an electric field. In particular, after the biofilm forms on the stainless steel surface, the microorganisms could accelerate the metal's corrosion, which can affect the stability of the BES (*62–64*). Therefore, in order to overcome these shortcomings, stainless steel needs to be modified to further improve its performance as a bioanode. Several improvement methods for stainless steel as bioanodes for BES are summarized in the following sections.

Modification on Stainless Steel Anode

Optimization of Stainless Steel Electrode Structure

The electrode distance, anode surface area, and mass-transfer efficiency can affect the performance of the BES greatly. It has been reported that the electrode distance has a major influence on the internal resistance, which would exert significant effects on BES performance, especially for large reactors (*65*, *66*). Moreover, a higher surface area provides a larger surface for microorganisms to attach and more sites for electrons to transfer between the exoelectrogens and the electrode, thereby demonstrating noteworthy improvements in power generation performance (*67*). In addition to the above factors, the anode arrangement and the electrolyte hydrodynamics also affect the BES performance.

Stainless steel has drawn increased interest as an electrode material in BESs due to its low cost, high mechanical strength, and electrical conductivity. Though the low biocompatibility of stainless steel limits its wide application in BESs, it is still a promising alternative to conventional carbon materials (*68*, *69*). Compared with the conventional carbon brush anode, stainless steel is easier to process into different shapes. Through the optimization of these stainless steel anode shapes, the electrodes can be made more unrestricted so that the internal resistance can be reduced and the mass transfer efficiency can be increased (*70*). For instance, stainless steel mesh was used as a bioanode in a dual-chamber MFC with a rectangular methacrylate reactor, the anode mesh structure ensures better hydrodynamics and provides high surface for microorganism attachment and growth (*71*). Some groups reported that the internal resistance of BES could decrease by filling the anode compartment with small stainless steel balls (*72*). The stainless steel brush and stainless steel wool can also be used as electrodes in a BES to improve its performance (*73*, *74*). Liu et al. used three-dimensional (3D) printing technology to fabricate 3D-structured stainless steel anodes, which are advantageous for good electrochemical performance for the BES (*75*). Table 2 summarizes the applications of different shapes of stainless steel electrodes in BESs.

Table 2. Comparison of Different Structures of Stainless Steel in BESs

Structures of Stainless Steel	*Reactor Configuration*	*Applications*	*Electrode Size*	*Maximum Power or Current Density*	*Ref No.*
Stainless steel balls	Dual-chamber BES	Packed bed anode	130 cm^2	0.06 mA cm^{-2}	(*72*)
Stainless steel mesh	Dual-chamber BES	Anode	25 cm^2	140 mW m^{-2}	(*71*)
Stainless steel foam	Single-chamber BES	Anode	2 cm^2	80 A m^{-2}	(*68*)
Stainless steel felt	Microbial electrolysis desalination and chemical-production cell	Anode	72 cm	24 A m^{-2}	(*69*)
Folded stainless steel mesh	Single-chamber BES	Anode	0.1 m^2	NA	(*65*)
3D printing stainless steel	Urine-powered MFCs	Anode	1 cm^3	0.93 W m^{-3}	(*75*)

Carbon Modification

In order to improve the biocompatibility and corrosion resistance of stainless steel, various methods can be adopted to modify stainless steel and regulate its properties. Carbon materials such as activated carbon (AC) (*76*), nanocarbon materials (*77*), graphene (*78*), and carbon black (CB) (*79*) are widely used to combine the advantages of carbon and stainless steel. AC is usually prepared by high temperature carbonization (*80*). Because the AC has a high specific surface area, it is widely used in adsorption (*81*), deodorization (*82*), and water purification (*83*). Coating-AC on a stainless steel surface could produce 3D microporous anodes, which could provide a large surface for attaching and growing microorganisms (*76*). The AC has excellent biocompatibility and chemical stability. As shown in Figure 1, the AC–coated stainless steel prepared in our lab had more microorganisms attached on its surface than in the bare stainless steel anode. These results demonstrated that the AC–modified stainless steel could be a promising high performance bioanode for BESs (*84*).

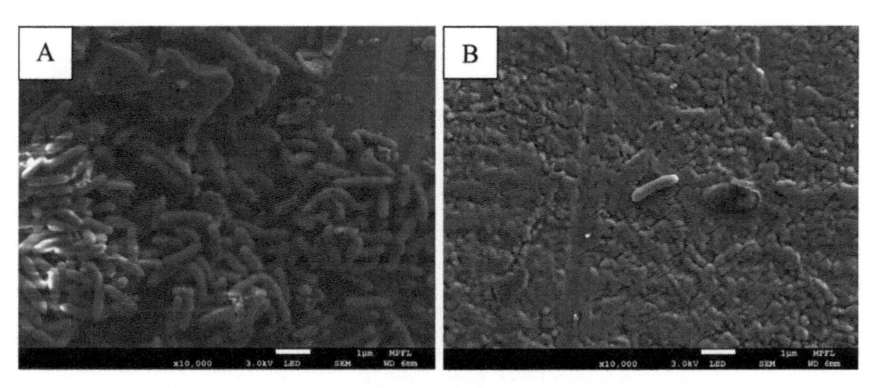

Figure 1. SEM images of microorganisms attached on different electrode materials (A. AC–coated stainless steel; B. Bare stainless steel).

Nanocarbon materials such as carbon nanofibers and carbon nanotubes (CNTs) with excellent physical and chemical properties are also widely used in bioanodes in BESs. Recently, the modification of stainless steel by various nanocarbon materials has been promising. For anode modification, nanocarbon materials have been reported to significantly improve the anode surface area and provide more sites for microorganism attachment and electron transfer, which resulted in a lower charge for transfer resistance and a higher power output (*85*). Yu et al. coated CNTs in a stainless steel mesh as bioanodes for BESs, which produced a maximum power density of 3360 mW m^{-2} that was 7.4 times higher than carbon cloth anodes (*86*). Chen et al. reported a self-connected bioanode of carbon nanofiber–stainless steel mesh prepared by direct growth of carbon nanofibers onto a stainless steel mesh via a chemical vapor deposition process (*87*). However, it has been reported that nanocarbon materials may be toxic to microorganisms because of their small size and easy access to microbial cells (*88*). Graphene has attracted a lot of research interest in recent years due to its excellent electrical conductivity and biocompatibility (*78*). Graphene can improve the surface area of bioanodes and decrease the electrode resistance (*89, 90*). Compared to traditional carbon cloth, graphene-coated stainless steel anodes had a better performance (*91*). The long-term stability of graphene-modified stainless steel anodes strengthens the idea of scaled-up BESs (*92*). In addition to the previously mentioned carbon materials, research into CB-modified stainless steel electrodes have been reported. The components of CBs are mainly aromatic hydrocarbons and other organic matters (*93–95*). CBs could improve the biocompatibility and electrical conductivity of stainless steel bioanodes. Carbon black–modified stainless steel (CB/SS) was developed as a high-performance BES bioanode. One of the CB-coating methods on stainless steel is the binder-free dipping and drying process (*79*). A CB coating greatly improved the attachment of microorganisms and facilitated electron transfer between microorganisms and the anode. The power generation performance of the MFC was greatly enhanced by CB modification (*96*). Another CB-coating method is in situ flame synthesis of CB on stainless steel. As shown in Figure 2, Singh et al. reported a CB-coated stainless steel mesh with an inexpensive flame-deposition method. CB nanoparticles generated from candle soot were used to modify the ultra-fine stainless steel mesh (*97, 98*). The results showed that the anode prepared by flame deposition is porous and mechanically and electrochemically stable. The fabricated electrodes were electroconductive and biocompatible. As shown in Figure 3, CB/SS anodes prepared in our lab showed the same porous structure. These structures can facilitate the growth of microorganisms and the formation of porous biofilm.

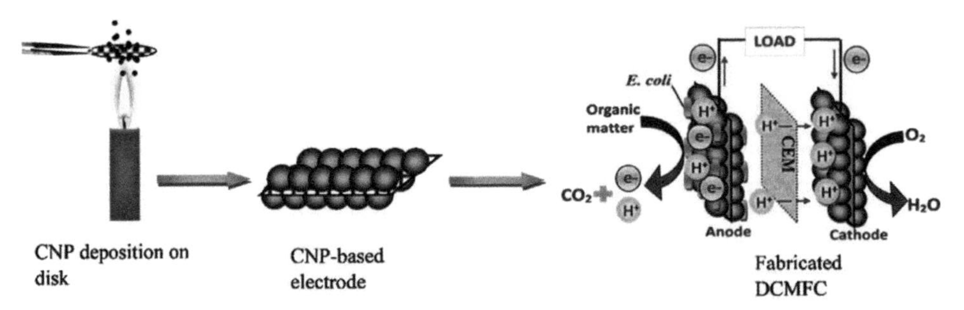

Figure 2. Schematic representation of the DCMFC fabrication (98). Reproduced with permission from reference (98). Copyright 2018 Elsevier.

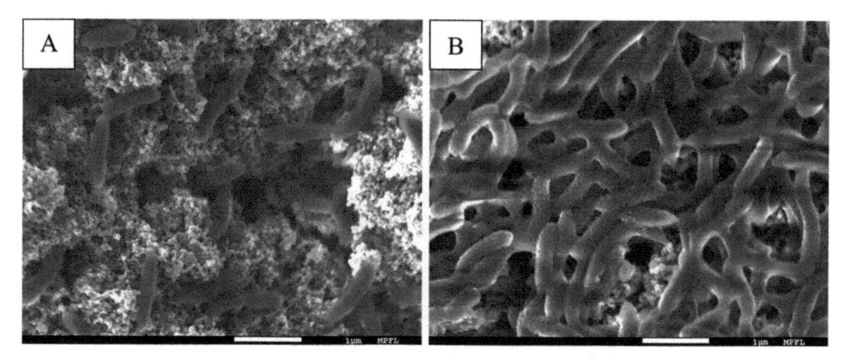

Figure 3. SEM images of microorganisms attached on a CB-coated stainless steel anode with different operating times (A. 7 d; B. 20 d).

The composites of carbon and other materials were also used to modify stainless steel to prepare high-performance composite bioanodes for MFCs. For example, the maximum power density of a BES with composites of AC and Fe_3O_4-modified stainless steel mesh anodes is 809 mW m^{-2}, which is 56 times higher than that of a bare stainless steel mesh anode (*99*). Moreover, carbon materials together with conductive polymers are used to modify stainless steel to further improve its performance. Li et al. used polypyrrole- and sargassum-activated carbon to modify stainless steel sponges (PPy/SAC/SS) to prepare a high-performance bioanode for BESs. The PPy/SAC layer effectively enhanced the corrosion resistance and electrical conductivity while also facilitating the formation of an evenly distributed biofilm. The maximum power density of a BES with a PPy/SAC/SS anode is 45.2 W m^{-3}, which is 2.9 times higher than a bare SS anode (*100*). Venkata Mohan et al. reported the polyaniline and carbon nanotube (PANI/CNT)–modified stainless steel anode. The results of bioelectrochemical characterization showed that the PANI/CNT layer modification improved the electrocatalytic activity of the bioanode. After the PANI/CNT modification, the maximum power density of BES increased to 48 mW m^{-3} (*101*).

Metal Oxides Modification

Coating metal oxides on a stainless steel surface can enhance the electron transfer efficiency, biocompatibility, and corrosion resistance (*102*). The common metal oxides for stainless steel modification include titanium dioxide (TiO_2), iron oxide, and manganese dioxide (MnO_2). It has been reported that TiO_2 has excellent biocompatibility, which can increase the electricity generation (*103*). Ying et al. electroplated TiO_2 onto the surface of stainless steel to improve the corrosion resistance and current output. Compared to the untreated stainless steel mesh, the TiO_2 modification increased the power-generation performance of the BES significantly (*104*). The iron oxides such as FeOOH, goethite, and Fe_2O_3 were also widely used in stainless steel modification. The iron oxide–coated stainless steel anode surface could promote extracellular electron transfer (*105*). Some nanostructured iron oxides could also improve the performance of photocatalytic biological anodes (*106*). The microstructure of the stainless steel surface could be changed greatly under heating at high temperatures. An iron oxide nanoparticle film could be formed on a stainless steel surface through flame oxidation or being heated in a muffle furnace (*107*, *108*). Figure 4 showed the typical microstructure of an iron oxide nanoparticle film formed by heating treatments in our lab. Guo et al. heated stainless steel felt in a 600 °C muffle furnace for 5 min to form an iron oxide layer on its surface. The iron oxide layer could enhance the biocompatibility and corrosion resistance of stainless steel, which helped to improve the power-generation performance of the BES (*109*). A flame-oxidized stainless steel anode was also used in BES-based biosensors to improve their performance (*110*).

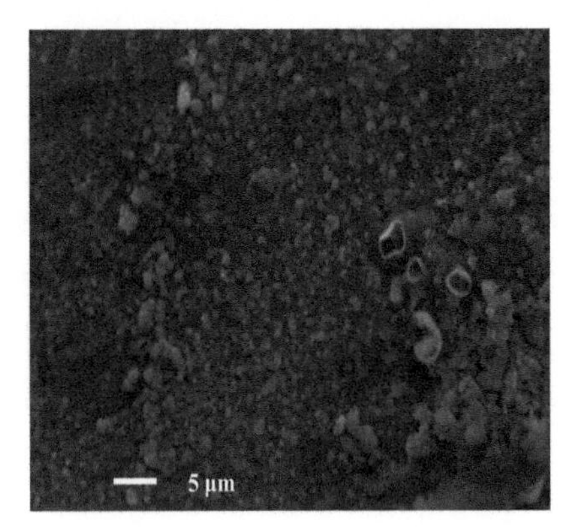

Figure 4. The iron oxide particle formation after heating treatment.

Moreover, these metal oxide materials often combine together with carbon and conductive polymer materials to modify stainless steel. For example, Yu et al. made composite anodes by rolling FeOOH and Fe_3O_4 into AC on stainless steel, respectively. The addition of FeOOH or Fe_3O_4 can enhance both the capacitance and diffusion efficiency of stainless steel bioanode and improve the anode performance of BES (*99*, *102*). Moreover, Liang et al. coated the reduced graphene oxide and MnO_2 on stainless steel anodes to promote the adhesion of microorganisms and electron transfer efficiency (*111*).

Conductive Polymer Modification

Besides the previously mentioned methods, conductive polymers such as polypyrrole (PPy) polyaniline (PANI) and polythiophene are also used to modify stainless steel electrodes because of their excellent electrical conductivity and biocompatibility. In our previous research, the PPy and Poly(3,4-ethylenedioxythiophene) (PEDOT) were electrochemically polymerized in situ to modify stainless steel plate electrodes. After modification, the corrosion resistance, electrical conductivity, and biocompatibility of the bioanode were enhanced. Based on the SEM images shown in Figure 5, the conductive polymer modification could improve the roughness of the anode surface, which can be helpful for the attachment and growth of microorganisms. A higher amount of microorganisms attached to the anode surface can significantly improve the power generation performance of a BES. The maximum power density of a BES with PPy/SS and PEDOT/SS anodes were 1190.9 mW/m^2 and 608.6 mW/m^2, respectively, which were 29 and 6 times higher than a BES with bare stainless steel plate anodes (*59*, *112*). In addition, the polymer film could prevent direct contact between the stainless steel and the electrolyte enhancing the corrosion resistance of the electrode (*113*). Moreover, conductive polymer-modified stainless steel anodes had lower internal resistance and higher capacitance than bare SS anodes. These ensured that the polymer modification could improve the bioanode performance making them a promising bioanode material for BES (*114*). The conductive polymer modification not only significantly improves the power–generation performance of the BES, but it also improve its start-up performance. The polymer film shortens the formation time of exoelectrogen film on an anode surface (*115, 116*). With the development of BESs, conductive polymer modification has been combined with other modification technologies to further improve the performance of bioanodes. Conductive polymer modification usually cooperated with metal oxide modification and carbon modification. As previously mentioned, the conductive polymer works together with AC (*100*), nanocarbon materials (*101*), and metal oxides like MnO_2 to modify the stainless steel to produce high-performance bioanodes for BESs. For instance, Korakot et al. electrodeposited MnO_2 on PPy-coated stainless steel anodes to enhance the anode's active surface area, electrical conductivity, and corrosion resistance. The maximum power density of a BES with MnO_2- and PPy-coated stainless steel anodes was around 440 mW m^{-2}, which was much higher a BES with bare stainless steel anodes (*117*). These composite anode materials are expected to further improve the performance of a BES to promote its development and application.

Figure 5. SEM images of different polymer modified stainless steel anodes. A. PPy/SS electrode (59). Reproduced with permission from reference (59). Copyright 2018 Elsevier. B. PEDOT/SS electrode (112). Reproduced with permission from reference (112). Copyright 2018 American Chemical Society.

Other Modification

There are additional modification methods for stainless steel electrodes such as nitrogen doping (*118*), methylene blue coating (*119*), surface hydrophilization (SDBS) (*120*), and pretreatment with NH_4Cl and NaOH (*121*). Godwin et al. immobilized methylene blue on a 304L stainless steel surface to get a high-performance bioanode for MFCs. The composite electrodes have shown improved performance compared to traditional graphite electrodes. The maximum power density of composite electrodes with pure cultures of *Escherichia coli K-12* was 39.35 mW m^{-2}, which showed a 6-fold increase in the maximum power density compared to graphite electrodes (*119*). A surfactant like cetyl trimethyl ammonium bromide (CTAB) could make the anode surface more hydrophilic, which can accelerate the anodic biofilm formation and enhance current output of BESs. The maximum current output with treated electrodes was 23.8% higher than those with untreated electrodes (*120*). In addition, many pretreatment methods can enhance the anode performance. Tang et al. used different methods to pretreat stainless steel and tested their performance as bioanodes in MFCs. The stainless steel was immersed into distilled water, NaOH, HCl, and NH_4Cl, respectively, for 24 h, and the results indicated that stainless steel pretreated by NH_4Cl showed the best power-generation performance in BESs (*121, 122*).

Table 3 summarized the applications of different modification methods on stainless steel anodes in BESs. The data shown in Table 3 demonstrated that carbon material modification of stainless steel could produce high-performance bioanodes for BESs. The highest power density was obtained in a BES with CNT–modified stainless steel anodes.

Table 3. Comparison of Modified Stainless Steel Anodes in BESs

Anode Materials	*Reactor Configuration*	*Anode Size*	*Maximum Power or Current Density*	*Ref No.*
AC/stainless steel fiber felts	Two-chamber MFC	3.2 cm^2	560 mW m^{-2}	(*76*)
Pyrolytic carbon/stainless steel felt	Multi-electrode BES	1 cm^2	3.65 mA cm^{-2}	(*84*)
Carbon nanotube/stainless steel	Single-chamber MFCs	7 cm^2	261 mW m^{-2}	(*85*)
Carbon nanohorn/stainless steel	Single-chamber MFCs	7 cm^2	327 mW m^{-2}	(*85*)
Multi-wall carbon nanotubes/ stainless steel meshes	Microfabricated 24-well MFC arrays	-	3360 mW m^{-2}	(*86*)
Carbon nanofibers/stainless steel	Microbial BES	1 cm^2	1.28 mA cm^{-2}	(*87*)
Graphite/stainless steel meshes	Novel annular single-chamber microbial fuel cell (ASCMFC)	126 cm^2	20.2 W m^{-3}	(*78*)
Graphene suspension/stainless steel meshes	Single-chamber MFCs	–	1.77 mW m^{-2}	(*90*)
Graphene/stainless steel meshes	MFCs	1 cm^2	2668 mW m^{-2}	(*91*)
Graphene/stainless steel fibers	Two-chamber MFCs	3.2 cm^2	2143 mW m^{-2}	(*91*)
CB/stainless steel meshes	MFCs	-	3215mW m^{-2}	(*79*)
CB/stainless steel meshes	MFCs	-	1.91 mA cm^{-2}	(*96*)
Carbon nanostructures/stainless steel meshes	Single-chamber MFCs	6.5 cm^2	187 mW m^{-2}	(*98*)
Carbon nanoparticles /stainless steel wire	Two-chamber MFCs	-	1650 mW m^{-2}	(*99*)
Fe_3o_4/AC/stainless steel meshes	Single-chamber MFCs	-	809 mW m^{-2}	(*99*)
PPy/sargassum-activated carbon/ stainless steel	Two-chamber MFCs	-	45.2 W m^{-3}	(*100*)
PANI/CNT/stainless steel mesh	Two-chamber MFCs	6 cm^2	48 mW m^{-2}	(*101*)
TiO_2 thin film-modified stainless steel mesh	Two-chamber MFCs	7 cm^2	2.87 W m^{-2}	(*104*)
Nanostructured a-Fe_2O_3/stainless steel	Two-chamber BES	12.6 cm^2	2.8 A m^{-2}	(*106*)

Table 3. (Continued). Comparison of Modified Stainless Steel Anodes in BESs

Anode Materials	*Reactor Configuration*	*Anode Size*	*Maximum Power or Current Density*	*Ref No.*
iron oxide nanoparticles/stainless steel felt	A custom-made BES reactor	2 cm^2	1.92 mA cm^{-2}	(*107*)
Fe_2O_3/stainless steel	Two-chambers MFCs	20.3 cm^2	827.25 mW m^{-2}	(*108*)
MnO_2/PPy/ stainless steel	Single-chamber MFCs	3 cm^2	440 mW m^{-2}	(*117*)
Graphene oxide/manganese oxide nanoparticles/stainless steel fibre felt	Two-chambers MFCs	7.1 cm^2	1045 mW m^{-2}	(*111*)
PPy/stainless steel	Single-chamber MFCs	4 cm^2	1190.94 mW m^{-2}	(*59*)
PEDOT/ stainless steel	Single-chamber MFCs	4 cm^2	608.6 mW m^{-2}	(*113*)
Polyaniline/stainless steel plates	Single-chamber MFCs	1.5 cm^2	0.078 mW cm^{-2}	(*114*)
PANI /stainless steel	Air-cathode MFCs	7.5 cm^2	0.288 mW cm^{-2}	(*116*)
PPY/stainless steel	air-cathode MFCs	7.5 cm^2	0.187 mW cm^{-2}	(*116*)
Polyaniline/stainless steel	MFCs	1.5 cm^2	10.5 A m^{-2}	(*117*)
PPY/stainless steel	MFCs	1.5 cm^2	5.0 A m^{-2}	(*117*)
Nitrided stainless steel	Proton exchange membrane fuel cell (PEMFC)	4.5 cm^2	-	(*118*)
Methylene blue/stainless steel	Dual-chamber biobatteries	-	39.35 mW m^{-2}	(*119*)
NH_4Cl/stainless steel	Single-chamber MFCs	-	130 mW m^{-3}	(*121*)
HCl/stainless steel	Single-chamber MFCs	-	110 mW m^{-3}	(*121*)
Distilled water/stainless steel	Single-chamber MFCs	-	59 mW m^{-3}	(*121*)
NaOH/stainless steel	Single-chamber MFCs	-	65 mW m^{-3}	(*121*)
Neutral-red steel Surface/carbon-coated stainless steel felt	Two-chamber reactor	1 cm^2	6.0 A m^{-2}	(*122*)
Sodium dodecyl benzene Sulfonate/ carbon-coated stainless steel	Two-chamber reactor	1 cm^2	4.5 A m^{-2}	(*122*)

Conclusions and Perspectives

Bioanode designs are still great challenges to BESs. Compared to carbon electrodes, stainless steel is cheaper and easier to process, especially since it can be easily combined with 3D printing technology to scale up to industrial scale with various designs. These advantages enable stainless steel to be widely used as anodes in BESs. Stainless steel–based bioanodes have shown promising applications in BESs. However, further research and modifications are needed. Until recently, many reported modification methods included carbon modification, metal oxide modification, and conductive polymer modification. Among these modifications, carbon modification combined the advantages of carbon and stainless steel, which showed the most promising performance. Conductive polymer modification has also attracted a lot of attention. Carbon modification and conductive polymer modification can significantly improve the biocompatibility and corrosion resistance of stainless steel, which is expected to be widely used in stainless steel anode modification. On the other hand, all these studies were conducted on a laboratory scale. There are still some issues to be addressed before these electrodes are scaled up for mass production, such as process optimization and equipment improvement. The electrode materials still have room for improvement. Further studies on more effective anode materials and optimization of the process are expected to address these challenges.

References

1. He, Z.; Minteer, S. D.; Angenent, L. T. Electricity Generation from Artificial Wastewater Using an Upflow Microbial Fuel Cell. *Environ. Sci. Technol.* **2005**, *39*, 5262–5267.
2. Cai, W.; Fang, X.; Xu, M.; Liu, X.; Wang, Y. Sequential Recovery of Copper and Nickel from Wastewater Without Net Energy Input. *Water Sci. Technol.* **2015**, *71*, 754–760.
3. Lu, N.; Zhou, S.; Zhuang, L.; Zhang, J.; Ni, J. Electricity Generation from Starch Processing Wastewater Using Microbial Fuel Cell Technology. *Biochemical Eng. J.* **2009**, *43*, 246–251.
4. Guo, K.; Hassett, D. J.; Gu, T. Microbial Fuel Cells: Electricity Generation from Organic Wastes by Microbes. In *Microbial Biotechnology: Energy and Environment*; Arora., R., Ed.; CABI: Oxon, United Kingdom, 2012; pp 162–189.
5. Wang, J.; Bi, F.; Ngo, H.; Guo, W.; Jia, H.; Zhang, H.; Zhang, X. Evaluation of Energy-Distribution of a Hybrid Microbial Fuel Cell–Membrane Bioreactor (MFC–MBR) for Cost-Effective Wastewater Treatment. *Bioresource Technol.* **2016**, *200*, 420–425.
6. Kang, Y. L.; Pichiah, S.; Ibrahim, S. Facile Reconstruction of Microbial Fuel Cell (MFC) Anode with Enhanced Exoelectrogens Selection for Intensified Electricity Generation. *Int. J. Hydrogen Energ.* **2017**, *42*, 1661–1671.
7. Pang, S.; Gao, Y.; Choi, S. Flexible and Stretchable Biobatteries: Monolithic Integration of Membrane-Free Microbial Fuel Cells in a Single Textile Layer. *Adv. Energ. Mater.* **2018**, *8*, 1–8.
8. Guo, K.; Freguia, S.; Dennis, P. G.; Chen, X.; Donose, B. C.; Keller, J.; Gooding, J. J.; Rabaey, K. Effects of Surface Charge and Hydrophobicity on Anodic Biofilm Formation, Community Composition, and Current Generation in Bioelectrochemical Systems. *Environ. Sci. Technol.* **2013**, *47*, 7563–7570.
9. Wei, J.; Liang, P.; Huang, X. Recent Progress in Electrodes for Microbial Fuel Cells. *Bioresource Technol.* **2011**, *102*, 9335–9344.

10. Sonawane, J. M.; Yadav, A.; Ghosh, P. C.; Adeloju, S. B. Recent Advances in the Development and Utilization of Modern Anode Materials for High Performance Microbial Fuel Cells. *Biosen. Bioelectron.* **2017**, *90*, 558–576.
11. Ahn, Y.; Logan, B. E. Effectiveness of Domestic Wastewater Treatment Using Microbial Fuel Cells at Ambient and Mesophilic Temperatures. *Bioresource Technol.* **2010**, *101*, 469–475.
12. He, Y.; Xiao, X.; Li, W.; Sheng, G.; Yan, F.; Yu, H.; Yuan, H.; Wu, L. Enhanced Electricity Production from Microbial Fuel Cells with Plasma-Modified Carbon Paper Anode. *Phys. Chem. Chem. Phys.* **2012**, *14*, 9966–9971.
13. Wang, X.; Feng, Y. J.; Lee, H. Electricity Production from Beer Brewery Wastewater Using Single Chamber Microbial Fuel Cell. *Water Sci. Technol.* **2008**, *57*, 1117–1121.
14. Chaudhuri, S. K.; Lovley, D. R. Electricity Generation by Direct Oxidation of Glucose in Mediatorless Microbial Fuel Cells. *Nat. Biotechnol.* **2003**, *21*, 1229–1232.
15. Khosravanipour Mostafazadeh, A.; Drogui, P.; Brar, S. K.; Tyagi, R. D.; Bihan, Y. L.; Buelna, G. Microbial Electrosynthesis of Solvents and Alcoholic Biofuels from Nutrient Waste: A Review. *J. Environ. Chem. Eng.* **2017**, *5*, 940–954.
16. Sophia, A. C.; Bhalambaal, V. M. Utilization of Coconut Shell Carbon in the Anode Compartment of Microbial Desalination Cell (MDC) for Enhanced Desalination and Bio-Electricity Production. *J. Environ. Chem. Eng.* **2015**, *3*, 2768–2776.
17. Logan, B.; Cheng, S.; Watson, V.; Estadt, G. Graphite Fiber Brush Anodes for Increased Power Production in Air-Cathode Microbial Fuel Cells. *Environ. Sci. Technol.* **2007**, *41*, 3341–3346.
18. Jiang, D.; Li, X.; Raymond, D.; Mooradain, J.; Li, B. Power Recovery with Multi-Anode/Cathode Microbial Fuel Cells Suitable for Future Large-Scale Applications. *Int. J. Hydrogen Energ.* **2010**, *35*, 8683–8689.
19. Jiang, D.; Curtis, M.; Troop, E.; Scheible, K.; McGrath, J.; Hu, B.; Suib, S.; Raymond, D.; Li, B. A Pilot-Scale Study on Utilizing Multi-Anode/Cathode Microbial Fuel Cells (MAC MFCs) to Enhance the Power Production in Wastewater Treatment. *Int. J. Hydrogen Energ.* **2011**, *36*, 876–884.
20. Santoro, C.; Kodali, M.; Shamoon, N.; Serov, A.; Soavi, F.; Merino-Jimenez, I.; Gajda, I.; Greenman, J.; Ieropoulos, I.; Atanassov, P. Increased Power Generation in Supercapacitive Microbial Fuel Cell Stack Using Fe-N-C Cathode Catalyst. *J. Power Sources* **2019**, *412*, 416–424.
21. Jiang, Y.; Liang, P.; Zhang, C.; Bian, Y.; Sun, X.; Zhang, H.; Yang, X.; Zhao, F.; Huang, X. Periodic Polarity Reversal for Stabilizing the pH in Two-Chamber Microbial Electrolysis Cells. *Appl. Energ.* **2016**, *165*, 670–675.
22. Kocatürk-Schumacher, N. P.; Madjarov, J.; Viwatthanasittiphong, P.; Kerzenmacher, S. Toward an Energy Efficient Wastewater Treatment: Combining a Microbial Fuel Cell/Electrolysis Cell Anode with an Anaerobic Cembrane Bioreactor. *Frontiers in Energy Res.* **2018**, *6*, 1–12.
23. Huang, L.; Li, X.; Ren, Y.; Wang, X. Preparation of Conductive Microfiltration Membrane and Its Performance in a Coupled Configuration of Membrane Bioreactor with Microbial Fuel Cell. *RSC Adv.* **2017**, *7*, 20824–20832.
24. Song, J.; Liu, L.; Yang, F.; Ren, N.; Crittenden, J. Enhanced Electricity Generation by Triclosan and Iron Anodes in the Three-Chambered Membrane Bio-Chemical Reactor (TC-MBCR). *Bioresource Technol.* **2013**, *147*, 409–415.

25. Nakhate, P. H.; Joshi, N. T.; Marathe, K. V. A Critical Review of Bioelectrochemical Membrane Reactor (BECMR) as Cutting-Edge Sustainable Wastewater Treatment. *Rev. Chem. Eng.* **2017**, *33*, 143–161.
26. Li, S.; Cheng, C.; Thomas, A. Carbon-Based Microbial-Fuel-Cell Electrodes: From Conductive Supports to Active Catalysts. *Adv. Mater.* **2017**, *29*, 1–30.
27. Baudler, A.; Schmidt, I.; Langner, M.; Greiner, A.; Schröder, U. Does It Have to Be Carbon? Metal Anodes in Microbial Fuel Cells and Related Bioelectrochemical Systems. *Energ. Environ. Sci.* **2015**, *8*, 2048–2055.
28. Karchiyappan, T. Study of Electrochemical Process Conditions For the Electricity Production in Microbial Fuel Cell. *Energy Sources, Part A* **2018**, *40*, 951–958.
29. Srikanth, S.; Pavani, T.; Sarma, P. N.; Venkata Mohan, S. Synergistic Interaction of Biocatalyst with Bio-Anode as a Function of Electrode Materials. *Int. J. Hydrogen Energ.* **2011**, *36*, 2271–2280.
30. Zhu, X.; Logan, B. E. Copper Anode Corrosion Affects Power Generation in Microbial Fuel Cells. *J. Chem. Technol. Biotechnol.* **2014**, *89*, 471–474.
31. Mehanna, M.; Basseguy, R.; Delia, M.; Gubner, R.; Sathirachinda, N.; Bergel, A. Geobacter Species Enhances Pit Depth on 304L Stainless Steel in a Medium Lacking with Electron Donor. *Electrochem. Commun.* **2009**, *11*, 1476–1481.
32. Dou, W.; Liu, J.; Cai, W.; Wang, D.; Jia, R.; Chen, S.; Gu, T. Electrochemical Investigation of Increased Carbon Steel Corrosion via Extracellular Electron Transfer by a Sulfate Reducing Bacterium under Carbon Source Starvation. *Corros. Sci.* **2019**, *150*, 258–267.
33. Mehanna, M.; Basséguy, R.; Délia, M.; Bergel, A. Effect of Geobacter Sulfurreducens on the Microbial Corrosion of Mild Steel, Ferritic and Austenitic Stainless Steels. *Corros. Sci.* **2009**, *51*, 2596–2604.
34. Little, B. J.; Lee, J. S. Microbiologically Influenced Corrosion: an Update. *Int. Mater. Rev.* **2014**, *59*, 384–393.
35. Huang, Y.; Zhou, E.; Jiang, C.; Jia, R.; Liu, S.; Xu, D.; Gu, T.; Wang, F. Endogenous Phenazine-1-Carboxamide Encoding Gene PhzH Regulated the Extracellular Electron Transfer in Biocorrosion of Stainless Steel by Marine Pseudomonas Aeruginosa. *Electrochem. Commun.* **2018**, *94*, 9–13.
36. Chen, S.; Patil, S. A.; Brown, R. K.; Schröder, U. Strategies for Optimizing the Power Output of Microbial Fuel Cells: Transitioning from Fundamental Studies to Practical Implementation. *Appl. Energ.* **2019**, *233–234*, 15–28.
37. Hoseinzadeh, E.; Rezaee, A.; Farzadkia, M. Nitrate Removal from Pharmaceutical Wastewater Using Microbial Electrochemical System Supplied Through Low Frequency-Low Voltage Alternating Electric Current. *Bioelectrochemistry* **2018**, *120*, 49–56.
38. Namour, P.; Jobin, L. Electrochemistry, a Tool to Enhance Self-Purification in Water Systems While Preventing the Emission of Noxious Gases (Greenhouse Gases, H_2S, NH_3). *Curr. Opin. Electrochem.* **2018**, *11*, 25–33.
39. Rinaldi, W.; Abubakar; Rahmi, R. F.; Silmina. Tofu Wastewater Treatment by Sediment Microbial Fuel Cells. In *IOP Conference Series: Materials Science and Engineering*; Proceedings of the 3rd International Conference on Chemical Engineering Sciences and Applications, Banda

Aceh, Indonesia, Sept 20–21, 2017; Yunardi, S. A., Rosnelly, C. M., Rinaldi, W., Eds.; IOP: Banda Aceh, Indonesia, 2017.

40. Foad Marashi, S. K.; Kariminia, H. Performance of a Single Chamber Microbial Fuel Cell at Different Organic Loads and PH Values Using Purified Terephthalic Acid Wastewater. *J. Environ. Health Sci. and Eng.* **2015**, *13*, 2–6.
41. Jadhav, G. S.; Ghangrekar, M. M. Improving Performance of MFC by Design Alteration and Adding Cathodic Electrolytes. *Appl. Biochem. Biotech.* **2008**, *151*, 319–332.
42. Guerrini, E.; Cristiani, P.; Grattieri, M.; Santoro, C.; Li, B.; Trasatti, S. Electrochemical Behavior of Stainless Steel Anodes in Membraneless Microbial Fuel Cells. *J. Electrochem. Soc.* **2013**, *161*, H62–H67.
43. Wang, J.; Song, X.; Wang, Y.; Abayneh, B.; Ding, Y.; Yan, D.; Bai, J. Microbial Community Structure of Different Electrode Materials in Constructed Wetland Incorporating Microbial Fuel Cell. *Bioresource Technol.* **2016**, *221*, 697–702.
44. Dumas, C.; Basseguy, R.; Bergel, A. Electrochemical Activity of Geobacter Sulfurreducens Biofilms on Stainless Steel Anodes. *Electrochim. Acta.* **2008**, *53*, 5235–5241.
45. Peixoto, L.; Parpot, P.; Martins, G. Assessment of Electron Transfer Mechanisms During a Long-Term Sediment Microbial Fuel Cell Operation. *Energies* **2019**, *12*, 1–13.
46. Dumas, C.; Mollica, A.; Féron, D.; Basséguy, R.; Etcheverry, L.; Bergel, A. Marine Microbial Fuel Cell: Use of Stainless Steel Electrodes as Anode and Cathode Materials. *Electrochim. Acta* **2007**, *53*, 468–473.
47. Sajana, T. K.; Ghangrekar, M. M.; Mitra, A. Influence of Electrode Material on Performance of Sediment Microbial Fuel Cell Remediating Aquaculture Water. *Environ. Eng. Manag. J.* **2017**, *16*, 421–429.
48. Erable, B.; Bergel, A. First Air-Tolerant Effective Stainless Steel Microbial Anode Obtained from a Natural Marine Biofilm. *Bioresource Technol.* **2009**, *100*, 3302–3307.
49. Arunasri, K.; Annie Modestra, J.; Yeruva, D. K.; Vamshi Krishna, K.; Venkata Mohan, S. Polarized Potential and Electrode Materials Implication on Electro-Fermentative Di-Hydrogen Production: Microbial Assemblages and Hydrogenase Gene Copy Variation. *Bioresource Technol.* **2016**, *200*, 691–698.
50. Cristiani, P.; Franzetti, A.; Gandolfi, I.; Guerrini, E.; Bestetti, G. Bacterial DGGE Fingerprints of Biofilms on Electrodes of Membraneless Microbial Fuel Cells. *Int. Biodeter. Biodegr.* **2013**, *84*, 211–219.
51. Parkhey, P.; Gupta, P. Improvisations in Structural Features of Microbial Electrolytic Cell and Process Parameters of Electrohydrogenesis for Efficient Biohydrogen Production: a Review. *Renew. Sust. Energ. Rev.* **2017**, *69*, 1085–1099.
52. Lu, M.; Qian, Y.; Huang, L.; Xie, X.; Huang, W. Improving the Performance of Microbial Fuel Cells through Anode Manipulation. *ChemPlusChem* **2015**, *80*, 1216–1225.
53. Lanas, V.; Ahn, Y.; Logan, B. E. Effects of Carbon Brush Anode Size and Loading on Microbial Fuel Cell Performance in Batch and Continuous Mode. *J. Power Sources* **2014**, *247*, 228–234.
54. Liu, J.; Liu, J.; He, W.; Qu, Y.; Ren, N.; Feng, Y. Enhanced Electricity Generation for Microbial Fuel Cell by Using Electrochemical Oxidation to Modify Carbon Cloth Anode. *J. Power Sources* **2014**, *265*, 391–396.

55. Larrosa-Guerrero, A.; Scott, K.; Katuri, K. P.; Godinez, C.; Head, I. M.; Curtis, T. Open Circuit Versus Closed Circuit Enrichment of Anodic Biofilms in MFC: Effect on Performance and Anodic Communities. *Appl. Microbiol. Biot.* **2010**, *87*, 1699–1713.
56. Dewan, A.; Beyenal, H.; Lewandowski, Z. Scaling Up Microbial Fuel Cells. *Environ. Sci. Technol.* **2008**, *42*, 7643–7648.
57. Aelterman, P.; Versichele, M.; Marzorati, M.; Boon, N.; Verstraete, W. Loading Rate and External Resistance Control the Electricity Generation of Microbial Fuel Cells with Different Three-Dimensional Anodes. *Bioresource Technol.* **2008**, *99*, 8895–8902.
58. Zhou, X.; Chen, X.; Li, H.; Xiong, J.; Li, X.; Li, W. Surface Oxygen-Rich Titanium as Anode for High Performance Microbial Fuel Cell. *Electrochim. Acta.* **2016**, *209*, 582–590.
59. Pu, K.; Ma, Q.; Cai, W.; Chen, Q.; Wang, Y.; Li, F. Polypyrrole Modified Stainless Steel as High Performance Anode of Microbial Fuel Cell. *Biochem. Eng. J.* **2018**, *132*, 255–261.
60. Abu Bakar, M. H.; Shamsuddin, R. A. A.; Yunus, R. M.; Wan Daud, W. R.; Md. Jahim, J.; Aqma, W. S. Stainless Steel Application as Metal Electrode in Bioelectrochemical System. *J. Kejuruteraan.* **2018**, *1*, 65–75.
61. Gu, Y.; Ying, K.; Shen, D.; Huang, L.; Ying, X.; Huang, H.; Cheng, K.; Chen, J.; Zhou, Y.; Chen, T.; Feng, H. Using Sewage Sludge Pyrolytic Gas to Modify Titanium Alloy to Obtain High-Performance Anodes in Bio-Electrochemical Systems. *J. Power Sources* **2017**, *372*, 38–45.
62. Lee, J. S.; Little, B. J. Technical Note: Electrochemical and Chemical Complications Resulting from Yeast Extract Addition to Stimulate Microbial Growth. *Corrosion* **2015**, *71*, 1434–1440.
63. Xu, F.; Duan, J.; Lin, C.; Hou, B. Influence of Marine Aerobic Biofilms on Corrosion of 316L Stainless Steel. *J. Iron Steel Res. Int.* **2015**, *22*, 715–720.
64. Zhang, P.; Xu, D.; Li, Y.; Yang, K.; Gu, T. Electron Mediators Accelerate the Microbiologically Influenced Corrosion of 304 Stainless Steel by the Desulfovibrio Vulgaris Biofilm. *Bioelectrochemistry* **2015**, *101*, 14–21.
65. Wang, H.; Cui, D.; Yang, L.; Ding, Y.; Cheng, H.; Wang, A. Increasing the Bio-Electrochemical System Performance in Azo Dye Wastewater Treatment: Reduced Electrode Spacing for Improved Hydrodynamics. *Bioresource Technol.* **2017**, *245*, 962–969.
66. Park, J.; Lee, B.; Shi, P.; Kim, Y.; Jun, H. Effects of Electrode Distance and Mixing Velocity on Current Density and Methane Production in an Anaerobic Digester Equipped with a Microbial Methanogenesis Cell. *Int. J. Hydrogen Energ.* **2017**, *42*, 27732–27740.
67. Bajracharya, S.; Sharma, M.; Mohanakrishna, G.; Dominguez Benneton, X.; Strik, D. P. B. T.; Sarma, P. M.; Pant, D. An Overview on Emerging Bioelectrochemical Systems (BESs): Technology for Sustainable Electricity, Waste Remediation, Resource Recovery, Chemical Production and Beyond. *Renew. Energ.* **2016**, *98*, 153–170.
68. Ketep, S. F.; Bergel, A.; Calmet, A.; Erable, B. Stainless Steel Foam Increases the Current Produced by Microbial Bioanodes in Bioelectrochemical Systems. *Energ. Environ. Sci.* **2014**, *7*, 1633–1637.
69. Lu, Y.; Luo, H.; Yang, K.; Liu, G.; Zhang, R.; Li, X.; Ye, B. Formic Acid Production Using a Microbial Electrolysis Desalination and Chemical-Production Cell. *Bioresource Technol.* **2017**, *243*, 118–125.

70. Liang, D.; Peng, S.; Lu, S.; Liu, Y.; Lan, F.; Xiang, Y. Enhancement of Hydrogen Production in a Single Chamber Microbial Electrolysis Cell Through Anode Arrangement Optimization. *Bioresource Technol.* **2011**, *102*, 10881–10885.
71. Vilà-Rovira, A.; Puig, S.; Balaguer, M. D.; Colprim, J. Anode Hydrodynamics in Bioelectrochemical Systems. *RSC Adv.* **2015**, *5*, 78994–79000.
72. Manohar, A. K.; Mansfeld, F. The Internal Resistance of a Microbial Fuel Cell and Its Dependence on Cell Design and Operating Conditions. *Electrochim. Acta.* **2009**, *54*, 1664–1670.
73. Kim, K.; Zikmund, E.; Logan, B. E. Impact of Catholyte Recirculation on Different 3-Dimensional Stainless Steel Cathodes in Microbial Electrolysis Cells. *Int. J. Hydrogen Energ.* **2017**, *42*, 29708–29715.
74. Call, D. F.; Merrill, M. D.; Logan, B. E. High Surface Area Stainless Steel Brushes as Cathodes in Microbial Electrolysis Cells. *Environ. Sci. Technol.* **2009**, *43*, 2179–2183.
75. Zhou, Y.; Tang, L.; Liu, Z.; Hou, J.; Chen, W.; Li, Y.; Sang, L. A Novel Anode Fabricated by Three-Dimensional Printing for Use in Urine-Powered Microbial Fuel Cell. *Biochem. Eng. J.* **2017**, *124*, 36–43.
76. Hou, J.; Liu, Z.; Yang, S.; Zhou, Y. Three-Dimensional Macroporous Anodes Based on Stainless Steel Fiber Felt for High-Performance Microbial Fuel Cells. *J. Power Sources* **2014**, *258*, 204–209.
77. Zhang, Y.; Liu, L.; Van der Bruggen, B.; Yang, F. Nanocarbon Based Composite Electrodes and Their Application in Microbial Fuel Cells. *J. Mater. Chem. A.* **2017**, *5*, 12673–12698.
78. Mahdi Mardanpour, M.; Nasr Esfahany, M.; Behzad, T.; Sedaqatvand, R. Single Chamber Microbial Fuel Cell with Spiral Anode for Dairy Wastewater Treatment. *Biosens. Bioelectron.* **2012**, *38*, 264–269.
79. Zheng, S.; Yang, F.; Chen, S.; Liu, L.; Xiong, Q.; Yu, T.; Zhao, F.; Schröder, U.; Hou, H. Binder-Free Carbon Black/Stainless Steel Mesh Composite Electrode for High-Performance Anode in Microbial Fuel Cells. *J. Power Sources* **2015**, *284*, 252–257.
80. Girgis, B. S.; Khalil, L. B.; Tawfik, T. Activated Carbon from Sugar-Cane Bagasse by Carbonization in the Presence of Inorganic Acids. *J. Chem. Technol. Biot.* **1994**, *61*, 87–92.
81. Park, J. H.; Hwang, R. H.; Yoon, H. C.; Yi, K. B. Effects of Metal Loading on Activated Carbon on Its Adsorption and Desorption Characteristics. *J. Ind. Eng. Chem.* **2019**, *74*, 199–207.
82. Przepiorski, J.; Yoshida, S.; Oya, A. Structure of K_2CO_3-Loaded Activated Carbon Fiber and Its Deodorization Ability Against H_2S Gas. *Carbon* **1999**, *37*, 1881–1890.
83. Spahn, H.; Schlunder, E. U. Scale-Up of Activated Carbon Columns for Water-Purification, Based on Results from Batch Tests .1. Theoretical and Experimental Determination of Adsorption Rates of Single Organic Solutes in Batch Tests. *Chem. Eng. Sci.* **1975**, *30*, 529–537.
84. Guo, K.; Hidalgo, D.; Tommasi, T.; Rabaey, K. Pyrolytic Carbon-Coated Stainless Steel Felt as a High-Performance Anode for Bioelectrochemical Systems. *Bioresource Technol.* **2016**, *211*, 664–668.
85. Yang, J.; Cheng, S.; Sun, Y.; Li, C. Improving the Power Generation of Microbial Fuel Cells by Modifying the Anode with Single-Wall Carbon Nanohorns. *Biotechnol Lett.* **2017**, *39*, 1515–1520.

86. Erbay, C.; Pu, X.; Choi, W.; Choi, M.; Ryu, Y.; Hou, H.; Lin, F.; de Figueiredo, P.; Yu, C.; Han, A. Control of Geometrical Properties of Carbon Nanotube Electrodes Towards High-Performance Microbial Fuel Cells. *J. Power Sources* **2015**, *280*, 347–354.
87. Wang, J.; Li, M.; Liu, F.; Chen, S. Stainless Steel Mesh Supported Carbon Nanofibers for Electrode in Bioelectrochemical System. *J. Nanomater.* **2016**, *2016*, 1–5.
88. Francis, A. P.; Devasena, T. Toxicity of Carbon Nanotubes: A Review. *Toxicol Ind Health* **2018**, *34*, 200–210.
89. Tsai, H.; Hsu, W.; Liao, Y. Effect of Electrode Coating with Graphene Suspension on Power Generation of Microbial Fuel Cells. *Coatings* **2018**, *8*, 1–11.
90. Zhang, Y.; Mo, G.; Li, X.; Zhang, W.; Zhang, J.; Ye, J.; Huang, X.; Yu, C. A Graphene Modified Anode to Improve the Performance of Microbial Fuel Cells. *J. Power Sources* **2011**, *196*, 5402–5407.
91. Hou, J.; Liu, Z.; Li, Y.; Yang, S.; Zhou, Y. A Comparative Study of Graphene-Coated Stainless Steel Fiber Felt and Carbon Cloth as Anodes in MFCs. *Bioproc. Biosyst. Eng.* **2015**, *38*, 881–888.
92. Champavert, J.; Ben Rejeb, S.; Innocent, C.; Pontié, M. Microbial Fuel Cell Based on Ni-tetra Sulfonated Phthalocyanine Cathode and Graphene Modified Bioanode. *J. Electroanal. Chem.* **2015**, *757*, 270–276.
93. Johansson, K. O.; Head-Gordon, M. P.; Schrader, P. E.; Wilson, K. R.; Michelsen, H. A. Resonance-Stabilized Hydrocarbon-Radical Chain Reactions May Explain Soot Inception and Growth. *Science* **2018**, *361*, 997–1000.
94. Rim, K.; Kim, S.; Han, J.; Kang, M.; Kim, J.; Yang, J. Effects of Carbon Black to Inflammation and Oxidative DNA Damages in Mouse Macrophages. *Mol. Cell Toxicol.* **2011**, *7*, 415–423.
95. Thomson, M.; Mitra, T. A Radical Approach to Soot Formation. *Science* **2018**, *361*, 978–979.
96. Peng, X.; Chen, S.; Liu, L.; Zheng, S.; Li, M. Modified Stainless Steel for High Performance and Stable Anode in Microbial Fuel Cells. *Electrochim. Acta.* **2016**, *194*, 246–252.
97. Lamp, J. L.; Guest, J. S.; Naha, S.; Radavich, K. A.; Love, N. G.; Ellis, M. W.; Puri, I. K. Flame Synthesis of Carbon Nanostructures on Stainless Steel Anodes for Use in Microbial Fuel Cells. *J. Power Sources* **2011**, *196*, 5829–5834.
98. Singh, S.; Bairagi, P. K.; Verma, N. Candle Soot-Derived Carbon Nanoparticles: An Inexpensive and Efficient Electrode for Microbial Fuel Cells. *Electrochim. Acta.* **2018**, *264*, 119–127.
99. Peng, X.; Yu, H.; Wang, X.; Zhou, Q.; Zhang, S.; Geng, L.; Sun, J.; Cai, Z. Enhanced Performance and Capacitance Behavior of Anode by Rolling Fe_3O_4 into Activated Carbon in Microbial Fuel Cells. *Bioresource Technol.* **2012**, *121*, 450–453.
100. Wu, G.; Bao, H.; Xia, Z.; Yang, B.; Lei, L.; Li, Z.; Liu, C. Polypyrrole/Sargassum Activated Carbon Modified Stainless-Steel Sponge as High-Performance and Low-Cost Bioanode for Microbial Fuel Cells. *J. Power Sources* **2018**, *384*, 86–92.
101. Yellappa, M.; Sravan, J. S.; Sarkar, O.; Reddy, Y. V. R.; Mohan, S. V. Modified Conductive Polyaniline-Carbon Nanotube Composite Electrodes for Bioelectricity Generation and Waste Remediation. *Bioresource Technol.* **2019**, *284*, 148–154.

102. Peng, X.; Yu, H.; Wang, X.; Gao, N.; Geng, L.; Ai, L. Enhanced Anode Performance of Microbial Fuel Cells by Adding Nanosemiconductor Goethite. *J. Power Sources* **2013**, *223*, 94–99.
103. Long, X.; Wang, H.; Wang, C.; Cao, X.; Li, X. Enhancement of Azo Dye Degradation and Power Generation in a Photoelectrocatalytic Microbial Fuel Cell by Simple Cathodic Reduction on Titania Nanotube Arrays Electrode. *J. Power Sources* **2019**, *415*, 145–153.
104. Ying, X.; Shen, D.; Wang, M.; Feng, H.; Gu, Y.; Chen, W. Titanium Dioxide Thin Film-Modified Stainless Steel Mesh for Enhanced Current-Generation in Microbial Fuel Cells. *Chem. Eng. J.* **2018**, *333*, 260–267.
105. Jadhav, D. A.; Ghadge, A. N.; Ghangrekar, M. M. Enhancing the Power Generation in Microbial Fuel Cells with Effective Utilization of Goethite Recovered from Mining Mud as Anodic Catalyst. *Bioresource Technol.* **2015**, *191*, 110–116.
106. Liang, Y.; Feng, H.; Shen, D.; Long, Y.; Li, N.; Zhou, Y.; Ying, X.; Gu, Y.; Wang, Y. Metal-Based Anode for High Performance Bioelectrochemical Systems Through Photo-Electrochemical Interaction. *J. Power Sources* **2016**, *324*, 26–32.
107. Guo, K.; Donose, B. C.; Soeriyadi, A. H.; Prévoteau, A.; Patil, S. A.; Freguia, S.; Gooding, J. J.; Rabaey, K. Flame Oxidation of Stainless Steel Felt Enhances Anodic Biofilm Formation and Current Output in Bioelectrochemical Systems. *Environ. Sci. Technol.* **2014**, *48*, 7151–7156.
108. Shamsuddin, R. A.; Wan Daud, W. R.; Kim, B. H.; Md. Jahim, J.; Abu Bakar, M. H.; Wan Mohd Noor, W. S. A.; Mohamad Yunus, R. Electrochemical Characterization of Heat-Treated Metal and Non-Metal Anodes Using Mud in Microbial Fuel Cell. *Sains Malays.* **2018**, *47*, 3043–3049.
109. Guo, K.; Soeriyadi, A. H.; Feng, H.; Prévoteau, A.; Patil, S. A.; Gooding, J. J.; Rabaey, K. Heat-Treated Stainless Steel Felt as Scalable Anode Material for Bioelectrochemical Systems. *Bioresource Technol.* **2015**, *195*, 46–50.
110. Liang, Q.; Yamashita, T.; Yamamoto-Ikemoto, R.; Yokoyama, H. Flame-Oxidized Stainless-Steel Anode as a Probe in Bioelectrochemical System-Based Biosensors to Monitor the Biochemical Oxygen Demand of Wastewater. *Sensors* **2018**, *18*, 1–6.
111. Liang, P.; Zhang, C.; Jiang, Y.; Bian, Y.; Zhang, H.; Sun, X.; Yang, X.; Zhang, X.; Huang, X. Performance Enhancement of Microbial Fuel Cell by Applying Transient-state Regulation. *Appl. Energ.* **2017**, *185*, 582–588.
112. Ma, Q.; Pu, K.; Cai, W.; Wang, Y.; Chen, Q.; Li, F. Characteristics of Poly(3,4-ethylenedioxythiophene) Modified Stainless Steel as Anode in Air-Cathode Microbial Fuel Cells. *Ind. Eng. Chem. Res.* **2018**, *57*, 6633–6638.
113. Reut, J.; Öpik, A.; Idla, K. Corrosion Behavior of Polypyrrole Coated Mild Steel. *Synthetic Met.* **1999**, *102*, 1392–1393.
114. Sonawane, J. M.; Al-Saadi, S.; Singh Raman, R. K.; Ghosh, P. C.; Adeloju, S. B. Exploring the Use of Polyaniline-Modified Stainless Steel Plates as Low-Cost, High-Performance Anodes for Microbial Fuel Cells. *Electrochim. Acta.* **2018**, *268*, 484–493.
115. Sonawane, J. M.; Patil, S. A.; Ghosh, P. C.; Adeloju, S. B. Low-Cost Stainless-Steel Wool Anodes Modified with Polyaniline and Polypyrrole for High-Performance Microbial Fuel Cells. *J. Power Sources* **2018**, *379*, 103–114.

116. Sonawane, J. M.; Ghosh, P. C.; Adeloju, S. B. Electrokinetic Behaviour of Conducting Polymer Modified Stainless Steel Anodes During the Enrichment Phase in Microbial Fuel Cells. *Electrochim. Acta.* **2018**, *287*, 96–105.
117. Phonsa, S.; Sreearunothai, P.; Charojrochkul, S.; Sombatmankhong, K. Electrodeposition of MnO_2 on Polypyrrole-Coated Stainless Steel to Enhance Electrochemical Activities in Microbial Fuel Cells. *Solid State Ionics* **2018**, *316*, 125–134.
118. Pugal Mani, S.; Rajendran, N. Corrosion and Interfacial Contact Resistance Behavior of Electrochemically Nitrided 316L SS Bipolar Plates for Proton Exchange Membrane Fuel Cells. *Energy* **2017**, *133*, 1050–1062.
119. Hoffman, A. B.; Suresh, S.; Evitts, R. W.; Kennell, G. F.; Godwin, J. M. Dual-Chambered Bio-Batteries Using Immobilized Mediator Electrodes. *J. Appl. Electrochem.* **2013**, *43*, 629–636.
120. Guo, K.; Soeriyadi, A. H.; Patil, S. A.; Prévoteau, A.; Freguia, S.; Gooding, J. J.; Rabaey, K. Surfactant Treatment of Carbon Felt Enhances Anodic Microbial Electrocatalysis in Bioelectrochemical Systems. *Electrochem. Commun.* **2014**, *39*, 1–4.
121. Tang, Y. L.; Bi, X. W.; Sun, H.; Fu, J. X.; Peng, M.; Zou, H. F. Effect of Anode with Pretreatment on the Electricity Generation of a Single Chamber Microbial Fuel Cell. *Adv. Mater. Research.* **2010**, *156-157*, 742–746.
122. Liang, Y.; Feng, H.; Shen, D.; Li, N.; Guo, K.; Zhou, Y.; Xu, J.; Chen, W.; Jia, Y.; Huang, B. Enhancement of Anodic Biofilm Formation and Current Output in Microbial Fuel Cells by Composite Modification of Stainless Steel Electrodes. *J. Power Sources* **2017**, *342*, 98–104.

Chapter 9

Studies on Controlled Protein Folding *versus* Direct Electron-Transfer Reaction of Cytochrome C on MWCNT/Nafion Modified Electrode Surface and Its Selective Bioelectrocatalytic H_2O_2 Reduction and Sensing Function

Annamalai Senthil Kumar,[*,1,2] Nandimalla Vishnu,[1] and Bose Dinesh[1]

[1]Nano and Bioelectrochemistry Research Laboratory, Department of Chemistry, School of Advanced Sciences, Vellore Institute of Technology, Vellore-632 014, India

[2]Carbon dioxide Research and Green Technology Centre, Vellore Institute of Technology, Vellore-632 014, India

[*]E-mail: askumarchem@yahoo.com; askumar@vit.ac.in.

A systematic study on the relationship between direct electron-transfer (DET) function and the folding effect of the Cytochrome c (Cytc) protein has been performed. The aging effect of the Cytc protein-pH 7 phosphate buffer solution (PBS) at a fixed temperature of 5 °C was investigated using spectroscopic and electrochemical techniques. A strong correlation between the aging and electron-transfer (ET) function activity of the Cytc protein was observed. UV-Vis spectroscopic characterization of the aged Cytc protein showed a significant shift in the absorption signal from 405 nm to 413 nm due to the geometric transformation of the heme complex from rhombohedral to tetrahedral symmetry. Similarly, the FTIR and fluorescence spectroscopic results reveal cleavage of the amide-I and II bonds and opening of tryptophan amino-acid residues after aging up to 18 days. These results are similar to the literature reports on denaturants assisted by folding and unfolding behaviors of Cytc protein. A Cytc-pH 7 PBS aged over 18 days was chemically modified as a bioelectrode by successive drop casting of MWCNT-ethanol suspension (1st layer), Cytc-pH PBS (2nd layer), and 1% Nafion (Nf) solution (3rd layer) on GCE (GCE/MWCNT@Cytc-Nf), and the DET activity was studied. The protein electrode showed a highly stable and well-defined redox peak at $E^{0\prime} = -0.31 \pm 0.02$ V vs Ag/AgCl and a Cytc active surface excess of 29.05 nmol cm^{-2} in N_2-purged pH 7 PBS. The redox peak is found to be a surface-confined and proton-coupled ET reaction in characteristic. Furthermore, the GCE/MWCNT@Cytc-Nf system showed a highly stable and efficient response for the bioelectrocatalytic reduction of H_2O_2 demonstrated by cyclic voltammetry (CV) and rotating disc electrode (RDE) techniques in N_2-

© 2020 American Chemical Society

purged pH 7 PBS. Bio-electrochemical sensing of H_2O_2 was tested by amperometric i-t and flow injection analysis (FIA) methods showing very stable sensing currents with a detection limit value of 1.6 μM. The following three factors are key for the successful development of an ET active Cytc-bioelectrode system: (i) Proper unfolding; (ii) a cooperative π-π, ionic, covalent, hydrophobic, hydrophilic, and hydrogen bondings with an underlying electrode; and (iii) proper orientation of the protein.

Introduction

Understanding the relationship between the structure and electron-transfer (ET) functionality of proteins and enzymes is a continued research interest in the interdisciplinary areas of chemistry, biochemistry, and biomedical systems. Cytochrome c (Cytc), a water-soluble protein (104 residues), functions as an ET chain associated with adenosine triphosphate synthesis in mitochondria (*1*). It has been reported that that ET reaction and the functionalities of Cytc are controlled by the protein-folding process (*2–8*), wherein the polypeptide chain of the protein alters its functional 3D structure (*1–4*). In general, the following structural and chemical changes happen while the Cytc protein is involved in the folding process: (i) Alterations in the central geometry of the heme complex where Methionine-80 and histidine-18 serve as axial ligands of the central Fe (i.e., geometric transformation from rhombohedral [asymmetric] to tetrahedral [symmetric] crystalline structure) (*5, 9, 10*); (ii) Formation of disulphide bonds, free thiol functional groups (-SH) of the amino acid reactions, and a new covalent link (-S-S-) (*11–13*); (iii) hydrophilic and hydrogen-bondings of water molecules interacting with the protein network (*14, 15*); (iv) hydrophobic (including π-π interaction) interactions (*16–19*); and (v) ionic interactions (each Cytc carries 9+ charges) (*20, 21*). The following are various folding forms of a protein: (i) the primary structure, which is a totally unfolded linear amino acid network; (ii) the secondary structure, which is a first step of the protein folding via the intramolecular hydrogen bonding process (α- and β-helix structures); (iii) the tertiary structure, which is a structure stabilized by hydrophobic and hydrophilic interactions; and (iv) the quaternary structure, which is an energetically lowered and completely folded structure (native form). Among these forms, the quaternary form of the Cytc protein structure has been determined to be an effective system for functional applications (*1, 2, 3, 5, 6, 9, 10, 14, 16*). Indeed, the quaternary structure is not non-amenable for molecular wiring and ET functionalities owing to the deeply shielded and insulated nature of the out shell protein structure. In fact, voltammetric studies of Cytc on solid electrodes like Au (*22*), Ag (*23*), SnO_2, In_2O_3, and Pt (*24*) have shown a poor redox feature of the cytc-heme-Fe(III)/Fe(II) system, which is due to bioincompatibility and the completely folded nature of the protein.

In order to improve the molecular wiring and ET property of Cytc protein, Cytc self-assembled monolayer systems are prepared using gold underlying electrodes that have been prefunctionalized with the following monolayers: 4-mercaptopyridine; 4,4′-dipyridyl disulphide (*25, 26*); N-acetylcysteine (*27*); alkanethiols terminated in trimethylammonium, sulfonate, methyl, amine, and carboxylic acid groups (*28*); 2-mercaptoethanesulfonate, 2-mercaptoethanol, and mercaptopropionic acid (*29*); and 11-mercapto-1-undecanoic acid (*30, 31*). In all of these cases, independent molecular interactions based on either hydrophobic (*25, 26, 28*), hydrophilic (*28*), ionic, or covalent (*28–30*) bonding were turned for the Cytc self-assembling. For instance,

$Si(CH_3)_{11}N^+(Me)_3$ (Standard electrode potential of $E^{o\prime} = 0.13$ V vs Ag/AgCl; surface excess of $\Gamma_{Cytc} = 16{\times}10^{-12}$ mol com^{-2}), $Si(CH_3)_{11}N^+(Me)_3$+$Si(CH_3)_{11}SO_3^-$ ($E^{o\prime} = -0.1$ V vs Ag/AgCl) (*28*), or $Si(CH_3)_{11}COOH$ ($E^o = 0.02$ V vs Ag/AgCl and $\Gamma_{Cytc} = 10{\times}10^{-12}$ mol com^{-2}) (*29*) functionalized Au systems in association with ionic and covalent bondings of Cytc protein were used. Meanwhile, Cytc protein was denatured using denaturants like urea (*31*), guanidine hydrochloride (GnHCl) (*32*), surfactants (*33*), acid and alkaline solutions (*30*), and thermal methods (*34*), which may rapture or collapse and unfold the basic network structure. For instance, the methionine link is removed from the heme prosthetic site by urea and GnHCl treatment (*20*), which have also been used for the ET studies.

There are several key questions relating to the structure, orientation, and ET of Cytc protein have been unanswered in the current literature. What is the true redox potential of Cytc protein when it is associated with a cooperative effect of π-π, covalent,and hydrogen bonding and ionic-interaction, hydrophobic, and hydrophilic bonding? Note that the redox potential of the Cytc has been studied as an open-circuit potential using the voltammetric titration method, where solid electrodes like Au are exposed to a dilute solution of Cytc (*35, 36*). The above experimental data could not be considered as a true value owing to the bioincompatible nature of the protein when interacting with metal electrodes. What is the relationship between protein folding and molecular wiring? What is the structural orientation of Cytc on the ET function? What is the influence of aging pure Cytc solution and its unfolding behaviour? Is it possible to develop a Cytc protein–stabilized bioelectrode with a cooperative effect of covalent, π-π, ionic, and hydrogen bonding with hydrophilic and hydrophobic interactions? In this research, we addressed all the above questions. A Cytc-protein chemically modified electrode that consists of a cooperative interactions based on ionic, π-π, hydrophobic, hydrophilic, covalent, and hydrogen bondings between the underlying electrode surface and Cytc protein has been successfully developed. For the first time in the literature report, a freshly prepared Cytc in a pH-7 phosphate buffer solution (PBS) was subjected to aging for different time lengths (up to 80 days) without any added surfactants and denaturing agents used as an optimal unfolded protein for the direct electron transfer (DET) studies.

Based on the experimental results, an optimal underlying electrode that is composed of an anionic exchanging polymer, Nafion (Nf), and a carboxylic acid–functionalized multiwalled carbon nanotube (MWCNT) modified on a glassy carbon electrode (GCE), designated as GCE/MWCNT-Nf, has been introduced for the effective molecular wiring and ET reaction of Cytc protein (Scheme 1). The GCE/MWCNT-Nf provides the following bio-mimicking support: (i) ionic interaction as the sulfonic acid in Nf interacts with the positive charge of the Cytc protein; (ii) covalent bonding as the carboxylic acid functional group of MWCNT links covalently with amino-terminal protein sites via amide-II linkage formation (*28, 29*); (iii) hydrogen bonding and hydrophilic interaction as the MWCNT-COOH and Nf-$SO_3^-.H_3O^+$ support for that; (iv) hydrophobic interaction as the -CF_2-CF_2- core of Nf and the basal plane of MWCNT can provide the support; and (v) π-π interaction as bonding occurs between sp^2 carbon of MWCNT and aromatic units of protein and amino acid sequences (phenylalanine, tryptophan, and tyrosine). Carbon nanomaterial–based chemically modified electrodes (CMEs) such as a single-walled carbon nanotube (SWCNT) (*37*), Au_{nano}-chitosan-CNT (*38*), chitosan-ionic liquid-CNT (*39*), Au_{nano}-ionic liquid-MWCNT (*40*), DNA-MWCNT (*40*), polyaniline-grafted MWCNT (*41*), graphene oxide-Au_{nano}-MWCNT-Nf (*42*), and pristine MWCNT (*43*) have been used for the Cytc protein DET reaction. In tthese cases, denatured Cytc proteins prepared and treated using a surfactant (*41*), alkaline solution (*40*), organic solvent

(*40*), electrochemical deposition condition (*40*), and ultrasonic treatment method have been used and showed electrochemical parameters of $E^{o\prime} = -0.25$ to 0 V vs Ag/AgCl with $\Gamma_{Cytc} = 2—10\times10^{-12}$ mol cm^{-2}. A new Cytc protein chemically modified electrode, GCE/MWCNT-Nf@Cytc, is introduced in this work and has shown a stable and well-defined redox peak of $E^{o\prime} = -0.31\pm0.02$ V vs Ag/AgCl with a $\Gamma_{Cytc} = 29.05$ nmol cm^{-2}, which is about 2–10,000 times higher than the previously reported Cytc-based electrochemical systems (*25–30*). As a functional model, bioelectrocatalytic reduction of hydrogen peroxide has been studied. By the end of this chapter, possible utility of this new Cytc bioelectrode as an electrochemical detector (ECD) for flow injection analysis (FIA) without any fouling problems will be demonstrated.

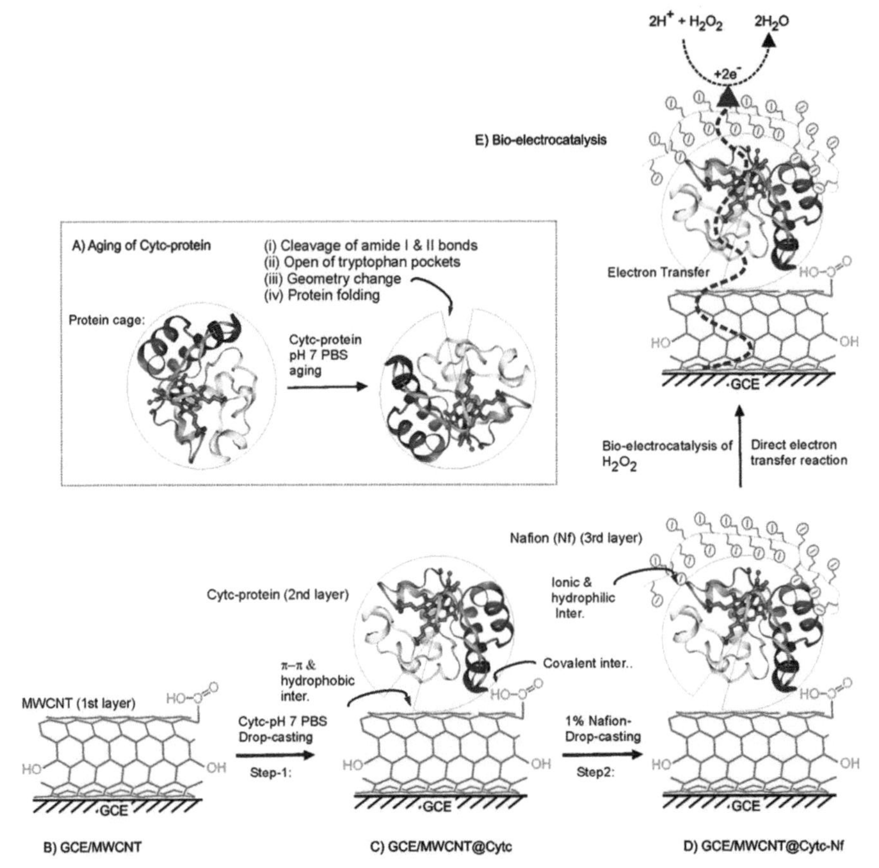

Scheme 1. Illustration for (A) aging of Cytc-protein; (B–D) preparation of GCE/MWCNT@Cytc-Nf bioelectrode by successive drop casting of MWCNT (1st layer), aged Cytc-protein solution (2nd layer), and 1% Nf solution (3rd layer); and (E) Bioelectrocatalysis of H_2O_2 reduction by DET reaction.

Experimental Section

Reagents and Materials

Cytc from a horse heart, a pristine MWCNT (90% carbon basics, outer diameter 10-15 nm and inner diameter 0.1-10 nm), and a 5-wt. % Nf-lower aliphatic alcohol mixture solution were purchased from Sigma-Aldrich (USA) and used as received. All reagents were analytical grade reagents used without any further purification. 0.1-M PBSs of varying pHs from 1 to 11 were prepared by mixing standard stock solutions of Na_2HPO_4 and NaH_2PO_4 and adjusting the pH by the addition of either 0.1 M of H_3PO_4 or 0.1 M of NaOH. Prior to each electrochemical experiment, all the solutions were deoxygenated with extra pure N_2 gas for about 15±2 min. Aqueous solutions were prepared using deionized and alkaline potassium permanganate distilled water (designated as DD water).

Apparatus

The electrochemical measurements were carried out using a CHI 440B workstation. A conventional three-electrode cell using a GCE as the working electrode (area of 0.0707 cm^2), saturated Ag/AgCl as a reference electrode, and Pt wire as a counter electrode was used. Hydrodynamic amperometric measurements were performed using a CHI 760D electrochemical workstation (Austin, TX, USA). The FIA system consisted of a Hitachi L-2130 pump delivery and a Rehodyne model 7125 sample injection valve (20 µL loop) with interconnecting Teflon tubes and a conventional electrochemical FIA cell (BAS, USA). For FT-IR, an IR-Afinity-1 (Shimadzu, Japan) instrument was used. For IR measurements, Cytc samples were dried in a glass plate and tested by the KBr pellet method. Raman spectroscopic analysis was performed using AZILTRON, PRO 532 (USA) with a 532-nm laser excitation source. An Ocean Optics instrument (JAZ-EL200XR1) was used for UV—Vis measurements.

Procedures

A carboxylic acid–functionalized MWCNT was prepared by treating a commercial MWCNT sample (pristine MWCNT) with 15 M of HNO_3 (*44*). A stock solution of MWCNT was prepared by dispersing 2 mg of a MWCNT in 500 µL of absolute ethanol followed by 5 min of sonication. 5 µL of the MWCNT-ethanolic suspension was drop casted on the cleaned GCE surface, and the resulting GCE/MWCNT was dried at room temperature (RT) (25 ± 2 °C) for 3(±0.5) min and cyclic voltammetry (CV) was run at the potential of −0.4 to 1.2 V in a pH of 7 N_2 saturated PBS to check the background current response. The electrode was then air dried for 2 min. A Cytc-pH 7 PBS stock solution was prepared by dissolving 5 mg of Cytc solid in 500 µL of pH 7 PBS and optimally stored for a month at RT. This solution was tested for aging studies. For chemically modified electrode preparation, 5 µL of Cytc-pH 7 PBS was drop casted over a GCE/MWCNT electrode where it was dried for 15 min. Finally, 5 µL of 1% Nf-ethanolic solution was casted on the surface and dried for 5(±1) min at RT. The resulting electrode was designated as GCE/MWCNT@Cytc-Nf (Scheme 1). All other control Cytc-immobilized electrodes were prepared in a similar way with 2 mg and 500 µL of ethanolic suspension of carbon nanomaterials. Prior to the electrochemical experiments, the freshly prepared GCE/MWCNT@Cytc-Nf was placed in 10 mL of a 0.1-M, pH-7 PBS (N_2 purged)–containing electrochemical cell and then potential cycled continuously for ten segments in a potential range of -0.7 V to +0.7 V vs Ag/AgCl at a scan rate (v) of 50 mV s^{-1}.

Results and Discussion

Aging Effect of Cytc Protein

This experiment was carried out by exposing a diluted 2-mg mL^{-1} Cytc-pH 7 PBS solution for varying days (1–80 days) in a refrigerator at a fixed temperature, T= 5±2 °C. A portion of the stock solution with proper dilution using a pH 7 PBS (N_2-purged) was subjected to various spectroscopic and electrochemical studies. Figure 1A is a typical UV-Vis response of Cytc-pH 7 PBS aged at different lengths (0–18 days). After four days, a strong absorption peak at $\lambda_{max} = 405$ nm was recorded; after that, a positive shift to $\lambda_{max} = 413$ nm was noticed. In addition, a weak absorption signal at $\lambda_{max} = 529$ nm was also observed. It was reported that the native form of Cytc protein has characteristic absorption bands at $\lambda_{max} \sim 408$ nm and 525 nm corresponding to the electronic transition of porphyrins $\pi \rightarrow \pi^*$ (a_{1u}, $a_{2u} \rightarrow e_g$; Soret band) and conformational change in the hydrophobic heme-binding region, respectively (*9, 45*). Based on the literature reports (*5, 9, 10, 45*), it has been determined that transformation in the planar geometry of the heme site (due to "heme complex-protein" interaction and loss of M80 ligand) from rhombohedral to tetrahedral is happened. This observation is responsible for the absorption change.

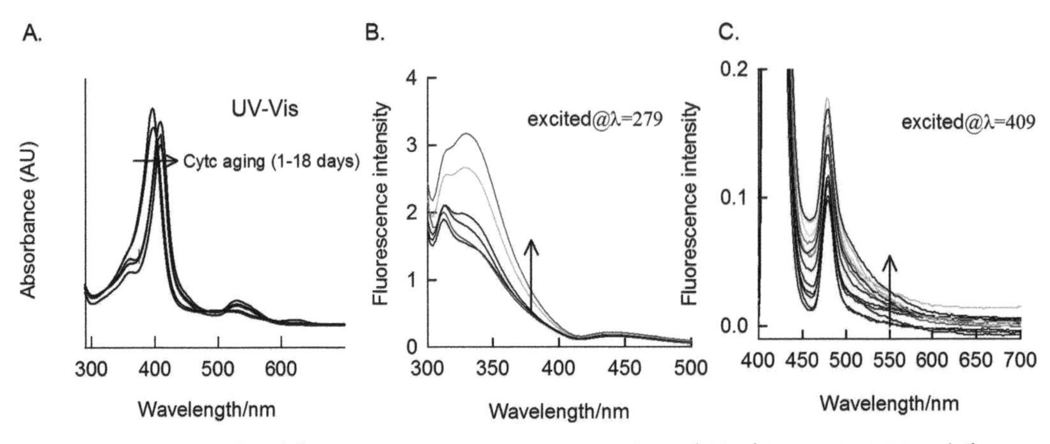

Figure 1. UV-Vis (A) and fluorescence spectroscopic responses (B and C) of Cytc-pH 7 PBS at different aging times (0–18 days). Fluorescence samples were excited at $\lambda_{max} = 279$ nm (B) and $\lambda_{max} = 409$ nm (C).

Figure 1B and 1C are typical fluorescence spectroscopic responses of Cytc protein-pH 7 PBS aged for diffent lengths of time (0–18 days) that have been excited at $\lambda = 279$ nm and 409 nm. A systematic increase in the fluorescence signal upon increasing the aging time was noticed. This observation corresponds with the literature report on Cytc protein denatured by urea (*31*), GdnHCl (*32*), and other denaturants (*33, 34*). In general, tryptophan amino acid residues in the Tyr 46, Tyr 48, Tyr 52, Tyr 62, Tyr 70, and Tyr 107 sites are responsible for the fluorescence activity of Cytc protein (*46*). It is noteworthy that the native form of Cytc protein is not fluorescence active in nature due to the excitation energy transfer of the heme complex. Thus, the observed fluorescence signal can be considered as proof of the unfolding of the Cytc protein.

To further understand the unfolding process, the aged Cytc protein solutions were subjected to FTIR studies. As can be seen in Figure 2, significant IR bands at 2950 cm^{-1} (Ar-CH=CH-), 1654 cm^{-1} (Amide-I, H_2N-CO-R), 1537 cm^{-1} (Amide II, -NH-CO-R), and 1410 and 1261 cm^{-1} (Aromatic-CH=CH-) due to characteristic functional groups of the Cytc protein were noticed (*36*). It can be seen that the amide I and II bands are getting weaker, whereas, the aromatic -CH=CH- signals are becoming stronger upon aging the Cytc-protein. Based on collective experimental results from the UV-Vis, fluorescence, and IR spectroscopic studies, the following conclusions can be drawn:

(i) There is a significant unfolding of the Cytc protein solution upon the aging process. Loss of M80 ligand (6th coordination site) and hence changes in the central geometric structure of the heme complex from rhombohedral to tetrahedral are probable reasons for the observation (*5, 9, 10*);
(ii) Cleavage of the amide I and II bonds upon aging; and
(iii) Opening of tryptophan pockets and exposure of hydrophobic aromatic sites (-CH=CH-) (*46*). Hydration of the protein is an accountable factor for the result.

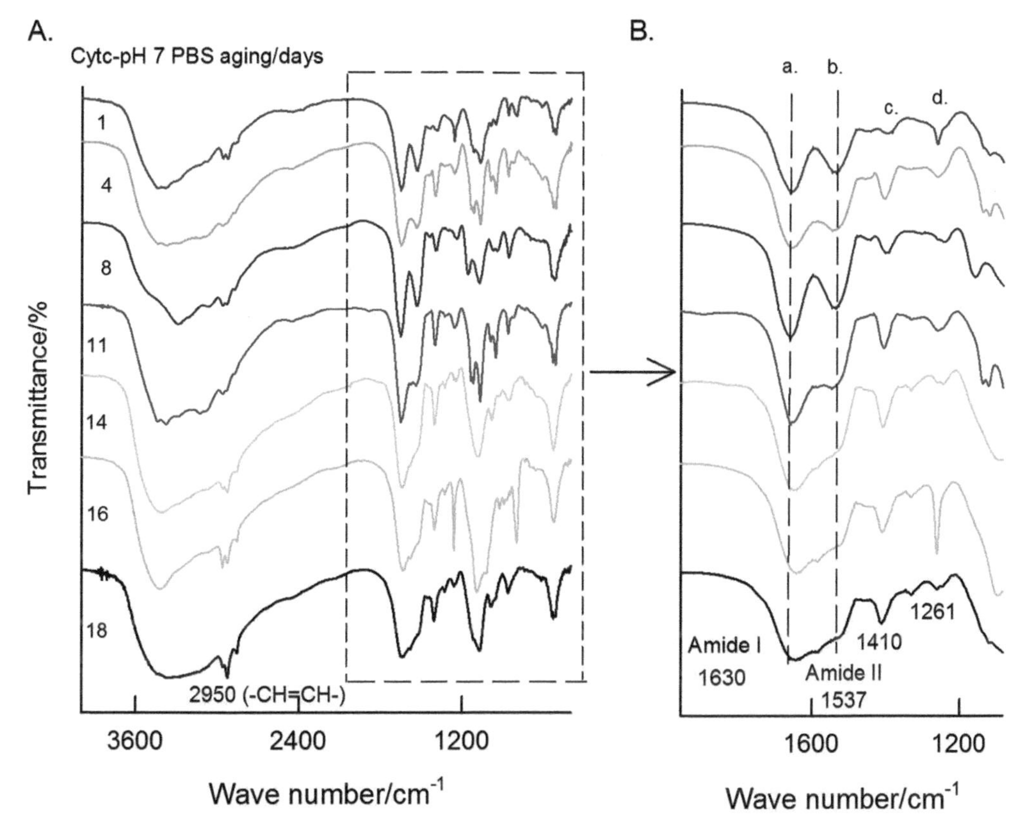

Figure 2. (A) FT-IR response of Cytc-pH 7 PBS at different aging days (as dried KBr pellet). (B) Expanded view of the FT-IR spectrum region between 1000 and 2200 cm^{-1}.

Overall, protonation of Cytc protein and the simultaneous opening of heme crevice and pockets (unfolding) make the central metal iron of heme readily accessible for ET and proton transfer reactions (Scheme 1).

DET Behavior of Cytc-Protein on MWCNT

The relationship between the Cytc-protein unfolding and its DET characteristics, wherein, the protein or enzyme is capable of shuttling its electrons to the electrode and electrolyte interface without any external transducer or mediator (which is similar to natural system), was investigated by CV technique. For this investigation, a Cytc-protein solution aged for 1–80 days was subjected to electrochemical investigation after modification as a MWCNT@Cytc-Nf chemically modified GCE (GCE/MWCNT@Cytc-Nf). Figure 3A is a typical CV response of GCE/MWCNT@Cytc-Nf in an N_2-purged pH 7 PBS. A regular increase in the redox peak current against an increase in the aging time was noticed. Figure 3B is a plot of anode and cathode peak currents against the aging time of the solutions. A steep rise in the peak current signal up to 18 days after that near plateau in the response was observed. It is interesting to notice that the observation is parallel to the response noticed with the spectroscopic characterization (Figure 1 and 2) and other literature reports on the denatured structure of the Cytc systems (*31–34*). Based on the results, it can be concluded that aging of Cytc resulted in a specific unfolding structure (a secondary or tertiary form) with exposure of the heme crevice and tryptophan pockets along with the opening of hydrophobic aromatic sites (-CH=CH-) (*5*, *9*, *10*, *46*). These structural alterations are important conditions for the DET reaction of Cytc protein on an MWCNT surface (Scheme 1).

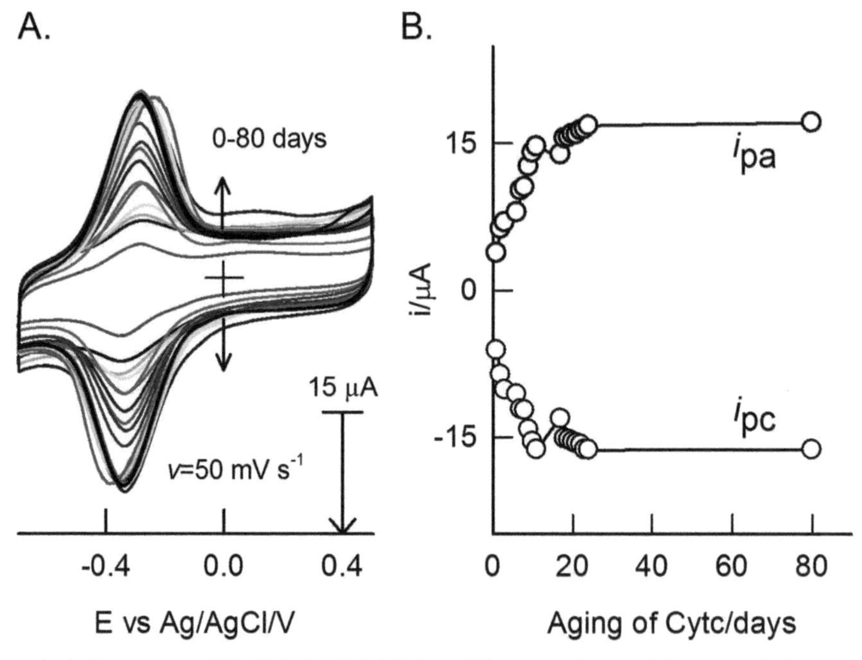

Figure 3. (A) CV response of GCE/MWCNT@Cytc-Nf prepared using different aged Cytc-pH 7 PBS samples in N_2-purged pH 7 PBS at a scan rate of 50 mV s^{-1} (B). Plot of peak current vs aging of Cytc protein (days).

To understand the nature of the ET reaction, the GCE/MWCNT@Cytc-Nf was subjected to a scan rate effect as shown in Figure 4A. The bioelectrode was prepared using a 20-day aged Cyt-pH 7 PBS. A regular increase in the i_{pa} and i_{pc} against incremental rise in the scan rate (v) was noticed. A plot of i_{pa} and i_{pc} vs the scan rate showed that a straight line starts from the origin indicating a perfect surface-confined ET feature of the Cytc-modified electrode. Furthermore, with the increase in scan rate, the oxidation peak shifts more positively, while the reduction peak shifts to more negative potentials. This follows the Laviron theory (*47*). The anodic (E_{pa}) and cathodic (E_{pc}) peak potentials are plotted against the logarithm of the scan rate (data not enclosed). A graph of E_{pa} and E_{pc} = f (log υ) yields a straight line with slope values of 68 and 81 mV dec^{-1}, respectively, for the anodic ($S_a = -2.3RT/\alpha nF$) and cathodic ($S_c = 2.3RT/[1-\alpha]nF$) portions. Based on the equation, $S_a/S_c = \alpha/(1-\alpha)$, wherein S_a and S_c are the anodic and cathodic slope values and is the transfer coefficient, the α is calculated as "0.51." This value indicates a symmetric energy barrier for the ET reaction. The ET rate constant, k_s, is calculated as 0.931 s^{-1} from the following Laviron model equation for a surface-controlled process, wherein peak-to-peak separation, $\Delta E_p > 0.200$ V is (*47*):

$$\log k_s = \alpha \log (1 - \alpha) + (1 - \alpha) \log\alpha - \log (RT/nFv) - \alpha(1 - \alpha) nF\Delta E_p/2.3RT \quad (1)$$

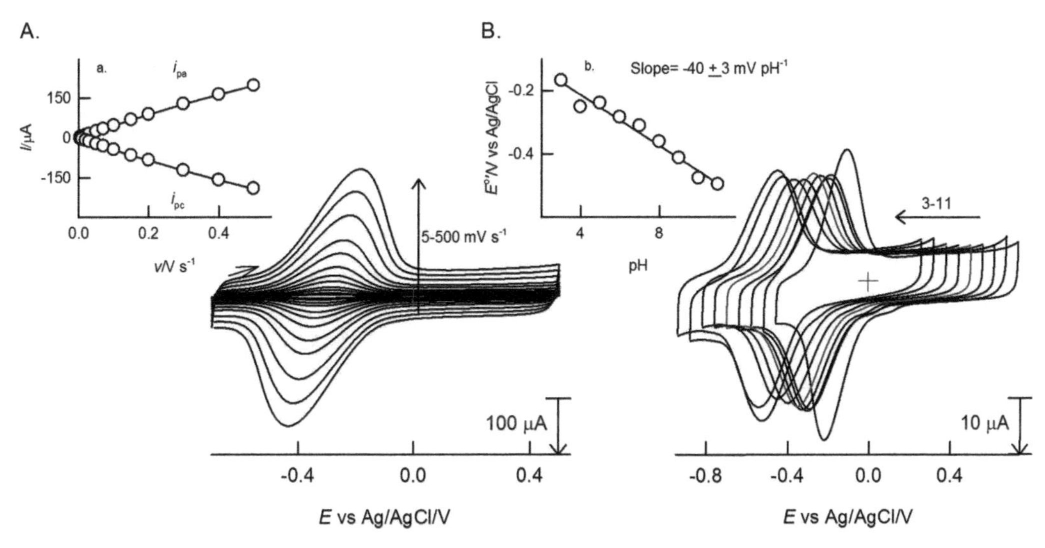

Figure 4. (A) Effect of scan rate (5-500 mV s^{-1}) in pH 7 PBS (N_2 purged); Inset: A plot of i_{pa} and i_{pc} vs scan rate. (B) Effect of solution pH (3–11) on GCE/MWCNT@Cytc-Nf; Inset: A plot of $E^{0\prime}$ vs pH.

The obtained k_s value is higher than that of the Cytc- and L-cysteine-modified electrode (0.28 s^{-1}) (*48*), the colloidal Cytc-immobilized Au-modified carbon paste electrode (1.21 s^{-1}) (*49*), and the Cytc-adsorbed amine-functionalized silica thin films (1.33 s^{-1}) (*50*). These results highlight the excellent DET features of Cytc on MWCNT/Nf.

In order to study the influence of pH on the ET reaction, the GCE/MWCNT@Cytc-Nf was subjected to CV studies in aqueous buffer solutions of various pHs in the range between 1 and 11 (N_2-purged solutions) at a fixed scan rate of 50 mV s^{-1} (Figure 4B). The redox peak responsible for the direct electrochemistry of Cytc is shifted negatively upon an increase in the solution's pH. A plot of formal potential ($E^{o\prime}$) against the pH is linear as exemplified in the inset to Figure 4B. This observation indicated that the ET reaction between the immobilized Cytc and the electrode is accompanied by proton-coupled ET behavior. A slope value, -40±2 mV pH^{-1}, which is less than the expected Nernstain value, -58 mV pH^{-1}, for a reversible proton-coupled ET reaction was observed. Such a low slope value was already reported with many other heme protein–modified electrodes, and it was attributed to the influence of the protonation states of trans ligands to the heme iron and amino acids around the heme and the protonation of the water molecule coordinated to the central position (*51*).

Physicochemical Characterizations

In order to understand the nature of covalent, π-π, ionic, hydrogen, hydrophobic, and hydrophilic bonding and structural orientation, Cytc-protein CMEs were prepared at various conditions. In this connection, different combinations of MWCNT (covalent, π-π, and hydrogen bonding) and Nf (ionic and hydrogen bonding) were combined with Cytc protein to prepare the CMEs. In first set of experiments, Cytc-modified electrodes without MWCNT or Nf (i.e., GCE/Cytc and GCE/Cytc-Nf) were prepared and studied independently in a pH 7 PBS (Figure 5A and B). Both electrodes showed an unstable and irreversible ($i_{pc}/i_{pa} \gg 1$, i_{pc} = cathodic peak and i_{pa} = anodic peak currents) reduction peak current response, which appeared at $E_{pc} = -0.38$ V versus Ag/AgCl, which is analogous to the irreversible reduction of the Cytc-heme-Fe(III)/Fe(II) site. The unstable CV peak current response indicates unfavorable orientation and bioincompatibility of the Cytc on bare GCE. In next set of experiments, mixtures of MWCNT+Cytc and MWCNT+Cytc+Nf suspensions were drop coated GCEs as GCE/MWCNT+Cytc and GCE/MWCNT+Cytc+Nf and subjected to CV examination. As seen in Figure 5C and D, a broad and irreversible peak at $E_{pc} = -0.38$ V was observed. Note that although all the bonding conditions are satisfactorily fulfilled by using MWCNT and Nf, there is no well-defined formation on the observed redox peak. It is likely that Cytc protein is not properly oriented in association with the various bondings. Furthermore, a slight alteration in the preparation condition, wherein the MWCNT+Cytc mixture (Figure 5E) and MWCNT (Figure 5F and G) were first drop casted on different GCEs followed by casting of Nf (Figure 5E), Cytc-Nf (Figure 5F), and Cytc (Figure 5G) as a second layer, were subjected to CV studies. Such a modification procedure is an example of a bilayer orientation for Cytc with the bonding matrix. Considerable improvement in the redox peak, but with a relatively less stable response, was noted. In a final trial experiment, MWCNT (1st layer), Cytc-pH 7 PBS (>18 days aged, 2nd layer), and an Nf solution (3rd layer) were successively drop casted on GCE as GCE/MWCNT@Cytc-Nf, and the CV responses were studied as shown in Figure 5H. The previously mentioned preparation procedure is a typical model for the sandwich arrangement of Cytc protein covering π-π, covalent, hydrophilic, hydrophobic, and hydrogen bonding interactions in the bottom layer using carboxylic-functionilized MWCNT and ionic, hydrophilic, hydrophobic, and hydrogen bonding interactions in the top layer using an Nf polymer. This time an excellent redox peak for the surface-confined ET feature of Cytc-Fe(III) and Fe(II) was noticed highlighting the optimal way to prepare the Cytc chemically modified electrode for bioelectroanalytical applications. These results reveal the importance of the protein folding, structural orientation, and cooperative bondings of Cytc with MWCNT and Nf for the facile ET reaction. Calculated apparent standard electrode potential, $E^{o\prime}$, and surface excess, Γ_{Cytc} values, are $-0.31(\pm0.02)$ V vs Ag/AgCl and $29.05(\pm2.1) \times 10^{-9}$ mol cm^{-2}, respectively. A relative standard deviation (RSD) of 10 successive CV cycles of the Cytc-modified electrode is 4.4%. Similarly, three independent preparations of the modified electrode resulted in a peak current signal alteration of RSD = 7.1%, which highlights the good stability and reproducibility of the GCE/MWCNT@Cytc-Nf.

Figures 6A–C are comparative Raman, UV-Vis, and IR spectroscopic characterizations of MWCNT@Cytc-Nf with its control samples. In the Raman spectroscopic analysis, both the systems showed qualitatively similar patterns (i.e., D and G bands) (*44*). Indeed, the ratio of peak signal, I_D/I_G, is found to be varied as 0.93 and 0.73, respectively. The increment in the I_D/I_G value indicates a reduction in the graphitic structure, which may be due to the conversion of the sp^2 to sp^3 sites on the carbon. Similarly, a marked shift in the soret band of Cytc in UV-Vis (405 →413 nm) and

IR (1648→1654 cm-1 and 1531→1537 cm^{-1}) were noticed. These observations support strong bonding between the Cytc protein's functional groups and the MWCNT and Nf via cooperative interactions (Scheme 1D).

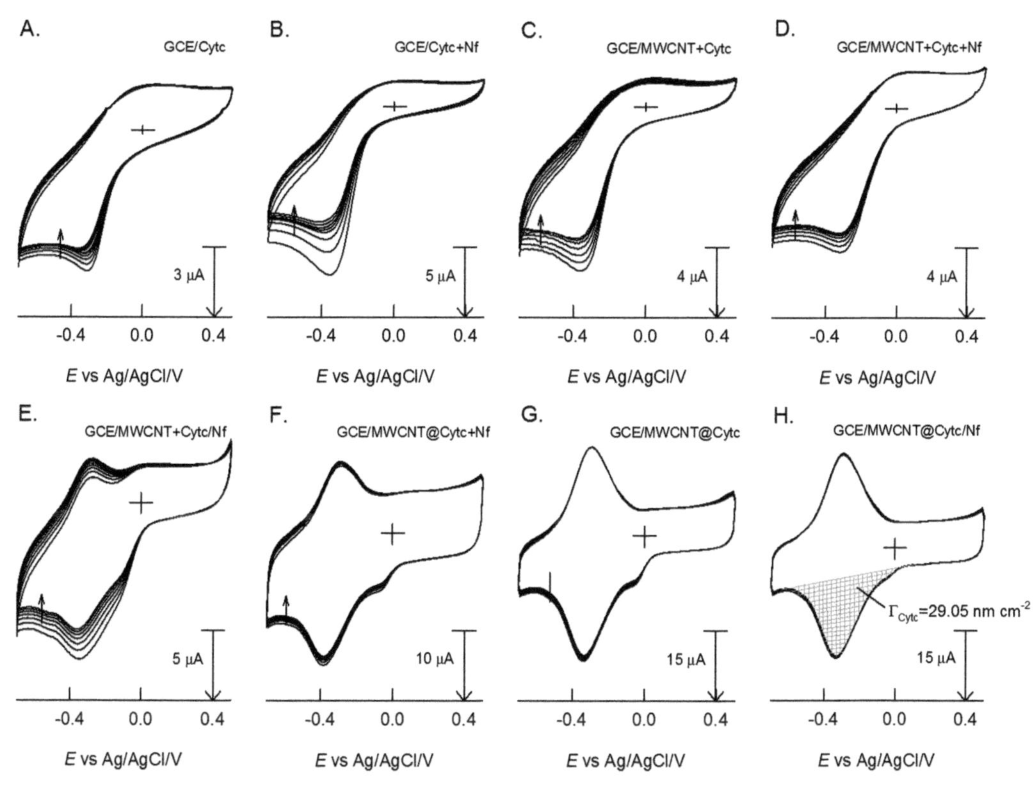

Figure 5. Continuous CV responses of various Cytc-immobilized GCE and MWCNT and Nf-modified electrodes in a pH-7 PBS (N_2 purged) at a scan rate of 50 mV s^{-1}. (A–D) are simple drop casting of Cytc or a mixture Cytc and Nf, MWCNT and Cytc, or MWCNT and Cytc and Nf, respectively. (E–G) are drop-castings of two layers and (H) is a three-layer coasting of the coating solutions.

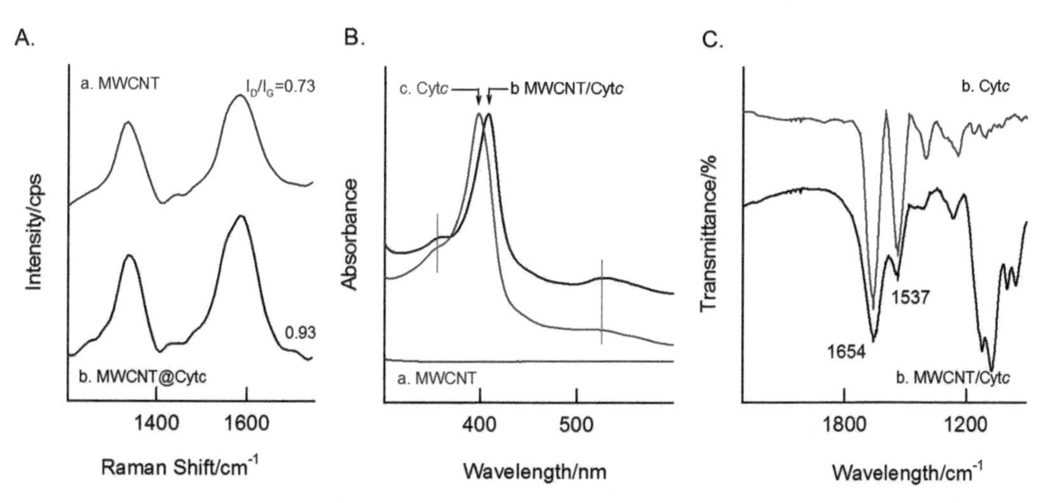

Figure 6. Comparative (A) Raman, (B) UV-Vis, and (C) FT-IR spectroscopic responses of MWCNT@Cytc with other controls.

Bioelectroacatalytic Activity towards Reduction of and Sensing of H_2O_2

Figure 7A shows the CVs of GCE/MWCNT@Cytc-Nf in pH-7 PBS (N_2 purged) without (a) and with the addition of 100 μM H_2O_2 at v = 10 mV s^{-1}. In the absence of H_2O_2, a characteristic redox peak of Cytc at $E^{0\prime}$= −0.31 V was noticed. Upon adding H_2O_2, the shape of the cyclic voltammogram changed dramatically with an increase of reduction peak current and a decrease of oxidation peak current. A further addition of H_2O_2 resulted in a strong increase in the peak catalytic current signal at −0.2 V (Figure 7B). In comparison with the control sample (GCE/MWCNT; data not enclosed), about 10-fold increases in the peak current signal and 400-mV reductions in the peak potential for the H_2O_2 reduction reaction were noticed. These observations indicate the bioelectrocatalytic function of the GCE/MWCNT@Cytc-Nf system toward H_2O_2 reduction reactions (Scheme 1E). Note that a pre-peak at $E^{0\prime}$= −0.1 V vs Ag/AgCl is also noticed with the modified electrode, which may be due to the Cytc protein adsorbed energetically at different sites of MWCNT (*50*). As a control sample, other interfering chemicals such as cysteine (CySH), ascorbic acid (AA), ibuprofen, histidine, reduced form of nicotinamide adenine dinucleotide (NADH), and dopamine (DA) were subjected to CV studies, but no alteration in the peak current and peak potential was noticed indicating the high selectivity of the present electrode for H_2O_2-sensing reaction. Figure 7B is a CV calibration plot for a linear increase in the concentration of H_2O_2 in a window of 0–1000 μM. The possible mechanism for the H_2O_2-reduction reaction through an ET step associated with chemical reaction (EC) pathway (Eqs. 2 and 3) is sketched as follows:

$$\text{MWCNT@Cytc-Fe(III)-Nf} + e^{-} + H^{+} \leftrightarrow \text{MWCNT@Cytc-Fe(II)-Nf} \quad (2)$$

$$\text{MWCNT@Cytc-Fe(II)-Nf} + 2H^{+} + H_2O_2 \rightarrow \text{MWCNT@Cytc-Fe(III)-Nf} + 2H_2O \quad (3)$$

Figure 8A is a typical rotating disc electrode (RDE) response of RDE-GCE/MWCNT@Cytc-Nf with and without 100 μM H_2O_2 at various rotation speeds in an N_2-purged pH-7 PBS solution. It can be seen that upon increasing the rotation speed, a regular increase in the H_2O_2 reduction peak current signal was noticed. The following Levich equation was applied to study the ET feature of the modified electrode toward an H_2O_2 reduction reaction (*52*):

$$i_L = [0.62nFAD^{2/3}v^{-1/6}C_{H2O2}]\omega^{1/2} \quad (4)$$

where i_L is the limiting current value for the H_2O_2 reduction reaction, ω is the angular velocity, and C_{H2O2} is the concentration of H_2O_2 and other symbols that have their own significance. Figure 8B is a plot of i_L measured at −0.4 V vs $\omega^{1/2}$ yielding a linear line starting from the origin. This observation indicates the diffusion-controlled ET reaction of the H_2O_2 reduction. It is interesting to note that there is no marked alteration in the redox peak current signal of the Cytc protein before and after the RDE studies (data not included), which indicates the good stability feature of the Cytc-modified electrode under a strong hydrodynamic condition.

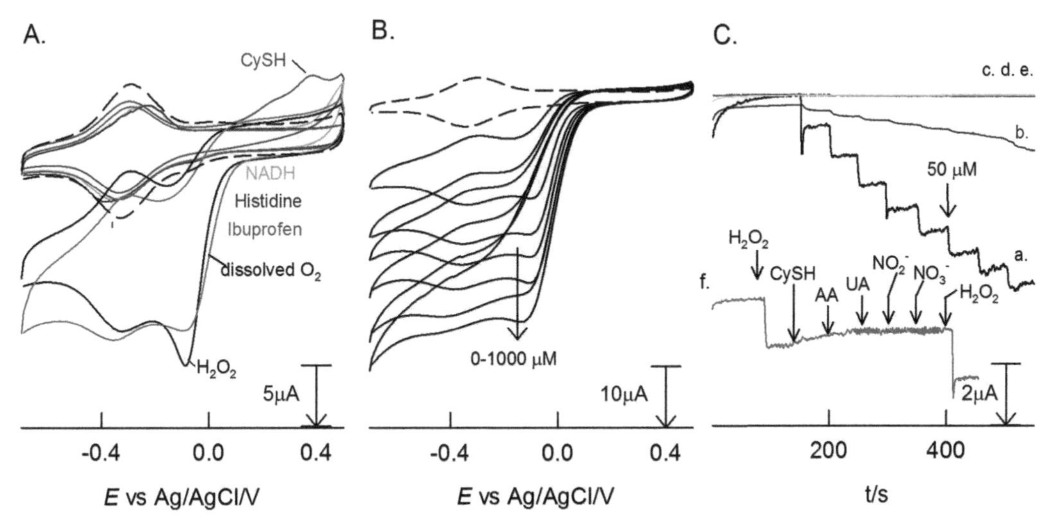

Figure 7. (A) Comparison of CV responses of GCE/MWCNT@Cytc-Nf with different analytes (100 μM) in a pH-7 PBS (N_2 purged) at v = 10 mV s^{-1}. (B) CV responses of GCE/MWCNT@Cytc-Nf with increasing concentrations of H_2O_2 (0-1000 μM). (C) Comparative amperometric i-t responses of GCE/MWCNT@Cytc-Nf (curve a), GCE/MWCNT (curve b), GCE/Nf (curve c), GCE/Cyt c (curve d), and GCE (curve e) for sensing of 50 μM of H_2O_2 in a stirred solution (RPM = 750) at E_{app} = 0 V vs Ag/AgCl. (f) An amperometric i-t response of GCE/MWCNT@Cytc-Nf toward H_2O_2 and various analytes (50 μM) including CySH, AA, UA, NO_2^-, and NO_3^- at E_{app} = 0.0 V vs Ag/AgCl in N_2-PBS.

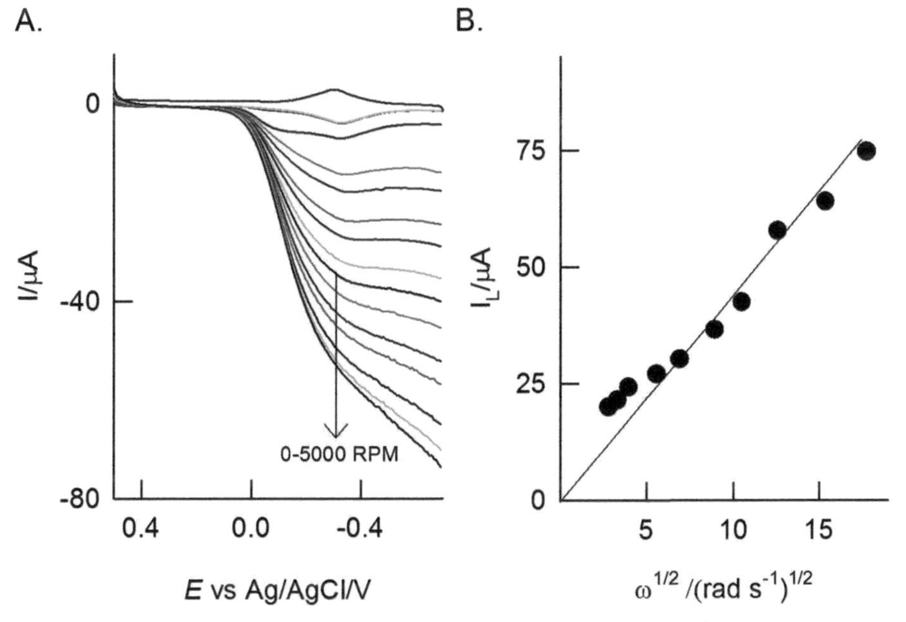

Figure 8. (A) RDE responses for the reduction of 100 μM of H_2O_2 on GCE/MWCNT@Cytc-Nf at different rotation speeds (50–5000 rpm) at v = 10 mV s^{-1}. (B) Typical Levich plot (i_L vs $\omega^{1/2}$) wherein the I_L value is obtained at E_L = -0.4 V vs Ag/AgCl.

Amperometric Detection and FIA of H_2O_2

Figure 7C shows comparative amperometric i–t responses of GCE/MWCNT@Cytc-Nf (a), GCE/MWCNT-Nf (b), GCE/Nf (c), GCE/Cytc (d), and GCE (e) for the continuous sensing of 50-μM spikes of the H_2O_2 in N_2-purged pH-7 PBS. A specific increase in the H_2O_2 steady-state current signal was noticed only with the optimal electrode, whereas the control electrodes like GCE/MWCNT, GCE/Nf, GCE/Cytc, and GCE showed feeble current signals. These observations support the efficient electrocatalytic reduction of the H_2O_2 by the Cytc heme-Fe(II)/Fe(III) active site, as mentioned in Eqs. 2 and 3. A corresponding calibration plot constructed using the amperometric i–t data was linear in the H_2O_2 concentration range (50–450 μM) with a current sensitivity and regression coefficient value of 19.43 (±0.2) nA/μM and 0.9949, respectively (data not shown). Eight continuous spikes of 50-μM H_2O_2 showed in an RSD value of 1.4%. The calculated detection limit (S/N = 3) for the H_2O_2 detection is 7.4 μM. The current sensitivity value obtained at 0.0 V vs Ag/AgCl is comparable with the literature values (Table 1) (*38, 39, 41–43, 49, 50, 53*).

The effect of interference from various biochemicals such as CySH, AA, uric acid (UA), nitrite (NO_2^-), and nitrate (NO_3^-) were examined at an applied potential of 0.0 V vs Ag/AgCl. As can be seen in Figure 7C, no marked alterations in the current signals were noticed upon the addition of these previously mentioned interfering agents, attributing the selective H_2O_2 biosensing ability of the GCE/i-MWCNT/Cytc electrode. It is surprising that the working electrode failed to show any current signal to NO_2^-, which is unusual because the Cytc-modified electrode was also reported to be effective for the NO_2^- reduction reaction (*54*). The anionic Nf overlayer coating on the working electrode can possibly ionically exclude the NO_2^- anion, hence the absence of any NO_2^- interference.

In order to detect the H_2O_2 in a low sample volume (20 μL), the GCE/MWCNT@Cytc-Nf system is extended to FIA, wherein FIA-GCE/MWCNT@Cytc-Nf was used as an ECD with a 0.1-M pH-7 PBS as a carrier system (Figure 9). Interrelated FIA parameters like flow rate (H_f) and applied potential (E_{app}) are individually optimized with 10 μM of H_2O_2. The optimal FIA condition for H_2O_2 detection in this work is: H_f= 0.8 ml min^{-1} and E_{app} = 0.0 V vs Ag/AgCl. Figure 8 is a typical FIA-ECD response of GCE/MWCNT@Cytc-Nf for H_2O_2 in a linear concentration range of 5 μM to 3 mM. The alculated current sensitivity and detection limit values are 1.5 nA μM^{-1} and 1.6 μM, respectively. Ten successive injections of 25 μM of H_2O_2 yielded an RSD value of 4.6% indicating the good stability of the ECD toward H_2O_2 detection. Furthermore, FIA-GCE/i-MWCNT@Cytc-Nf was also subjected to the detection of various interfering substances like CySH, AA, UA, NO_2^-, and NO_3^- (Figure 8B). No interference was noticed with the optimal electrode. These results reveal the selective bioelectrochemical behavior of GCE/MWCNT@Cytc-Nf for H_2O_2 reduction and sensing reaction.

Table 1. Literature Comparison of H_2O_2 Sensor with Some Represenative Cytc Chemically Modified Electrodes

	Cytc CMEs	*Denaturants*	$E_{1/2}$ *(mV)*	ΔE_p *(mV)*	Γ *(nM cm⁻²)*	*LOD (μM)*	E_{app} *(V)*	*Sensitivity (μA μM⁻¹)*	*Ref*
1	GCE/MWCNT/Chit/GNPs/ Cytc	Chitosan- cysteine	−160	60	-	0.90	−0.2	0.092	(38)
2	GCE/MWCNT/RTIL/GNP/ Cytc	RTIL	−129	36	-	3.0	−0.2	-	(39)
3	ITO/MWNT-g-PANI(O)/Cytc	CTAB and HCl	−20	250	-	0.3	-	0.0322	(41)
4	GCE/AuNP/GO-MWCNT/ Cytc	Ionic-polymer	−360	60	1.82	2.7×10^{-4}	−0.2	-	(42)
5	GCE/MWCNT/Cytc	Sonoprope	−338	156	0.19	0.15	−0.1	0.04×10^{-3}	(43)
6	CPE/Colloidal Au/Cytc	Surfactants	−35	60	0.034	-	-	-	(49)
7	ITO/Silica/Cytc	APTES	163	121	-	5	−2.0		(50)
8	Au/4-ATP/SWCNT/Cytc	GA	11.3	~50	0.060	3800	-0.11	3×10^{-6}	(53)
9	GCE/MWCNT @Cytc-Nf	Aging	−310	32	29.05	1.6	0.0	1.5×10^{-3}	This work

GNP: graphene nanoparticle: Chit: chitosan; RTIL: room temperature ionic liquid; ITO: indium tin oxide; PANI: polyaniline; AuNP: Gold nanoparticle; GO: graphene oxide; CPE: carbon paste electrode; CTAB: cetyltrimethylammonium bromide; APTES: (3-Aminopropyl)triethoxysilane: GA: glutaraldehyde; 4-ATP: 4-amino thiolphenol: SWCNT: Single walled carbon nanotube.

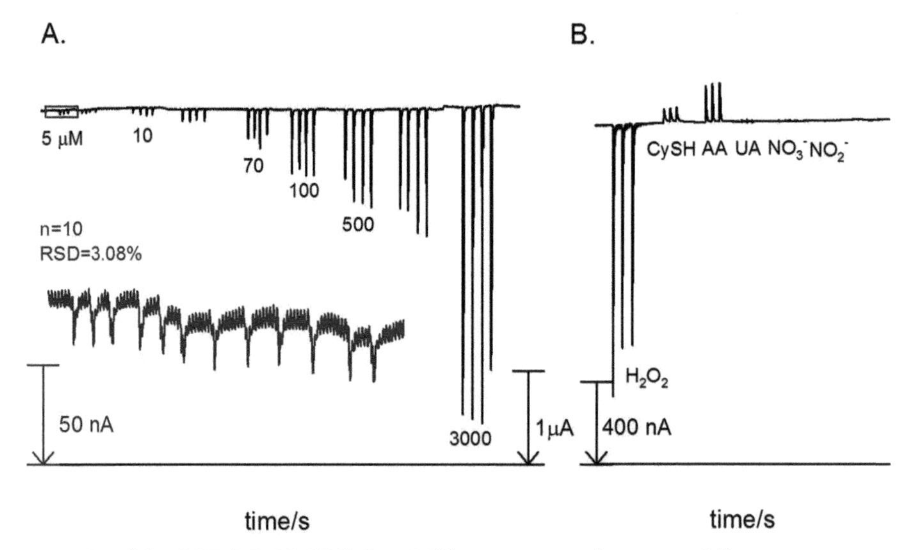

Figure 9. FIA of the GCE/MWCNT@Cytc-Nf detector system for sensing different concentrations of H_2O_2 (A) and other interference chemicals like CySH, AA, UA, NO_3^-, and NO_2^- of concentrations of 50 μM each. Other FIA conditions are: E_{app} = 0 V vs Ag/AgCl; hydrodynamic flow rate (H_f) = 800 μL min^{-1}; and carrier buffer solution = pH-7 N_2 PBS.

Figure 10 is a CV response of GCE/MWCNT@Cytc-Nf examined randomly at different days (up to 90 days) to determine the long-term stability of the bioelectrode. Note that Figure 3 was carried out by testing the CVs of Cytc proteins aged for different lengths of time that have been prepared as GCE/MWCNT@Cytc-Nf. However, for Figure 10, the Cytc protein aged for 20 days (a completely unfolded protein) and modified as GCE/MWCNT@Cytc-Nf was examined. Our aims are different in different cases. As can be seen, there is no marked alteration in the peak potential and peak current at up to 90 days of testing time.

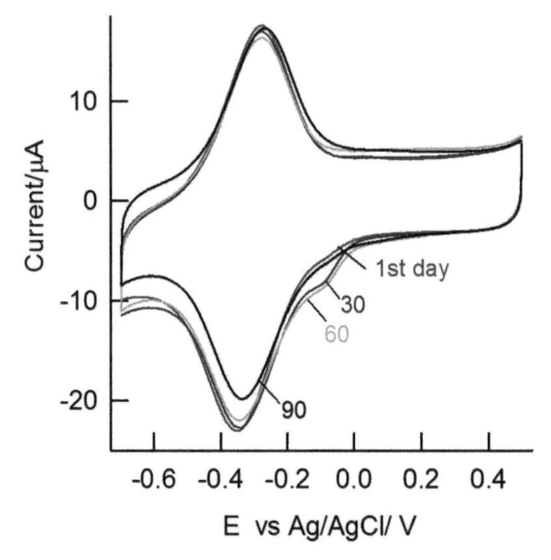

Figure 10. CV responses of a GCE/MWCNT@Cytc-Nf examined at different days in pH-7 PBS (N_2 purged) at scan rate = 50 mV s^{-1}.

Conclusions

In conclusion, a novel approach has been introduced for the entrapment and DET study of Cytc protein on the MWCNT surface. The aging effect of a Cytc-pH 7 PBS solution was studied systematically by spectroscopic and electrochemical techniques. A Cytc-pH 7 PBS sample aged up to 18 days at T = 5 °C showed a red shift in the Soret band absorption signal in UV-Vis, weakening of Amide-I and II bands in the FTIR, and increasing of fluorescence signals of tryptophan amino acid residues in fluorescence spectroscopic characterization indicating the denaturation-based unfolding of the Cytc protein. A MWCNT@Cytc-Nf chemically modified electrode prepared by a successive overlay method using the MWCNT, Cytc-pH 7 PBS, and Nf solution has shown a high redox active cyclic voltammetric signal at $E^{o\prime} = 0.31$ V vs Ag/AgCl with $\Gamma_{Cytc} = 29.05$ nmol cm^{-2}. The Cytc redox reaction in a blank pH-7 PBS is a typical surface-confined process following a proton-coupled, ET mechanism. The modified electrode showed a highly efficient and stable H_2O_2 bioelectrocatalytic response tested by CV and RDE techniques. Bioelectrochemical sensing of H_2O_2 was tested using amperometric i-t technique. The new bioelectrode is highly selective to H_2O_2 and showed nil interference effects on other biochemicals like CySH, AA, UA, NO_2^-, and NO_3^-. Under poised conditions, the modified electrode used as an ECD for H_2O_2 sensing by the FIA method yielded a RSD of 3.08%, sensitivity of 1.5 nA μM^{-1}, linear range of 5 μM to 3 mM, and a detection limit of 1.6 μM. The Cytc bioelectrode exhibited a stable response up to 90 days without any marked alteration in the peak potential and peak current signals.

ORCID

Annamalai Senthil Kumar: 0000-0001-8800-4038

Notes

The authors declare no competing financial interest.

Author Contributions

The manuscript was written through contributions of all authors. All authors have given approval to the final version of the manuscript.

Abbreviations

MWCNT: multiwalled carbon nanotube; Nf: nafion; Cytc: cytochrome c; GCE: glassy carbon electrode; PBS: phosphate buffer solution; RSD: relative standard deviation; FIA: flow injection analysis.

Acknowledgments

The authors acknowledge the Department of Science and Technology, Science and Engineering Research Board, India (DST-SERB-EMR/2016/002818), Scheme.

References

1. Wikstrom, M.; Krab, K.; Sharma, V. Oxygen Activation and Energy Conservation by Cytochrome c Oxidase. *Chem. Rev.* **2018**, *118*, 2469–2490.
2. Meyer, Y. P. Confirmations of cytochromes. *J. Biol. Chem.* **1971**, *246*, 1241–1248.
3. Dikiy, A.; Carpentier, W.; Vandenberghe, I.; Borsari, M.; Safarov, N.; Dikaya, E.; Beeumen, J. V.; Ciurli, S. Structural Basis for the Molecular Properties of Cytochrome C6. *Biochem.* **2002**, *41*, 14689–14699.
4. Hua, W.; Kana, Z. Y.; Maynea, L.; Englander, S. W. Cytochrome c Folds Through Foldon-Dependent Native-Like Intermediates in an Ordered Pathway. *Pro. Nat. Acad. Sci.* **2016**, *113*, 3809–3814.
5. Dolla, A.; Blanchard, L.; Guerlesquin, F.; Bruschi, M. The Protein Moiety Modulates the Redox Potential in Cytochromes c. *Biochimie* **1994**, *76*, 471–479.
6. Rafferty, S. P.; Guillemette, J. G.; Berghuis, A. M.; Smith, M.; Brayer, G. D.; Mauk, A. G. Mechanistic and Structural Contributions of Critical Surface and Internal Residues to Cytochrome c Electron Transfer Reactivity. *Biochem.* **1996**, *35*, 10784–10792.
7. Bortolotti, C. A.; Borsari, M.; Sola, M.; Chertkova, V.; Dolgikh, D.; Kotlyar, A.; Facci, P. Orientation-Dependent Kinetics of Heterogeneous Electron Transfer for Cytochrome c Immobilized on Gold: Electrochemical Determination and Theoretical Prediction. *J. Phys. Chem. C* **2007**, *111*, 12100–12105.
8. Armstrong, F. A.; Bond, A. M.; Hill, H. A. O.; Oliver, B. N.; Psalti, I. S. M. Electrochemistry of Cytochrome c, Plastocyanin, and Ferredoxin at Edge- and Basal-Plane Graphite Electrodes Interpreted via a Model Based on Electron Transfer at Electroactive Sites of Microscopic Dimensions in Size. *J. Am. Chem. Soc.* **1989**, *26*, 9185–9189.
9. Bhuyan, A. K.; Udgaonkar, J. B. Folding of Horse Cytochrome c in the Reduced State. *J. Mol. Biol.* **2001**, *312*, 1135–1160.
10. Salemme, F. R.; Freer, S. T.; Xuong, N. H.; Alden, R. A.; Kraut, J. The Structure of Oxidized Cytochrome c2 of Rhodospirillum Rubrum. *J. Biol.Chem.* **1973**, *248*, 3910–3921.
11. Betz, S. F.; Pielak, G. J. Introduction of a Disulfide Bond into Cytochrome c Stabilizes a Compact Denatured State. *Biochem.* **1992**, *31*, 12337–12344.
12. Mason, J. M.; Bendall, D. S.; Howe, C. J.; Worrall, J. A. R. The Role of a Disulfide Bridge in the Stability and Folding Kinetics of Arabidopsis Thaliana Cytochrome c6A. *Biochim. Biophys. Acta* **2012**, *1824*, 311–318.
13. Qina, M.; Wang, W.; Thirumalai, D. Protein Folding Guides Disulfide Bond Formation. *Proc. Nat. Acad. Sci.* **2015**, *112*, 11241–11246.
14. Yeh, S-R.; Han, S.; Rousseau, D. L. Cytochrome c Folding and Unfolding: A Biphasic Mechanism. *Acc. Chem. Res.* **1998**, *31*, 727–736.
15. Yeh, S-R.; Rousseau, D. L. Ligand Exchange During the Unfolding of Cytochrome c. *J. Biol. Chem.* **1999**, *274*, 17853–17859.
16. Nagasawa, T.; Fujimori, H. N.; Heinrich, P. C. Hydrophobic Interactions of Cytochrome c Oxidase. *Eur. J. Biochem.* **1979**, *94*, 31–39.
17. Reilly, N. J. O.; Magner, E. Electrochemistry of Cytochrome c in Aqueous and Mixed Solvent Solutions: Thermodynamics, Kinetics, and the Effect of Solvent Dielectric Constant. *Langmuir* **2005**, *21*, 1009–1014.

18. Parui, P. P.; Deshpande, M. S.; Nagao, S.; Kamikubo, H.; Komori, H.; Higuchi, Y.; Kataoka, M.; Hirota, S. Formation of Oligomeric Cytochrome c During Folding by Intermolecular Hydrophobic Interaction Between N- and C -Terminal α-Helices. *Biochem.* **2013**, *52*, 8732–8744.
19. Sun, J.; Ruchmann, J.; Pallier, A.; Jullien, L.; Desmadril, M.; Tribet, C. Unfolding of Cytochrome c Upon Interaction with Azobenzene Modified Copolymers. *Biomacromol.* **2012,** *13*, 3736–3746.
20. Dick, L. A.; Haes, A. J.; Duyne, R. P. V. Distance and Orientation Dependence of Heterogeneous Electron Transfer: A Surface-Enhanced Resonance Raman Scattering Study of Cytochrome c Bound to Carboxylic Acid Terminated Alkanethiols Adsorbed on Silver Electrodes. *J. Phys. Chem. B* **2000**, *104*, 11752–11762.
21. Gong, J.; Yao, P.; Duan, H.; Jiang, M.; Gu, S.; Chunyu, L. Structural Transformation of Cytochrome c and Apo Cytochrome c Induced by Sulfonated Polystyrene. *Biomacromol.* **2003,** *4*, 1293–1300.
22. Hinnen, C.; Parsons, R. Electrochemical and Spectroreflectance Studies of the Adsorbed Horse Heart Cytochrome c and Cytochrome c3 from *d. Vulgaris*, Miyazaki Strain, at Gold Electrode. *J. Electroanal. Chem.* **1983**, *147*, 329–337.
23. Reed, D. E.; Hawkridge, F. M. Direct Electron Transfer Reactions of Cytochrome c at Silver Electrodes. *Anal. Chem.* **1987**, *59*, 2339–2344.
24. Bowden, E. F.; Hawkridge, F. M.; Blount, H. N. Interfacial Electrochemistry of Cytochrome c at Tin Oxide, Indium Oxide, Gold, and Platinum Electrodes. *J. Electroanal. Chem. Inter. Electrochem.* **1984,** *161*, 355–376.
25. Lamp, B. D.; Hobara, D.; Porter, M. D.; Niki, K.; Cotton, T. M. Correlation of the Structural Decomposition and Performance of Pyridinethiolate Surface Modifiers at Gold Electrodes for the Facilitation of Cytochrome c Heterogeneous Electron-Transfer Reactions. *Langmuir* **1997**, *13*, 736–741.
26. Reilly, N. J. O.; Magner, E. Electrochemistry of Cytochrome c in Aqueous and Mixed Solvent Solutions: Thermodynamics, Kinetics, and the Effect of Solvent Dielectric Constant. *Langmuir* **2005**, *21*, 1009–1014.
27. Ruzgas, T.; Wong, L.; Gaigalas, A. K.; Vilker, V. L. Electron Transfer Between Surface-Confined Cytochrome c and an N-Acetylcysteine-Modified Gold Electrode. *Langmuir* **1998,** *14*, 7298–7305.
28. Chen, X.; Ferrigno, R.; Yang, J.; Whitesides, G. M. Redox Properties of Cytochrome c Adsorbed on Self-Assembled Monolayers: A Probe for Protein Conformation and Orientation. *Langmuir* **2002,** *18*, 7009–7015.
29. Imabayashi, S. I.; Mita, T.; Kakiuchi, T. Effect of the Electrostatic Interaction on the Redox Reaction of Positively Charged Cytochrome c Adsorbed on the Negatively Charged Surfaces of Acid-Terminated Alkanethiol Monolayers on a Au(111) Electrode. *Langmuir* **2005**, *21*, 1470–1474.
30. Millo, D.; Bonifacio, A.; Ranieri, A.; Borsari, M.; Gooijer, C.; Zwan, G. V. D. pH-Induced Changes in Adsorbed Cytochrome c. Voltammetric and Surface-Enhanced Resonance Raman Characterization Performed Simultaneously at Chemically Modified Silver Electrodes. *Langmuir* **2007**, *23*, 9898–9904.

31. Monari, S.; Ranieria, A.; Bortolotti, C. A.; Peressini, S.; Tavagnacco, C.; Borsari, M. Unfolding of Cytochrome c Immobilized on Self-Assembled Monolayers. An Electrochemical Study. *Electrochim. Acta* **2011**, *56*, 6925–6931.
32. Ferri, T.; Poscia, A.; Ascoli, F.; Santucci, R. Direct Electrochemical Evidence for an Equilibrium Intermediate in the Guanidine-Induced Unfolding of Cytochrome c. *Biochim. Biophys. Acta* **1996**, *1298*, 102–108.
33. Wu, Y.; Hu, S. Voltammetric Investigation of Cytochrome c on Gold Coated with a Self-Assembled Glutathione Monolayer. *Bioelectrochem.* **2006**, *68*, 105–112.
34. Behera, R. K.; Nakajima, H.; Rajbongshi, J.; Watanabe, Y.; Mazumdar, S. Thermodynamic Effects of the Alteration of the Axial Ligand on the Unfolding of Thermostable Cytochrome c. *Biochem.* **2013**, *52*, 1373–1384.
35. Dolla, A.; Blanchard, L.; Guerlesquin, F.; Bruschi, M. The Protein Moiety Modulates the Redox Potential in Cytochromes c. *Biochimie* **1994**, *76*, 471–479.
36. Hellwig, P.; Grzybek, S.; Behr, J.; Ludwig, B.; Michel, H.; Mäntele, W. Electrochemical and Ultraviolet/Visible/Infrared Spectroscopic Analysis of Heme a and a3 Redox Reactions in the Cytochrome c Oxidase from Paracoccus denitrificans: Separation of Heme a and a3 Contributions and Assignment of Vibrational Modes. *Biochem.* **1999**, *38*, 1685–1694.
37. Wang, J.; Li, M.; Shi, Z.; Li, N.; Gu, Z. Direct Electrochemistry of Cytochrome c at a Glassy Carbon Electrode Modified with Single-Wall Carbon Nanotubes. *Anal. Chem.* **2002**, *74*, 1993–1997.
38. Xiang, C.; Zou, Y.; Sun, L. X.; Xu, F. Direct Electrochemistry and Electrocatalysis of Cytochrome c Immobilized on Gold Nanoparticles–Chitosan–Carbon Nanotubes-Modified Electrode. *Talanta* **2007**, *74*, 206–211.
39. Xiang, C.; Zoua, Y.; Sun, L. X.; Xu, F. Direct Electron Transfer of Cytochrome c and its Biosensor Based on Gold Nanoparticles/Room Temperature Ionic Liquid/Carbon Nanotubes Composite Film. *Electrochem. Commun.* **2008**, *10*, 38–41.
40. Shie, J. W.; Yogeswaran, U.; Chen, S. M. Electroanalytical Properties of Cytochrome c by Direct Electrochemistry on Multi-Walled Carbon Nanotubes Incorporated with DNA Biocomposite Film. *Talanta* **2008**, *74*, 1659–1669.
41. Lee, K-P.; Gopalan, A. I.; Komathi, S. Direct Electrochemistry of Cytochrome c and Biosensing for Hydrogen Peroxide on Polyaniline Grafted Multi-Walled Carbon Nanotube Electrode. *Sens. Actuators, B* **2009**, *141*, 518–525.
42. Dinesh, B.; Mani, V.; Saraswathi, R.; Chen, S. M. Direct Electrochemistry of Cytochrome c Immobilized on a Graphene Oxide–Carbon Nanotube Composite for Picomolar Detection of Hydrogen Peroxide. *RSC Adv.* **2014**, *4*, 28229–28237.
43. Eguílaz, M.; Gutiérrez, A.; Rivas, G. Non-Covalent Functionalization of Multi-Walled Carbon Nanotubes with Cytochrome c: Enhanced Direct Electron Transfer and Analytical Applications. *Sens. Actuators, B* **2016**, *225*, 74–80.
44. Prakasam, G.; Kumar, A. S. An Iron Impurity in Multiwalled Carbon Nanotube Complexes with Chitosan that Biomimics the Heme-Peroxidase Function. *Chem. Eur. J.* **2013**, *19*, 17103–17112.
45. Wiederkehr, R. S.; Hoops, G. C.; Aslan, M. M.; Byard, C. L.; Mendes, S. B. Investigations on the Q and CT Bands of Cytochrome c Submonolayer Adsorbed on an Alumina Surface using

Broadband Spectroscopy with Single-Mode Integrated Optical Waveguides. *J. Phys. Chem. C* **2009**, *113*, 8306–8312.

46. Salemme, F. R.; Freer, S. T.; Xuong, N. H.; Alden, R. A.; Kraut, J. The Structure of Oxidized Cytochrome c_2 of *Rhodospirillum rubrum*. *J. Biol. Chem.* **1973**, *248*, 3910–3921.
47. Laviron, E. General Expression of the Linear Potential Sweep Voltammogram in the Case of Diffusionless Electrochemical Systems. *J. Electroanal. Chem.* **1979**, *101*, 19–28.
48. Liu, Y. C.; Cui, S. Q.; Zhao, J.; Yang, Z. S. Direct Electrochemistry Behaviour of Cytochrome c/L-Cysteine Modified Electrode and its Electrocatalytic Oxidation to Nitric Oxide. *Bioelectrochem.* **2007**, *7*, 416–420.
49. Ju, H. X.; Liu, Q. S.; Ge, X. B.; Lisdat, F.; Scheller, W. F. Electrochemistry of Cytochrome c Immobilized on Colloidal Gold Modified Carbon Paste Electrodes and its Electrocatalytic Activity. *Electroanal.* **2002**, *14*, 141–147.
50. Zhang, X.; Wang, J.; Wu, W.; Qian, S.; Man, Y. Immobilization and Electrochemistry of Cytochrome c on Amino-Functionalized Mesoporous Silica Thin Films. *Electrochem. Commun.* **2007**, *9*, 2098–2104.
51. Amreen, K.; Kumar, A. S.; Mani, V.; Huang, S. T. Axial Coordination Site-Turned Surface Confinement, Electron Transfer, and Bio-Electrocatalytic Applications of a Hemin Complex on Graphitic Carbon Nanomaterial-Modified Electrodes. *ACS Omega* **2018**, *3*, 5435–5444.
52. Bard, A. J.; Faulkner, R. L. *Electrochemical Methods: Fundamentals and Applications*, 2nd ed.; Wiley: New York, 2001.
53. Nagaraju, D. H.; Pandey, R. K.; Lakshminarayanan, V. Electrocatalytic Studies of Cytochrome c Functionalized Single Walled Carbon Nanotubes on Self-Assembled Monolayer of 4-ATP on Gold. *J. Electroanal. Chem.* **2009**, *627*, 63–68.
54. Chen, Q.; Ai, S.; Zhu, X.; Yin, H.; Ma, Q.; Qiu, Y. A Nitrite Biosensor Based on the Immobilization of Cytochrome c on Multi-Walled Carbon Nanotubes–PAMAM–Chitosan Nanocomposite Modified Glass Carbon Electrode. *Biosens. Bioelectron.* **2009**, *24*, 2991–2996.

Chapter 10

Novel Bioelectrocatalytic Strategies Based on Immobilized Redox Metalloenzymes on Tailored Electrodes

Gabriel García-Molina, Marcos Pita, and Antonio L. De Lacey*

Instituto de Catálisis y Petroleoquímica, CSIC, c/ Marie Curie 2, L10, 28049 Madrid, Spain

*E-mail: alopez@icp.csic.es.

The attachment of redox metalloenzymes to electroactive surfaces is of great importance for both fundamental and applied studies. Oriented covalent immobilization of a metalloenzyme on the electrode allows for the use of electrochemical methods to study its catalytic mechanism by direct electron transfer. Furthermore, redox metalloenzymes specifically catalyze many reactions of great interest in the energy conversion process, such as H_2 production or oxidation, O_2 evolution or reaction, and generation of biochemical power within living cells. Therefore, their optimized immobilization on electrodes has potential applications in biofuel cell development, water splitting or CO_2 valorization using electrochemical and photoelectrochemical cells, and electroenzymatic systems for cofactor regeneration. This chapter reviews novel strategies developed in recent years for tailoring electrodes with the aim of specific, stable, and oriented binding of redox metalloenzymes. This review is focused on the strategies designed for three different applications: (1) development of efficient oxygen reduction biocathodes; (2) production of hybrid semiconductor and enzyme photoelectrocatalysts; and (3) biomimetic reconstitution of membrane enzymes over electrodes.

Introduction

The coupling of enzymatic activity to electrodes is a very versatile strategy for many applications, including biosensing, cofactor regeneration, and biofuel cells among others. The reason is that this coupling combines the interesting catalytic properties of enzymes, such as specificity, high turnover, and mild operational conditions, with the high sensitivity and thermodynamic control of electrochemical methods.

In particular, the attachment of redox metalloenzymes to electroactive surfaces is of great importance for both fundamental and applied studies. These types of enzymes have metals in their active sites that change their redox state during the catalytic cycle. They often contain redox relays

© 2020 American Chemical Society

within their protein structure that transport electrons between the active site and the enzyme's redox partner. These redox relays are either aligned iron-sulphur clusters, heme groups, or Cu complexes, which act as electronic wires to and from the enzymatic active site (*1, 2*). Therefore, adequate immobilization of a metalloenzyme on the electrode allows for the use of a panoply of electrochemical methods to study its catalytic mechanism by direct electron transfer (*3*). A requisite for direct electron transfer is that the redox center located on the enzyme surface is at a short distance (less than 20 Å) from the electrode surface. This condition can be fulfilled by oriented immobilization of the enzyme on a planar electrode surface or by its incorporation into a nanostructured electrode (*4*). Redox metalloenzymes specifically catalyze many reactions of great interest in the energy conversion process, such as H_2 production and oxidation, O_2 evolution and reaction, and generation of biochemical power within living cells. Thus, their optimized immobilization on electrodes has many applications in biofuel cell development (*5*), electrochemical and photoelectrochemical cells for water splitting (*6*), CO_2 valorization (*7*), and electroenzymatic systems for cofactor regeneration (*8*).

An optimal coupling between enzyme and electrode is key for all these applications. Optimal coupling has two requirements: the enzyme maintains its native configuration and catalytic activity after the immobilization process on the electrode surface, and fast transfer of electrons between electrode and enzyme, either by DET or by redox mediators incorporated in the interface, is established. Both conditions can be achieved by adequately tailoring the electrode surface. This can be done by chemical functionalization, introduction of nanostructure, deposition of conductive or redox polymers, and, in the case of membrane enzymes, supporting phospholipid bilayers on the electrode.

This chapter reviews novel strategies that have been developed in the last 10 years by our group, as well by others, for optimizing the interaction of redox metalloenzymes with electrodes for several applications. In particular, we focus on the development of efficient oxygen reduction biocathodes, hybrid semiconductor or enzyme photoelectrocatalysts, and reconstituted membrane enzymes over electrodes.

Oxygen Reduction Biocathodes Based on Multicopper Oxidases Immobilized on Nanostructured Electrodes

Enzymatic biofuel cells have been a focus of research to increase their power output by adjusting the biocatalysts' performance and aiming the selective oxidation of organic matter in the anode (*9*). The most common comburent used in the cathode is O_2, which is available in the atmosphere and in equilibrium with an aqueous solution. However, its electrochemical reduction requires high overpotentials, causing a great loss of energy and power. Biocatalysis offers tools to improve these limitations by performing the reduction process closer to the reaction thermodynamic value.

The use of biocatalysts for this process has to meet several requirements. First of all, the selected enzyme must reduce O_2. There are many enzymes capable of O_2 reduction, which are named oxidases. Most oxidases only use two electrons from O_2 to produce H_2O_2, which is not desirable as it can generate reactive oxygen species that inactivate the enzymes or damage the setup, thus lowering the electrochemical efficiency at the cathode. There are, however, some particular oxidases known as multicopper oxidases (MCOs) that can accomplish the selective reduction of O_2 to H_2O in a four-electron process (*10*). Therefore, a successful combination of MCOs with electrodes may avoid unwanted byproducts and additionally yield an onset potential for O_2 reduction cathode close to the thermodynamical value.

MCOs are metalloenzymes characterized by having two active sites containing copper cations. One of the active sites comprises one Cu cation coordinated to three amino acids and is responsible for substrate oxidation. The other active site has a Cu_3 cluster and is in charge of O_2 reduction to water. Both active sites are connected via an intramolecular electron path formed by histidine and cysteine amino acids. The surrounding amino acids change the copper cations' EPR signal, which determines the most common nomenclature for the MCOs' active sites: The oxidizing Cu is known as a T1 copper site, and the trinuclear cluster is known as a T2 or T3 copper site (*11*). The two most popular classes of MCOs are polyphenol oxidases (commonly known as laccases) and bilirubin oxidases (BOxs). Laccases are obtained from either plants, trees, or fungi. Normally, the plant laccases have a T1 redox potential that is lower than the fungal ones. This fact has created another classification: low redox potential laccases or high redox potential laccases. The interesting ones for developing biocathodes are the high redox potential ones, as they are the only ones that would allow a positive-enough O_2 reduction potential to develop an efficient biofuel cell (*12*). Another characteristic of laccases, especially for fungal laccases, is that they are only active in acidic environments. This is not the case for BOx, as their major difference from laccases is that they maintain activity at neutral pH (*12*). Most research on the development of O_2-reducing biocathodes has been performed using either laccases or BOxs as biocatalysts (*12*).

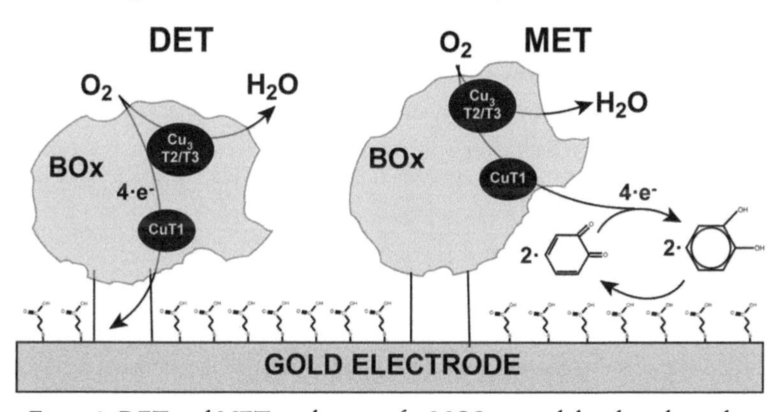

Figure 1. DET and MET mechanisms for MCOs immobilized on electrodes.

One of the major challenges in developing an efficient bioelectrode is how to electrically connect the enzyme with the electrodes. There are two strategies: the use of a redox compound for mediated electron transfer (MET), or a DET connection between the enzyme and electrode (Figure 1). MET allows for a faster connection but includes an extra step in the potential cascade, which diminishes the onset potential from the value of the T1 copper site. On the other hand, the use of redox hydrogels as the MET system allows for stable entrapment of MCOs and measuring high current densities (*13*, *14*). Another example of a laccase biocathode connected via MET to an electrode was developed by attaching an Os-bipyridine derivative complex to a polyelectrolyte, which was appended on an electrode surface and used as a switchable electrode (Figure 2) (*15*). In principle, DET allows for obtaining the maximum onset potential; however, the steric hindrances inherent to the proteic tertiary structure of the MCOs are a kinetic barrier to reach such onset potential. This section will focus on different strategies of MCO immobilization on different electrodes for optimizing DET.

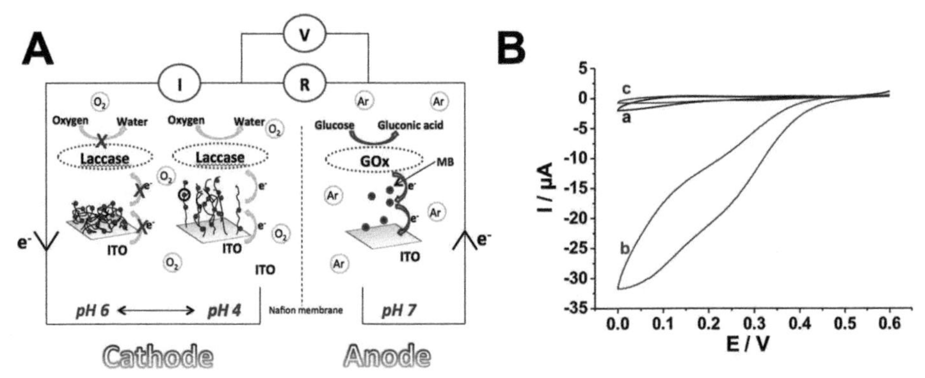

Figure 2. (A) Biofuel cell composed of a pH-switchable, logically controlled biocatalytic cathode and glucose-oxidizing anode. (B) Response of Trametes versicolor laccase connected via MET to an ITO electrode: (a) and (c) are closed state, (b) is open state (pH 4.2, room temperature, $E'[O_2/H_2O] = 0.78$ V). Redox potentials are vs. Ag and AgCl (3 M KCl) reference electrode. Reproduced with permission from Reference (15). Copyright 2009 American Chemical Society.

The key to obtaining a fast DET between an enzyme and an electroactive surface is a successful connection of the active site to the electrode. One factor is the orientation of the enzyme. In general, the active site and its substrate access channel should be oriented facing the electrode. In this scenario, the distance between the electrode and the catalytic site is minimal, facilitating the electron hopping. There are cases where this strategy is not successful (i.e., enzymes where there is only one access point to the active site, and it is blocked by the electrode itself) (*16*). In these cases, the MET often yields better results because DET can only be achieved with a flexible binding between the enzyme and the electrode, which allows some protein movement. There are other cases where the enzymes' metallic active site is buried in the proteic structure and is not directly accessible, but there is an electronic path inside the enzyme itself. This is the case in many hydrogenases, nitrogenases, and even some formate dehydrogenases among other enzymes, where a series of iron-sulfur clusters are contained and connected from the enzyme's surface to the active site. In these cases, the orientation facing the electrode should be, of course, the outer Fe_4S_4 cluster (*17*). The case of MCOs is particularly advantageous for attempting DET due to their exposed T1 Cu sites. When the enzyme is immobilized with the T1 Cu site facing the electrode, it is possible to develop a high redox potential biocathode. In this case, the intramolecular pathway of MCOs dodges the blocking of substrate access to the active site due to immobilization on the electrode because the O_2 reduction takes place in a different region of the enzyme: the T2 and T3 site. Moreover, it is crucial to avoid immobilization with the T2 and T3 site facing the electrode, as this would cause short circuiting of the intramolecular electron path and biocathode performance at a lower redox potential. Another advantage of MCOs, and particularly laccases, is their natural affinity for polyaromatic derivatives, which happen to be the substrates in their native activity for degrading lignin (*10, 18*). Modification of graphite-based electrodes with polyaromatic groups by covalent bonds (such as anthracene) (*19*) or by non-covalent π-π pyrene stacking (*20*) facilitate the orientation of laccase, which leads to efficient DET.

Trametes hirsuta laccase (ThLc) is a high redox potential laccase with a standard redox potential of 780 mV compared to the standard hydrogen electrode (SHE) for the Cu T1 site, which is close to the thermodynamical value for O_2 reduction (*11*). Although ThLc gave fair cathodic currents for O_2 reduction by simple physical absorption on a graphite electrode (*11*), an increase of the operational stability was obtained via covalent immobilization on the electrode's surface (*21*). Low density graphite (LDG) electrodes were modified by electrochemical grafting of 4-nitroaryl diazonium salt, which were further reduced to aminoaryl groups. Carbodiimide (EDC) activation of the enzyme's carboxylates allowed formation of amide bonds with the electrode's surface groups. However, electrocatalytic currents of O_2 reduction by DET were only half of those obtained by MET, which suggests a mixture of orientations of the immobilized laccase (*21*). On the other hand, DET-based electrocatalysis was shown to be insensitive to chloride inhibition, whereas MET was not. It was argued that Cl^- probably binds near the Cu T1 site, thus affecting electron exchange with a redox mediator but not with the electrode (*21*).

An improved strategy for oriented immobilization of ThLc was developed by taking advantage of the glycosylated structure of the enzyme, as two of its three sugar residues are located near its Cu T1 site (*22*). First, the hydroxyls of the sugar residues were oxidized to aldehydes. Then, a two-step immobilization process was done by forming imine bonds between the aldehydes of the sugar residues and the electrode's amino groups. Once the enzyme was attached, EDC coupling of the enzyme's carboxylate groups to the electrode was performed to increase binding stability (*23, 24*).

In order to increase the current density of the biocathodes, nanostructured electrodes were tested (*24*). Carbon nanotubes–carbon microcloth hierarchically nanostructured electrodes were synthesized, and three ThLc immobilization strategies were compared: aminoaryl diazonium modification followed by direct EDC coupling, aminoaryl diazonium modification followed by the two-step imino plus EDC coupling, and π-π stacking of pyrene pentanoic acid with further EDC coupling. All three strategies improved the DET performance when compared with LDG, but the best of them was based on grafting the amino aryl derivatives via diazonium reduction, development of sugar-oriented imino bonds, and further EDC covalent bonding. In the best conditions (O_2-enriched solution and electrode rotation at 1000 rpm), the current density reached a maximum of 1.4 $mA \cdot cm^{-2}$. The only drawback was a reduction in the chloride resistance, which yielded a 16% current decrease in presence of 140 mM Cl^- concentration (*24*).

Much more challenging has been the development of laccase biocathodes on gold electrodes. Gold presents interesting advantages over graphite for both general and in vivo applications in terms of chemical inertness, stability, resistance to corrosion, and biocompatibility. However, some drawbacks need to be addressed in order to grant a reasonable bioelectrochemical performance. Unlike unmodified graphite, bare gold strongly adsorbs the amino acid residues of the peptidic chain of the protein, which often causes a multipunctual link of the protein to the electrode that yields loss of its tertiary structure and activity. Therefore the functionalization of the gold surface becomes a critical issue for enzyme immobilization in bioelectrochemistry. The first attempts were performed by Atanassov's group in 2004 with *Rhus vernicifera* and *Coriolus hirsutus* laccases (*25*), and by Gorton's group in 2006 with ThLc (*26*). The gold electrodes were modified with thiol derivatives to build self-assembled monolayers (SAMs). Carboxylic- and hydroxyl-terminated thiol SAMs did not yield positive results, whereas 4-aminothiophenol (4-ATP) SAMs did. ThLc covalent bonding to Au-4-ATP by the previously mentioned two-step strategy yielded a very mild electrocatalytic response starting at the T1 site redox potential, a higher response at the T2 and T3 redox potential, and a higher response at lower potentials. Experiments were performed to check for the production of the H_2O_2 byproduct, a marker for enzyme partial denaturation or inadequate orientation, which appeared only at the lower applied potentials (*26*).

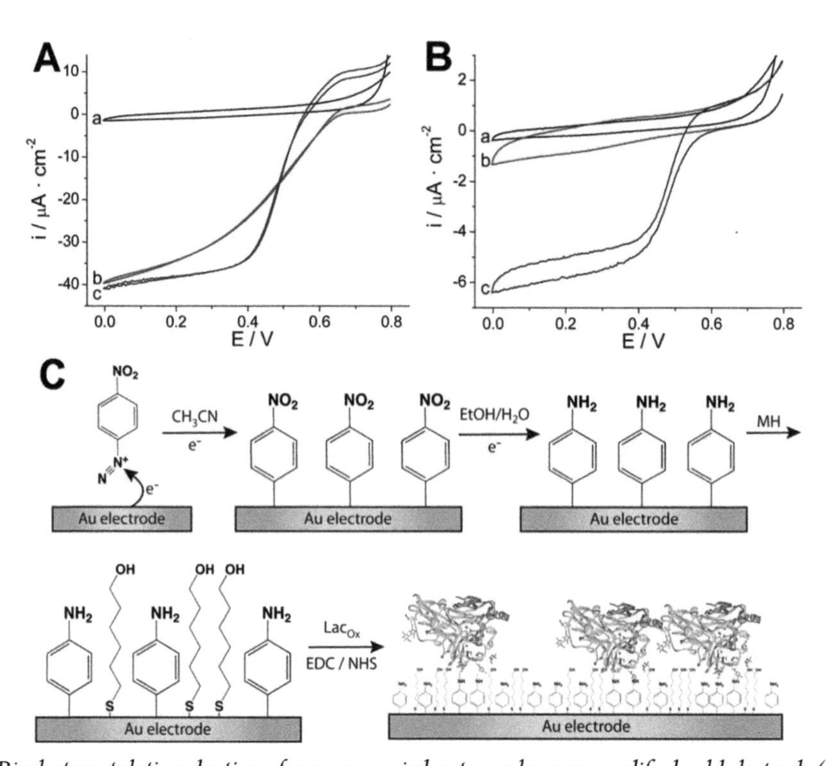

Figure 3. Bioelectrocatalytic reduction of oxygen carried out on a laccase-modified gold electrode (A) using a two-step—Schiff base and NHS-EDC—immobilization strategy, and (B) direct reduction just using NHS-EDC covalent linkage. (a) Background current provided by the electrode before laccase immobilization. (b) Direct electron current measured at 1000 rpm under oxygen saturation. (c) Mediated electron current measured with 0.1 mM ABTS at 1000 rpm under oxygen saturation. Temperature was 27 °C, pH of 4.2, and redox potential was compared to. Ag and AgCl (3 M KCl). Thus, $E'(O_2/H_2O) = 0.77$ V. (C) Strategy for the modification of gold with an aryl diazonium salt and MH and further laccase immobilization. Reproduced with permission from reference (23). Copyright 2011 American Chemical Society.

The SAM strategy for the modification of gold was not a very successful one compared with the modified carbon electrodes. An improvement was achieved when using diazonium salt derivatives for the gold modification, changing the thiol SAMs for gold-carbon bonds (Figure 3) (23). The electrochemical reduction of diazonium derivatives produces aromatic radicals that are prone to react with the already deposited aromatic rings. This undesirable effect may yield multilayer growth in the aryl groups. The solution presented was to form incomplete monolayers where there is little chance for multilayer growth and, in a second step, incubate a thiol to block the naked gold and avoid possible denaturation of the ThLc. Surprisingly, the most suitable thiol was 6-mercapto-1-hexanol and not 4-ATP, which was not really needed for the enzyme immobilization as there were 4-aminophenyl molecules directly attached to the gold surface. The resulting biocathodes, after modification with ThLc, gave DET current densities for O_2 reduction (23). However, the chloride inhibition could not be properly shielded in the same way as using LDG. This effect was attributed to Cl^- adsorption on the gold electrode, which affected DET at the Cu T1 site. The upgrade in the performance was deeply studied by STM-SEIRAS microscopy (27). After an intensive study of the behavior of pure 4-ATP SAMs and the mixed 4-aminophenyl:4-ATP monolayer, it could be seen that the carbon-gold bonds acted as anchors for the architecture. They provided higher stability and lesser mobility, whereas the SAMs showed a significant mobility around the surface, providing blurry images (*27*).

Additional attempts to improve the DET performance of ThLc focused on adding gold nanostructures to LDG electrodes. Three different nanostructures were studied: 16-nm diameter Au nanoparticles (AuNPs), 5-nm diameter AuNPs (*28*), and 30x8-nm gold nanorods (*29*). Nanostructures were immobilized in the following ways. The LDG was modified with aryldiazonium derivatives as had been done before (*21*), incubated with a $NaNO_2$ solution, and finally incubated in a gold nanostructure dispersion during different times for up to three days (Figure 4). Nitrite causes the diazotization of the amino-terminated modified LDG, which reacted with the gold nanoparticles and tethered them to the surface. The Au-LDG electrodes were later modified with ThLc following the strategy developed for gold electrodes. The most significant result was only obtained with the small nanoparticles of 5-nm diameter. In this case, two electrochemical behaviors were observed instead of one. The cyclic voltammograms (CVs) comprised two overlapping processes: the typical one where the rate-limiting step is DET depending on the distance between the enzyme and the electrode surface, and another one in which the enzyme active site is efficiently wired to the AuNPs-modified electrode. In the latter case, a Nernstian-dependent current response is obtained, where the rate-limiting step is the enzymatic reaction (Figure 4) . As this result only occurred with the smallest AuNPs, the hypothesis was that the smallest AuNPs had a higher orbital coupling with the T1 active site of the laccase that allowed for faster electron exchange (*28*). The chloride inhibition measured for these systems was for the 5-nm AuNPs similar to LDG electrodes, whereas the 16-nm AuNPs and nanorods (*29*) suffered higher inhibition, getting closer to gold planar electrodes.

Laccases are, together with BOxs, the most used redox enzymes for the development of biocathodes and have been extensively tested in both MET and DET regimes. They are able to promote O_2 reduction at very high potentials and close to the thermodynamic potential and can produce large current densities by using nanostructured electrodes, air-breathing cathodes, or redox polymer-based electrodes (*12*, *13*, *30*). Although laccases are not naturally reversible, another research direction has aimed to artificially force the enzymes to oxidize water to O_2, either electrochemically (*31*) or by the assistance of light energy (*32*, *33*), which presents possible alternatives to PSII and serves as an inspiration for biomimetic synthesis.

Hybrid Inorganic and Biological Photocatalysts Based on Redox Metalloenzymes Attached to Semiconductors for Water Electrolysis

Photoelectrochemical devices are currently considered as an emerging technology for sustainable production of fuels, either for water splitting to produce molecular hydrogen or for carbon dioxide reduction to formate or methanol. The energetic gain obtained by visible light absorption during the photoechemical process allows for a considerable decrease in the the overpotential for these reduction reactions (*34*). In particular, the attachment of enzymes that specifically catalyze these reactions to semiconductor particles is a promising strategy, as it avoids using expensive and non-sustainable noble metals as cocatalysts.

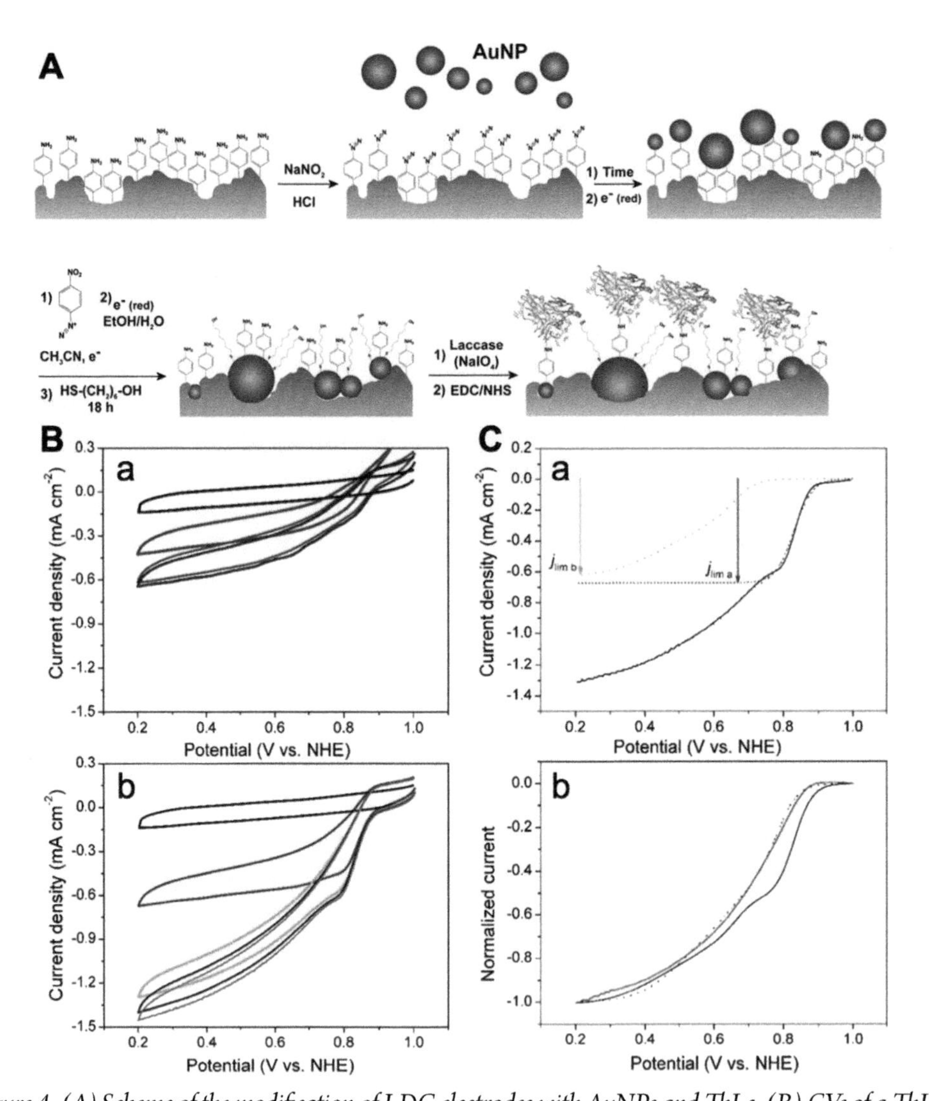

Figure 4. (A) Scheme of the modification of LDG electrodes with AuNPs and ThLc. (B) CVs of a ThLc-AuNP(5 nm)-LDG electrode at different rotation rates under air (a) or under 1 atm O_2 (b). Scan rate was 10 mV/s. Rotation rates were 0 (red), 500 (green), 1000 (blue), or 1500 rpm (gray). The black line corresponds to the CV measured in the presence of 30 mM NaF. (C) Mathematical fittings for the blank-subtracted forward scan of the CVs recorded at 1500 rpm under 1 atm of O_2. (a) ThLc- AuNP(5 nm)-LDG electrode: Black line is the experimental CV; red dotted line is the theoretical curve of a one-electron Nernstian process; green dashed line is the theoretical curve for a reductive electrocatalytic process; blue dashed line corresponds to the addition of the red and green curves. (b) Comparison of ThLc-AuNP(5 nm)-LDG and ThLc-LDG electrodes: Black and red lines are the experimental CVs of ThLc-AuNP(5 nm)-LDG and ThLc-LDG electrodes, respectively; blue dotted line is the theoretical curve for a reductive electrocatalytic process. pH of 4.2, 27 °C, $E'(O_2/H_2O) = 0.98$ V vs NHE. Reproduced with permission from reference (28) . Copyright 2012 American Chemical Society.

Early research already showed that hydrogenases, which are enzymes that reversibly catalyze reduction of protons to molecular hydrogen, could directly receive excited electrons from the conductive band of TiO_2 under UV irradiation (35, 36). A decade ago, Reisner and coworkers showed that the NiFeSe-hydrogenase (NiFeSe-Hase) from *Desulfomicrobium baculatum* favorably adsorbed onto the surface of Ru dye-sensitized TiO_2 nanoparticles. In this way, the hybrid photocatalyst could absorb visible light, which produces H_2 with a maximal turnover frequency (TOF) of 50 s^{-1} in the presence of triethanolamine as a sacrificial agent and under standard conditions (25 °C, pH 7.0) (37). A few years, later King and coworkers were able to improve the significantly charged transfer between a visible-light excited semiconductor and a hydrogenase. CdS nanorods were modified with an SAM of 3-mercaptopropionic acid for promoting oriented adsorption of *Clostridium acetobutylicum* FeFe-hydrogenase I (FeFe-Hase) by electrostatic interactions between the negatively charged carboxylate groups on the semiconductor surface and the positively charged patch that surrounds the entry point for the enzymatic electron transport chain (*38, 39*). This strategy yielded a TOF of H_2 production up to 900 s^{-1} and 20% photon conversion under illumination at 405 nm (*38*). The CdS-hydrogenase complexes were operationally stable for 4 h before losing completely their photocatalytic activity (*38*).

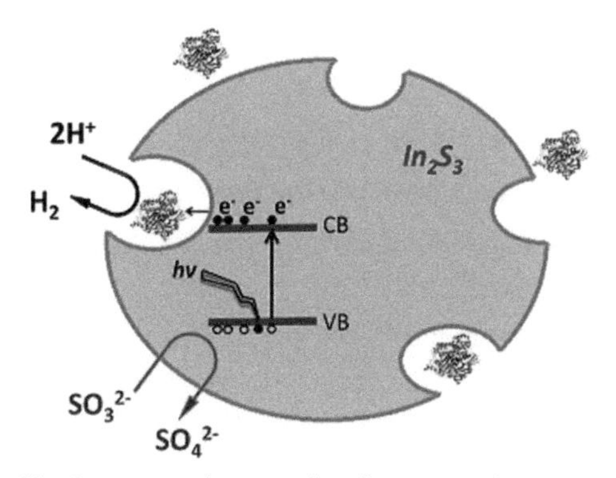

Figure 5. Hybrid In_2S_3 and hydrogenase photocatalyst for H_2 production. Reproduced with permission from reference (40). Copyright 2016 American Chemical Society.

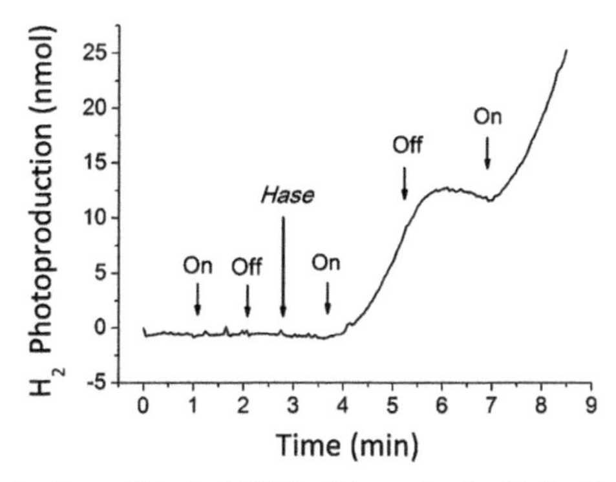

Figure 6. Photocatalytic production of H_2 by NiFeSe-Hase mixed with In_2S_3 particles monitored by mass spectrometry. The measurements were performed at 37 °C in 50 mM of Tris-HCl and 0.2 M of sodium sulfite at pH of 7. The arrows mark the times at which the lamp was turned on or off and of Hase injection into the reactor vessel. Reproduced with permission from reference (40). Copyright 2016 American Chemical Society.

In our group, we have studied the attachment of the NiFeSe-Hase from *Desulfovibrio vulgaris* Hildenborough to In_2S_3 (Figure 5). The advantage of In_2S_3 is that it is less toxic than CdS, while having a similar band gap energy (approximately 2.3 eV) to CdS, a conductive band potential of about -0.8 V versus H^+ and H_2 at a pH 0, and it can be easily synthesized by a solvothermal method. Therefore, this semiconductor is well suited for photochemical H_2 production. We have studied the attachment of the hydrogenase molecules to In_2S_3 microparticles and measured the rates of H_2 photoproduction under visible light irradiation using a membrane-inlet mass spectrometer connected to the photoreactor (Figure 5) (*40*).

This experimental setup allowed for optimization of the rate of interfacial charge exchange from the conduction band of the semiconductor to the redox relay of the hydrogenase. The mass spectrometry measurements showed that H_2 photoproduction under visible light irradiation using sodium sulfite as the sacrificial reagent was only detected after hydrogenase was added to the In_2S_3 microparticle suspension (Figure 6).

Increased rates of H_2 photoenzymatic production were obtained if the hydrogenase and semiconductor microparticles were incubated under mild stirring for several hours at 4 °C. The optimal rates of H_2 photoproduction were obtained after 6 h of incubation, reaching a turnover frequency (TOF) of 986 s^{-1} (Figure 7). Moreover, 94% of photocatalytic efficiency was achieved when compared to the H_2-production activity of the hydrogenase using reduced methyl viologen as an electron donor. It was shown that 89% of the photocatalytic activity was due to the enzyme being attached to the semiconductor particles (*40*). Thus, it was concluded that the long incubation period was required to get the hydrogenase molecules within the pores of the semiconductor microparticles. Indeed, the pore analysis of the In_2S_3 indicated an average diameter of 16.5 nm, which is a suitable pore size for accomodating hydrogenase molecules with a diameter of approximately 5 nm. In this way, the hydrogenase molecules introduced into the pores would have the semiconductor surface very near their most exposed redox site, therefore allowing fast transfer of excited electrons from the conductor band to the enzyme (Figure 5).

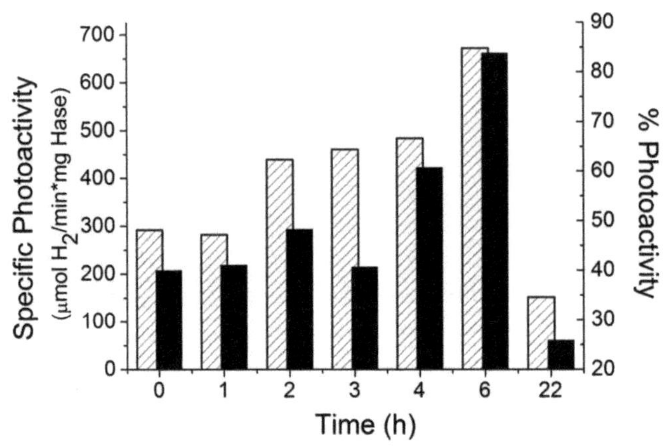

Figure 7. Photocatalytic production of H_2 by NiFeSe-Hase in combination with In_2S_3 particles monitored by mass spectrometry. Striped column bars represent the specific activity of H_2 photoproduction by Hase after different incubation times. Black column bars represent the percentage of photoactivity of Hase compared to the specific activity of the sample measured with reduced MV as an electron donor. Reproduced with permission from reference (40). Copyright 2016 American Chemical Society.

Photoelectrochemical production of H_2 requires replacing the sacrificial agent by an electrode as an electron donor to reduce protons. However, the above mentioned semiconductors that have been interfaced successfully with hydrogenases have the inconvenience of being n-types. This means that the holes generated in the valence bands after light irradiation are transported slowly within these materials toward the interface with the electrode, where they should be refilled with electrons. In order to overcome this kinetic limitation, the direct adsorption of hydrogenase on a p-type semiconductor or one covered with a layer of protecting n-type semiconductor, which form a p-n heterojunction, have been studied. Reisner and coworkers have reported the immobilization of *D. baculatum* NiFeSe-Hase on TiO_2-protected p-Si. The role of the n-type TiO_2 was to protect the p-type Si from oxidation in aqueous solutions, as well as to provide biocompatible support for the hydrogenase attachment. In this photocathode configuration, photoelectrochemical H_2 production was measured with a faradaic yield of approximately 95% (*41*). Alternatively, Zhao and coworkers reported that FeFe-Hase adsorbed onto a nanoporous black Si photocathode resulted in photocurrents above 1 mA/cm^2 (*42*). These very high photocurrents can be attributed to the efficiency of hole transport across the black Si and to the high rate of charge transfer between the nanostructured semiconductor and the hydrogenase. The main drawback of this photobiocathode was the low stability of the hydrogenase caused either by its desorption or its inactivation (*42*). Photoelectrochemical production of H_2 has also been reported with another FeFe-Hase adsorbed to a CuO_2-ZnO photocathode (*43*). Cu_2O is a visible light–absorbing p-type semiconductor with valence and conducting bands that match those of the n-type ZnO, thus allowing an efficient p-n heterojunction between both materials. The ZnO layer protected Cu_2O from its photoreaction in water and served as support for the enzyme immobilization (*43*).

Photoelectrochemical water splitting requires combining an H_2-evolution photocathode with an O_2-evolution photoanode. In fact, the latter reaction is kinetically more difficult because it demands a transfer of four electrons and four protons combined with the formation of an O-O bond. Therefore, an efficient cocatalyst has to be matched with an n-type semiconductor that generates sufficiently high potential holes for water oxidation and quickly transfers the excited electrons to the anode. The most efficient catalysts for water oxidation are oxides of expensive and scarce elements, such as RuO_2 and IrO_2 (*44*). Therefore, biological catalysts have been studied as an alternative for photoelectrochemical O_2 evolution. Photosystem II is the enzyme complex responsible for catalyzing O_2 evolution in natural photosynthesis (*45*). Its attachment to electrodes for photoelectrochemical evolution of O_2 has been thoroughly studied by Reisner and coworkers (*46–48*). The main drawback of attaching photosystem II to electrodes relies on the complex structure of this multisubunit enzyme, which propduces low operational stability on the electrode and requires MET for achieving high photocurrents (*48*). In order to increase the operational stability of the photobiocatalyst on the anode, researchers have studied the attachment of thylakoid membranes (TMs) on nanostructured electrodes. These membranes isolated from spinach contain photosystem II in their natural environment, and it has been shown that they can be electrically wired by DET to anodes modified with multiwalled carbon nanotubes (*49, 50*) or indium tin oxide nanoparticles (*51*). We showed that higher anodic photocurrents could be measured (up to 5 $\mu A/cm^2$) when the TMs were deposited over an electroreduced graphene oxide support, which was subsequently modified with 4-aminophenyl groups by electrochemical reduction of the diazonium salt derivative (Figure 8) (*52*). The generated aminoaryl groups introduced positive charges in the electrode surface for favoring the immobilization of the negatively-charged TMs and to eliminate oxygen-containing functional groups that have been shown to contribute to oxidative stress of biological membranes. The operational stability of the TMs on the graphene-modified electrodes was also higher than those on other supports, retaining 64% of initial activity after 500 s and 18% after 1 h (52). Nevertheless, both photocurrent density and operational stability achieved with TM anodes are still too low for practical applications.

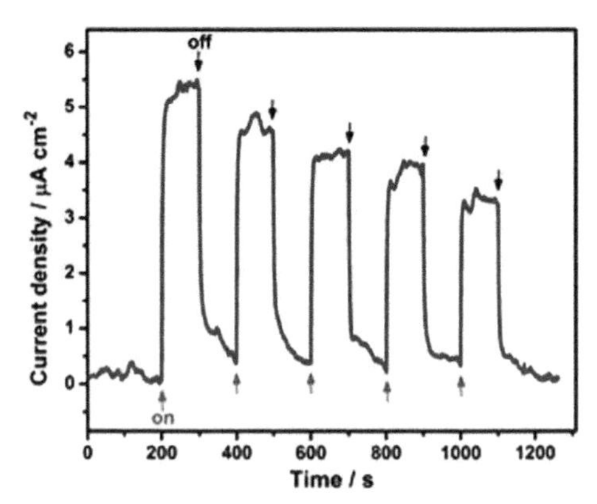

Figure 8. Representative background-subtracted amperogram for TMs immobilized on an electroreduced graphene oxide electrode under "light on" and "light off" conditions at an applied potential of 0.6 V vs. SHE. Reproduced with permission from reference (52). Copyright 2018 American Chemical Society.

In our group, we have studied the covalent attachment of ThLc to p-type sulfide semiconductors as an alternative photobioelectrocalyst for the oxygen evolution reaction (*32, 33*). These semiconductors are adequate for developing photoanodes because they quickly transport excited electrons to the electrode, and some of them have a valence band with a redox potential higher than the OH^- and O_2 pair. Although the native activity of laccases is the reduction of O_2 to H_2O, we previously reported that some native and mutant laccases bound to graphite electrodes can catalyze the reverse reaction at a neutral pH when a very high redox potential is applied (*31*). With this result in mind, we studied the functionalization of In_2S_3 deposited on transparent FTO electrodes with amino phenyl groups generated by electrochemical reduction of the diazonium salt derivative. The laccase molecules were then bound to the functionalized semiconductor surface by the two-step process described in the previous section (*24*), which favors their correct orientation on the surface for DET (Figure 9).

CV experiments under visible light irradiation, combined with an in situ O_2 microsensor, showed that photoelectrochemical O_2 evolution only happens if the photocathode contains both In_2S_3 and laccase (Figure 10) (*32*).

Chronoamperometric (CA) measurements under light illumination indicated that the maximal faradaic yield (45%) for O_2 evolution was obtained when the photobioanode was polarized at 1 V vs. SHE; although at 0.8 V, O_2 formation was already detected (*32*). This onset potential for photoelectrochemical O_2 evolution is at 0.4 V of lower overpotential than that required for the electrocatalytic system with the same laccase bond to a graphite electrode (*31*). This result reflects the energetic gain obtained by the visible light absorption event. However, the operational stability of the photobioelectrocatalyst proved to be low, mostly due to the inactivation of the laccase (*32*).

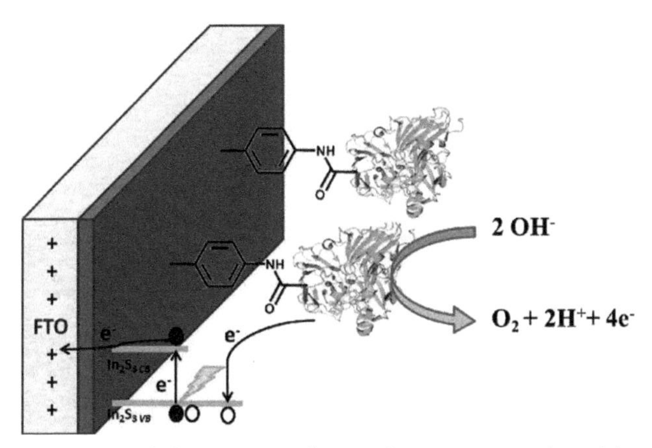

Figure 9. Covalent attachment of ThLc to In_2S_3 deposited on a FTO anode and functionalized with 4-aminophenyl groups. Reproduced with permission from reference (32). Copyright 2017 American Chemical Society.

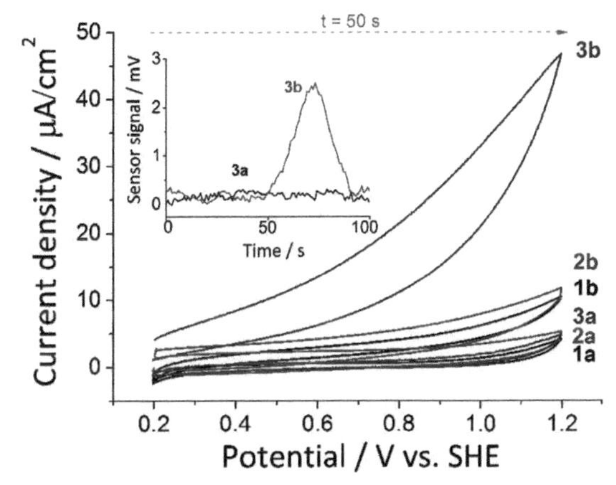

Figure 10. CVs measured in dark (a) and under illumination conditions (b) of FTO electrodes modified with covalently bound ThLc (1), modified with In_2S_3 functionalized with amino phenyl groups (2), and modified with In_2S_3 plus covalently bound ThLc (3). Scan rate was 20 mV/s. Inset: Potentiometric signal recorded by the oxygen sensor while scanning 3a and 3b CVs. Reproduced with permission from reference (32). Copyright 2017 American Chemical Society.

We have also studied the immobilization of the ThLc by the same methodology using SnS_2 instead of In_2S_3. The reason for this study is that the former sulfide has a higher stability compared to photocorrosion, and the redox potential of its valence band is thermodynamically more adequate for oxidizing water to molecular oxygen. Indium tin oxide nanoparticles were deposited on the photocathode in order to improve the rates of charge transfer from the semiconductor to the transparent electrode. The measured photocurrents of O_2 evolution were significantly smaller when SnS_2 was used as a semiconductor, but the operational stability of the photobiocathode was higher (Figure 11) and the best faradaic yield was 75% (33). Furthermore, it was importantly noted that a 0.6–V lower overpotential was required with this photoanode compared to the FTO, In_2S_3, and laccase ones. This latter result can be explained by the higher redox potential of the valence band of SnS_2 compared to In_2S_3.

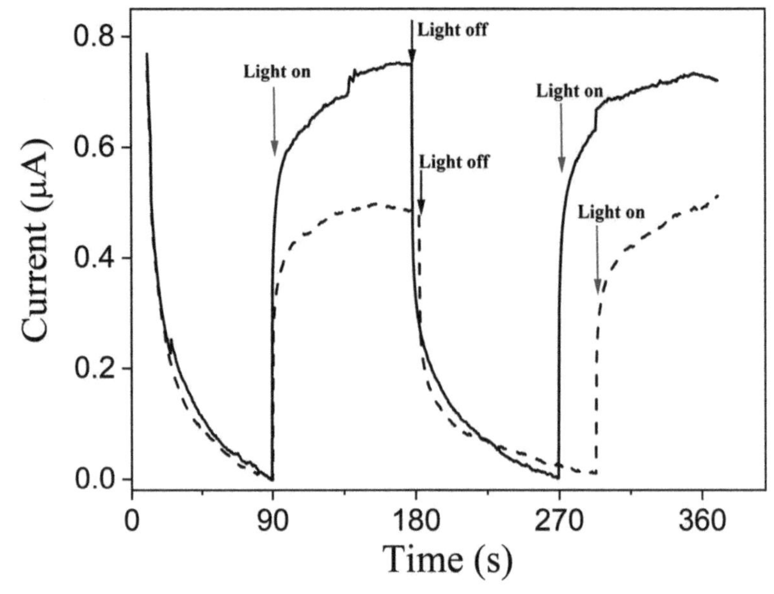

Figure 11. CA response of a FTO-SnS_2-ThLc-ITONP electrode at 1 V vs. SHE under dark and illumination cycles of 90 s applying 140 W (solid line) and 81.4 W (dashed line) of power to the lamp.

Bioelectrocatalytic Systems Based on Membrane Enzymes Reconstituted on Supported Phospholipid Bilayers

The main problem presented by membrane enzymes immobilized on electrodes is the partial or total loss of their catalytic activity due to aggregation by their hydrophobic domains, if not complete dissociation between hydrophilic and hydrophobic subunits (*53*). By reconstituting these enzymes in lipid bilayers, the goal is to mimic their natural environment and thereby optimize their activity and stability on the electrode (*54*). In our group, we have developed several strategies for this purpose in recent years.

In 2011, we reported the immobilization of the membrane form of the NiFeSe-Hase from *D. vulgaris* on planar gold electrodes by keeping its maximum activity (*55*). To accomplish this, a strategy was developed for covalent and oriented immobilization of an enzyme layer on the electrode surface and formation of a phospholipid bilayer on top using its lipid tail as scaffold (Figure 12A). A SAM of 4-ATP introduced positive charges to the electrode surface at a pH of 6. The negative surface around the distal cluster 4Fe4S of the hydrogenase was oriented by electrostatic charges toward the positively charged amino groups of the SAM (*17*). This cluster connects the surface of the enzyme with the Ni-Fe active site of the hydogenase through an electron transport chain. The *D. vulgaris* NiFe-Hase has a post-transcriptional modification during maturation that adds a lipid tail at the N-terminal end of the large subunit, which allows its attachment to the periplasmic membrane of cells (*56*). Therefore, the covalent immobilization of the membrane hydrogenase in the presence of phospholipids and CALBIOSORB BioBeads for detergent removal allowed for the stabilization of the enzyme by inserting its lipid tail into the PhBL (*55*). The electrochemical tests showed the oxidation of H_2 catalyzed by DET of the hydrogenase at a very low overpotential (Figure 12B).

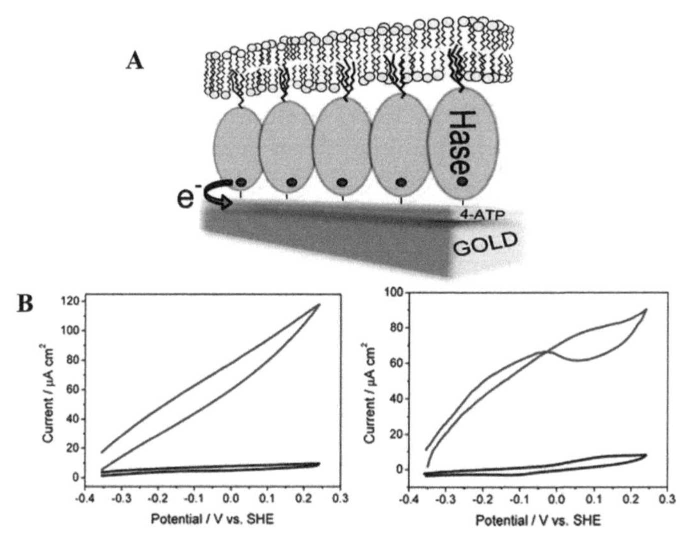

Figure 12. (A) Scheme representing the covalent and oriented immobilization of hydrogenase molecules with the subsequent formation of a membrane in the upper part of the proteins. The DET between the active center of the enzyme and the electrode is represented. (B) CVs of (left) a gold plate–modified with 4-ATP and (right) a gold wire to which the membrane hydrogenase was covalently immobilized in the presence of phospholipids and the CALBIOSORB adsorbent. The measurements were made in a 0.1-M phosphate buffer at a pH of 6 under 1 atm of N_2 (black lines) or 1 atm of H_2 (red lines) at a scan rate of 20 mV/s^{-1} and 40 ° C. Reproduced with permission from reference (55). Copyright 2011 American Chemical Society.

Following a similar strategy, our 2015 report showed that with the Au, Hase, and PhBL electrode system the energy obtained by the oxidation of H_2 could be stored to generate a proton gradient across the phospholipid bilayer (57). The immobilization method of the enzyme differed slightly from the previous one. First, the oriented covalent bonding of the NiFeSe-Hase was done, and the PhBL was formed over the monolayer of enzyme in a second step. To verify the accumulated energy in the form of protons in the electrode and PhBL interface, the local pH was measured using the SAM on the gold electrode as a redox probe. For this, differential pulse voltammetry (DPV), which is a more sensitive electrochemical technique for measuring peak potential shifts than CV, was used to measure the displacement of the redox potential of the SAM after its oxidation and subsequent dimerization. These dimers were formed after subjecting the monolayer of 4-aminothiophenol to a CV of 15 cycles between 0.25 V and 0.70 V compared SHE. The formed imino quinone groups have a two electron and two proton reversible redox process (58). By generating a calibration curve, the pH changes at the electrode and PhBL interface could be measured when H_2 oxidation catalyzed by the immobilized Hase was initiated by applying a redox potential of 0.18 V against SHE (57).

In nature, energy transduction mechanisms (such as photosynthesis and cellular respiration) store light and chemical energy in the form of an electrochemical gradient created through a lipid bilayer (59). In order to mimic this biochemical process, research has focused on whether the accumulated energy obtained with the proton gradient generated by the electroenzymatic oxidation of H_2 could drive the synthesis of adenosine triphosphate (ATP) by inserting ATP synthase into the PhBL over the hydrogenase layer. In 2016, we reported the reconstitution of the ATP synthase F1F0 (ATPase) from *Escherichia coli* to the Au, Hase, and PhBL system (60). For this goal, pores were opened in the liposome membrane by exposing them to a nonionic detergent, n-dodecyl β-D-maltoside (DDM). Later, they were incubated with the ATPase and CALBIOSORB BioBeads in order to eliminate the detergent and obtain proteoliposomes where the orientation of the ATPase was mainly with the catalytic head toward the exterior of the proteoliposomes (*61*, *62*). The steps for the Au, Hase, PhBL, and ATPase electrode formation were similar to those of the previous research (*60*). The final result was a system with two membrane enzymes: the Hase that oxidized H_2 for generating the proton gradient and the ATPase that used the proton gradient to synthesize ATP from adenosine diphosphate and inorganic phosphate that are present in the bulk electrolyte solution (*60*). The formation process of the floating PhBL was studied using a quartz microbalance (QCM). The synthesis of ATP was carried out in an anaerobic chamber by applying a potential of 0.2 V compared to SCE at the Au, Hase, PhBL, and ATPase electrode for the electroenzymatic oxidation of H_2, while an adenosine diphosphate concentration of 500 μM in phosphate buffer medium was the electrolyte solution. The ATP production in the bulk solution was measured by luciferase assay (*60*).

In another example of reconstituting the membrane enzymes on gold electrodes, the respiratory complex I (CI) was for the first time completely and functionally reconstituted in a biomimetic membrane over a modified gold electrode (*63*). CI oxidizes NADH to NAD+ while translocating protons across the cellular membrane, which generates a proton gradient that drives the synthesis of ATP (*64*). Previously, reconstitution of CI was attempted using fractions of cell membranes containing CI or by reconstituting the isolated protein in liposomes (*65–67*).

The strategy that followed for reconstituting CI on Au electrodes was very similar to the previous examples. The gold electrodes (in this case, gold wires) were modified with a monolayer of 4-aminothiophenol and, on it, a floating PhBL containing the CI was formed. In this work, the bacterial CI isolated from *Rhodothermus marinus* was used, which is formed by 14 subunits. The negative net charge in the lipid composition helped the formation of the bilayer by interacting with the positive charges of the amino groups at the SAM on the gold electrode. In addition to these lipids, 2,3-dimethyl-1,4-naphthoquinone (DMN) was included in the bilayer as an electrochemical mediator for the oxidation of NADH (Figure 13) (*63*).

Atomic force microscopy and faradaic impedance measurements showed the biomimetic construction over the electrode, while electrochemical measurements showed the CI functionality (*63*). To study the NADH oxidation and proton translocation, electrochemical tests were performed. In the case of NADH oxidation, CVs between -0.3 V and 0.6 V compared to SHE showed that the electrocatalytic effect was very dependent on the scan rate, which indicates that the electroenzymatic process was slow. CAs were also performed at a constant potential of +530 mV compared to SHE while adding increasing concentrations of NADH (Figure 14A). Both in the CV and CA measurements, it was observed that the oxidation current decreased over time despite the increased concentration of NADH in the medium. This effect was attributed to reduced naphtoquinone leaking from the PhBL due to its higher hydrophilicity. Negative controls were also performed with the system without DMN in the bilayer, and no electrocatalytic NADH oxidation was observed (*63*).

In the case of proton translocation, as in the cases of Au, Hase, and PhBL electrodes (*57*), we used DPV to measure the change in the oxidation peak potential of the redox pair that appears after the dimerization of the SAM monolayer (Figure 14B).

Importantly, Pelster and Minteer reported in 2016 that the combined reconstitution of complexes I, III, and IV from mitochondria (supercomplex) on a PhBL over an electrode (Figure 15) (*68*). The efficiency of the electron transport chain is considered to be caused by the synergy between complexes I, III, and IV in the membrane. Tests were done with the substrates of the enzymes that form the supercomplex and with their respective inhibitors, which concluded that the enzymes reconstituted over the electrode conserved their activity (*68*).

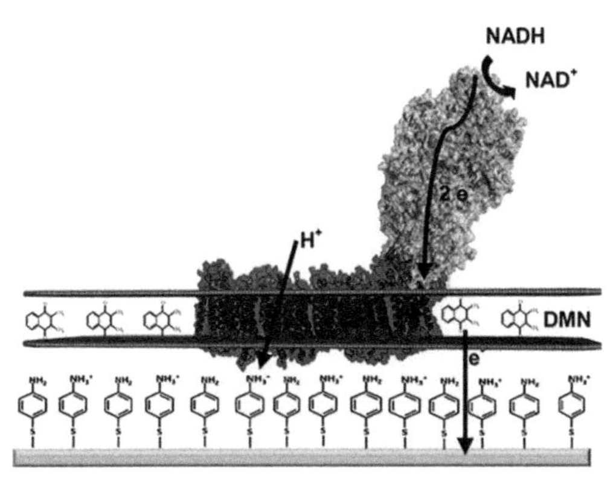

Figure 13. Schematic representation of CI reconstituted in a PhBL containing DMN over a gold electrode modified with a 4-ATP SAM. CI oxidized NADH, transferring the electrons to the DMN present in the PhBL, which acts as a redox mediator with the electrode. The electroenzymatic process was coupled to the translocation of protons through the biomimetic membrane. Reproduced with permission from reference (63). Copyright 2014 American Chemical Society.

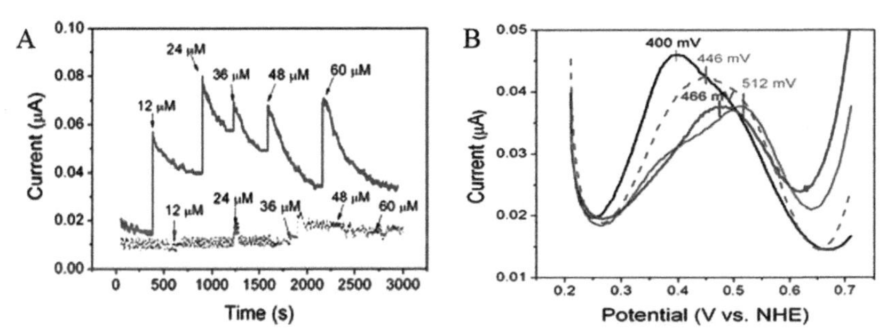

Figure 14. (A) CA recorded at 0.51 V vs. NHE in 0.1 M phosphate buffer, pH 7.0 with sequential additions of NADH at Au, 4-ATP, and PhBL+CI+DMN electrodes (red line) and Au, 4-ATP, and PhBL+DMN electrodes (black line). (B) DPVs of Au, 4-ATP, and PhBL+CI+ DMN in a 1-mM phosphate buffer, pH 7.0, 20 mM of Na_2S (red line); after the addition of 60 μM of NADH (green line); after the addition of 60 μM NADH and 70 μM of 2,4-dinitrophenol (dotted orange line). The black line corresponds to the DPV of Au, 4-ATP, and PhBL+ DMN in the presence of 60 μM of NADH. Reproduced with permission from reference (63). Copyright 2014 American Chemical Society.

The fractions of the isolated supercomplex were reconstituted on a PhBL simulating the composition of the mitochondria's inner membrane in order to preserve their maximum activity. Cardiolipin and the coenzyme ubiquinone were essential within the composition of the PhBL for supercomplex activity. When cardiolipin was eliminated, 98% of the activity was lost. Electrochemical studies demonstrated the interdependence of the activity of the different enzymes for the supercomplex electron transport chain. The gold-disk electrodes were modified with a mixed monolayer of β-mercaptoethanol and cysteamine to make the gold surface hydrophilic and provide an amine functional group for PEGylated cholesterol N-hydroxysuccinimide. Metabolon proteoliposomes were immobilized on the electrode by drop-casting them onto the electrode surface. Cytochrome c was the electron shuttle between complexes III and IV. Cytochrome c could move freely in the PhBL and electrode due to the small defects that the PhBL presents during its formation. Electrochemically, this can be seen by a reversible redox signal at 0.035 V compared to Ag and AgCl reference electrodes. In the electrochemical tests of the supercomplex, electrocatalytic waves are only observed in the presence of cytochrome c and increase with its concentration (Figure 16) (*68*).

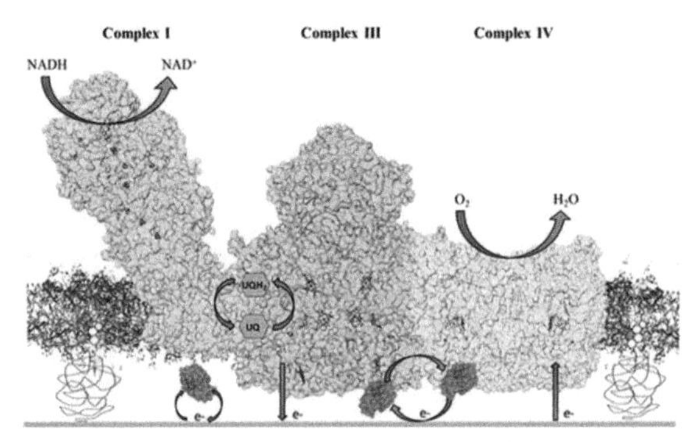

Figure 15. The scheme depicts the preferred orientation of the metabolon reconstituted into a tethered PhBL on an electrode. NADH oxidation and oxygen reduction take place in the bulk solution while cytochrome c oxidation and reduction take place in the hydrated space of the tethered PhBL facing the electrode surface. Reproduced from reference (68). Copyright American Chemical Society 2016.

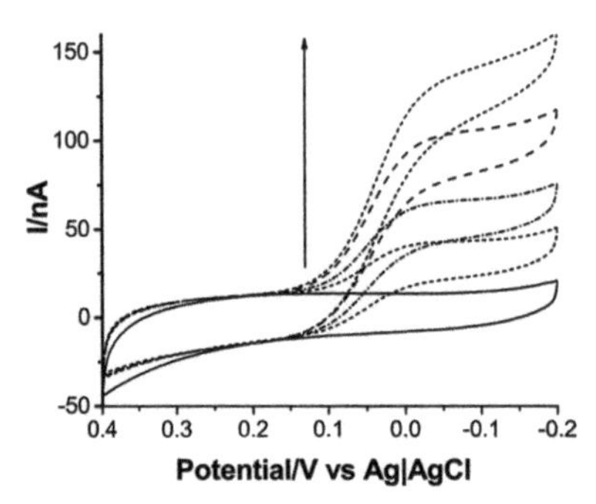

Figure 16. Representative CVs of the reconstituted metabolon with increasing oxidized cytochrome c concentrations (0–5 μM). The arrow indicates the increasing catalytic cathodic current with increasing concentrations of cytochrome c. Experiments were performed in a nitrogen-purged, 10-mM potassium phosphate buffer, pH of 7.5, and 70 mM potassium chloride at a scan rate of 5 m V s^{-1}. Reproduced with permission from reference (68). Copyright 2016 American Chemical Society.

References

1. Page, C. C.; Moser, C. C.; Chen, X.; Dutton, L. Natural Engineering Principles of Electron Tunneling in Biological Oxidation-Reduction. *Nature* **1999**, *402*, 47–52.
2. Leger, C.; Lederer, F.; Guigliarelli, B.; Bertrand, P. Electron Flow In Multicenter Enzymes: Theory, Applications and Consequences on the Natural Design of Redox Chains. *J. Am.Chem. Soc.* **2006**, *128*, 180–187.
3. Leger, C.; Bertrand, P. Direct Electrochemistry of Redox Enzymes as a Tool for Mechanistic Studies. *Chem. Rev.* **2008**, *108*, 2379–2438.
4. Milton, R. D.; Minteer, S. D. Direct Enzymatic Bioelectrocatalysis: Differentiating Between Myth and Reality. *J. Royal Soc. Interfaces* **2017**, *14*, 20170253.
5. Falk, M.; Blum, Z.; Shleev, S. Direct Electron Transfer Based Enzymatic Fuel Cells. *Electrochim. Acta* **2012**, *82*, 191–202.
6. Sokol, K. P.; Robinson, W. E.; Warnan, J.; Kornienko, N.; Nowaczyk, M. M.; Ruff, A.; Zhang, J. Z.; Reisner, E. Bias-Free Photoelectrochemical Water Splitting with Photosystem II on a Dye-Sensitized Photoanode Wired to Hydrogenase. *Nat. Energy* **2018**, *3*, 944–951.
7. Yuan, M.; Sahin, S.; Cai, R.; Abdellaoui, S.; Hickey, D. P.; Minteer, S. D.; Milton, R. D. Creating a Low-Potential Redox Polymer for Efficient Electroenzymatic CO_2 Reduction. *Angew. Chem. Int. Ed.* **2018**, *57*, 6582–6586.
8. Siritanaratkul, B.; Megarity, C.; Roberts, T.; Samuels, T.; Winkler, M.; Warner, J.; Happe, T.; Armstrong, F. Transfer of Photosynthetic NADP+/NADPH Recycling Activity to a Porous Metal Oxide for Highly Specific, Electrochemically-Driven Organic Synthesis. *Chem. Sci.* **2017**, *8*, 4579–4586.
9. Falk, M.; Alcalde, M.; Bartlett, P.; De Lacey, A. L.; Gorton, L.; Gutierrez-Sanchez, C.; Haddad, R.; Leech, D.; Ludwig, R.; Magner, E.; Mate, D. M.; Ó Conghaile, P.; Ortiz, R.; Pita, M.; Pöller, S.; Ruzgas, T.; Salaj-Kosla, U.; Schuhmann, W.; Sebelius, F.; Shao, M.; Stoica, L.; Tilly, J.; Toscano, M. D.; Vivekananthan, J.; Wright, E.; Shleev, S. Self-Powered Wireless Carbohydrate/Oxygen Sensitive Biodevice Based on Radio Signal Transmission. *Plos ONE* **2014**, *9*, E109104.
10. Solomon, E. I; Augustine, A. J.; Yoon, Y. O_2 Reduction to H_2O by the Multicopper Oxidases. *Dalton Transactions* **2008**, 3921–3932.
11. Shleev, S.; Christenson, A.; Serezhenkov, V.; Burbaev, D.; Yaropolov, A.; Gorrton, L.; Ruzgas, T. Electrochemical Redox Transformations of T1 And T2 Copper Sites in Native *Trametes hirsuta* Laccase at Gold Electrode. *Biochem. J.* **2005**, *385*, 745–754.
12. Mano, N.; de Poulpiquet, A. Oxygen Reduction in Enzymatic Biofuel Cells. *Chem. Rev.* **2018**, *118*, 2392–2468.
13. Soukharev, V.; Mano, N.; Heller, A. A Four-Electron O_2-Electroreduction Biocatalyst Superior to Platinum and a Biofuel Cell Operating at 0.88 V. *J. Am. Chem. Soc.* **2004**, *123*, 8368–8369.
14. Ackermann, Y.; Guschin, D. A.; Eckhard, K.; Shleev, S.; Schuhmann, W. Design of a Bioelectrocatalytic Electrode Interface for Oxygen Reduction in Biofuel Cells Based on a Specifically Adapted Os-Complex Containing Redox Polymer with Entrapped *Trametes hirsuta* Laccase. *Electrochem. Commun.* **2010**, *12*, 640–643.

15. Amir, L.; Kin Tam, T.; Pita, M.; Meijler, M. M.; Alfonta, L.; Katz, E. Biofuel Cell Controlled by Enzyme Logic Systems. *J. Am. Chem. Soc.* **2009**, *131*, 826–832.
16. Abad, J. M.; Gass, M.; Bleloch, A.; Schiffrin, D. J. Direct Electron Transfer to a Metalloenzyme Redox Center Coordinated to a Monolayer-Protected Cluster. *J. Am. Chem. Soc.* **2009**, *131*, 10229–10236.
17. Rüdiger, O.; Abad, J. M.; Hatchikian, E. C.; Fernandez, V. M.; De Lacey, A. L. Oriented Immobilization Of *Desulfovibrio gigas* Hydrogenase onto Carbon Electrodes by Covalent Bonds for Non-Mediated Oxidation of H_2. *J. Am. Chem. Soc.* **2005**, *127*, 16008–16009.
18. Solomon, E. I; Sundaram, U. M.; Machonkin, T. E. Multicopper Oxidases and Oxygenases. *Chem. Rev.* **1996**, *96* (7), 2563–2606.
19. Blanford, C. F.; Heath, R. S.; Armstrong, F. A. A Stable Electrode for High-Potential, Electrocatalytic O_2 Reduction Based on Rational Attachment of a Blue Copper Oxidase to a Graphite Surface. *Chem. Commun.* **2007**, 1710–1712.
20. Lalaoui, N.; Elouarzaki, K.; Le Goff, A.; Holzinger, M.; Cosnier, S. Efficient Direct Oxygen Reduction by Laccases Attached and Oriented on Pyrene-Functionalized Polypyrrole/Carbon Nanotube Electrodes. *Chem. Commun.* **2013**, *49*, 9281–9283.
21. Vaz-Dominguez, C.; Campuzano, S.; Rüdiger, O.; Pita, M.; Gorbacheva, M.; Shleev, S.; Fernandez, V. M.; De Lacey, A. L. Laccase Electrode for Direct Electrocatalytic Reduction of O_2 to H_2O with High-Operational Stability and Resistance to Chloride Inhibition. *Biosens. Bioelectron.* **2008**, *24*, 531–537.
22. Polyakov, K. M.; Federova, T. V.; Stepanova, T. V.; Cherkashin, E. A.; Kurzeev, S. A.; Strokopytov, B. V.; Lamzin, V. S.; Koroleva, O. V. Structure of Native Laccase from *Trametes hirsuta* at 1.8 Angstrom Resolution. *Acta Crystallogr.* **2009**, *D65*, 611–617.
23. Pita, M.; Gutierrez-Sanchez, C.; Olea, D.; Velez, M.; García-Diego, C.; Shleev, S.; Fernandez, V. M.; De Lacey, A. L. High-Redox Potential Cathode Based on Laccase Covalently Attached to Gold Electrode. *J. Phys. Chem. C* **2011**, *115*, 13420–13428.
24. Gutiérrez-Sánchez, C.; Jia, W.; Beyl, Y.; Pita, M.; Schuhmann, W.; De Lacey, A. L.; Stoica, L. Enhanced Direct Electron Transfer Between Laccase and Hierarchical Carbon Microfibers/Carbon Nanotubes Composite Electrodes. Comparison of Three Enzyme Immobilization Methods. *Electrochim. Acta* **2012**, *82*, 218–223.
25. Gupta, G.; Rajendran, V.; Atanassov, P. Bioelectrocatalysis of Oxygen Reduction Reaction by Laccase on Gold Electrodes. *Electroanalysis* **2004**, *16*, 1182–1185.
26. Pita, M.; Shleev, S.; Ruzgas, T.; Fernández, V. M.; Yaropolov, A. I.; Gorton, L. Direct Heterogeneous Electron Transfer Reactions of Fungal Laccases at Bare and Thiol-Modified Gold Electrodes. *Electrochem. Comm.* **2006**, *8*, 747–753.
27. Vaz-Domínguez, C.; Pita, M.; De Lacey, A. L.; Shleev, S.; Cuesta, A. Combined ATR-SEIRAS and EC-STM Study of the Immobilization of Laccase on Chemically Modified Au Electrodes. *J. Phys. Chem. C* **2012**, *116*, 16532–16540.
28. Gutierrez-Sanchez, C.; Pita, M.; Vaz-Domínguez, C.; Shleev, S.; De Lacey, A. L. Gold Nanoparticles as Electronic Bridges for Laccase-Based Biocathodes. *J. Am. Chem. Soc.* **2012**, *134*, 17212–17220.
29. Di Bari, C.; Shleev, S.; De Lacey, A. L.; Pita, M. Laccase-Modified Gold Nanorods for Electrocatalytic Reduction of Oxygen. *Bioelectrochemistry* **2016**, *107*, 30–36.

30. Szczesny, J.; Marković, N.; Conzuelo, F.; Zacarias, S.; Pereira, I. A. C.; Lubitz, N.; Plumeré, N.; Schuhmann, W.; Ruff, A. A Gas Breathing Hydrogen/Air Biofuel Cell Comprising a Redox Polymer/Hydrogenase-Based Bioanode. *Nat. Commun.* **2018**, *9*, 4715.
31. Pita, M.; Mate, D. M.; Gonzalez-Perez, D.; Shleev, S.; Fernandez, V. M.; Alcalde, M.; De Lacey, A. L. Bioelectrochemical Oxidation of Water. *J. Am. Chem. Soc.* **2014**, *136*, 5892–5895.
32. Tapia, C.; Shleev, S.; Conesa, J. C.; De Lacey, A. L.; Pita, M. Laccase-Catalyzed Bioelectrochemical Oxidation of Wáter Assisted with Visible Light. *ACS Catal.* **2017**, *7*, 4881–4889.
33. Jarne, C.; Paul, L.; Conesa, J. C.; Shleev, S.; De Lacey, A. L.; Pita, M. Underpotential Photoelectrooxidation of Water by Sns_2– Laccase Co-Catalysts on Nanostructured Electrodes with Only Visible-Light Irradiation. *Chemelectrochem* **2019**, *6*, 2755–2761.
34. Grätzel, M. Photoelectrochemical Cells. *Nature* **2001**, *414*, 338–344.
35. Cuendet, P.; Rao, K.; Grätzel; Hall, D. O. Light Induced H_2 Evolution in a Hydrogenase-$Ti0_2$ Particle System by Direct Electron Transfer or via Rhodium Complexes. *Biochimie* **1986**, *68*, 217–221.
36. Nikandrov, V. V.; Shlyk, M. A.; Zorin, N. A.; Gogotov, I. N.; Krasnovsky, A. A. Efficient Photoinduced Electron-Transfer from Inorganic Semiconductor TiO_2 to Bacterial Hydrogenase. *FEBS Lett.* **1988**, *234*, 111–114.
37. Reisner, E.; Powell, D.; Cavazza, C.; Fontecilla-Camps, J. C.; Armstrong, F. A. Visible Light-Driven H_2 Production by Hydrogenases Attached to Dye-Sensitized Tio2 Nanoparticles. *J. Am. Chem. Soc.* **2009**, *131*, 18457–18466.
38. Brown, K. A.; Wilker, M. B.; Boehm, M.; Dukovic, G.; King, P. W. Characterization of Photochemical Processes for H_2 Production by CdS Nanorod-[Fefe] Hydrogenase Complexes. *J. Am. Chem. Soc.* **2012**, *134*, 5627–5636.
39. Wilker, M. B.; Shinopoulos, K. E.; Brown, K. A.; Mulder, D. W.; King, P. W.; Dukovic, G. Electron Transfer Kinetics in CdS Nanorod-FeFe-Hydrogenase Complexes and Implications for Photochemical H_2 Generation. *J. Am. Chem. Soc.* **2014**, *136*, 4316–4324.
40. Tapia, C.; Zacarias, S.; Pereira, I. A. C.; Conesa, J. C.; Pita, M.; De Lacey, A. L. In Situ Determination of Photobioproduction of H_2 by In_2S_3-[NiFeSe] Hydrogenase from *Desulfovibrio vulgaris* Using Only Visible Light. *ACS Catal.* **2016**, *6*, 5691–5698.
41. Lee, C. H.; Park, H. S.; Fontecilla-Camps, J. C.; Reisner, E. Photoelectrochemical H_2 Evolution with a Hydrogenase Immobilized on a TiO_2-Protected Silicon Electrode. *Angew. Chem. Int. Ed.* **2016**, *55*, 5971–5974.
42. Zhao, Y.; Anderson, N.; Ratzloff, M. W.; Mulder, D. W.; Zhu, K.; Turner, J. A.; Neale, N.; King, P. W.; Branz, H. Proton Reduction Using a Hydrogenase-Modified Nanoporous Black Silicon Photoelectrode. *ACS Appl. Mater. Interfaces* **2016**, *8*, 14481–14487.
43. Tian, L.; Nemeth, B.; Berggren, G.; Tian, H. N. Hydrogen Evolution by a Photoelectrochemical Cell Based on a Cu_2O-Zno-FeFe Hydrogenase Electrode. *J. Photochem. Photobiol. A: Chem.* **2018**, *366*, 27–33.
44. Matheu, R.; Francas, L.; Chernev, P.; Ertem, M.; Batista, V.; Haumann, M.; Sala, X.; Llobet, A. Behavior of the Ru-Bda Water Oxidation Catalyst Covalently Anchored on Glassy Carbon Electrodes. *ACS Catal.* **2015**, *5*, 3422–3429.

45. Najafpour, M.; Ghobadi, M.; Larkum, A.; Shen, J.; Allahkhverdiev, S. The Biological Water-Oxidizing Complex at the Nano-Bio Interface. *Trends Plant Sci.* **2015**, *20*, 559–5668.
46. Mersch, D.; Lee, C. H.; Zhang, J. C.; Brinkert, K.; Fontecilla-Camps, J. C.; Rutherford, A. W.; Reisner, E. Wiring of Photosystem II to Hydrogenase for Photoelectrochemical Water Splitting. *J. Am. Chem. Soc.* **2015**, *137*, 8541–8549.
47. Kato, M.; Cardona, T.; Rutherford, A.; Reisner, E. Covalent Immobilization of Oriented Photosystem II on a Nanostructured Electrode for Solar Water Oxidation. *J. Am. Chem. Soc.* **2013**, *135*, 10610–10613.
48. Sokol, K.; Mersch, D.; Hartmann, V.; Zhang, J.; Nowaczyk, M.; Rögner, M.; Ruff, A.; Schuhmann, W.; Plumere, N.; Reisner, E. Rational Wiring of Photosystem II to Hierarchical Indium Tin Oxide Electrodes Using Redox Polymers. *Energy Environ. Sci.* **2016**, *9*, 3698–3709.
49. Calkins, J. O.; Umasankar, Y.; O'Neill, H.; Ramasamy, R. P. High Photo-Electrochemical Activity of Thylakoid-Carbon Nanotube Composites for Photosynthetic Energy Conversion. *Energy Environ. Sci.* **2013**, *6*, 1891–1900.
50. Pankratov, D.; Pankratova, G.; Dyachkova, T. P.; Falkman, P.; Åkerlund, H.-E.; Tascano, M. D.; Chi, Q.; Gorton, L. Supercapacitive Biosolar Cell Driven by Direct Electron Transfer Between Photosynthetic Membranes and CNT Network with Enhanced Performance. *ACS Energy Lett.* **2017**, *2*, 2635–2639.
51. González-Arribas, E.; Aleksejeva, O.; Bobrowski, T.; Toscano, M. D.; Gorton, L.; Schuhmann, W.; Shleev, S. Solar Biosupercapacitor. *Electrochem. Commun.* **2017**, *74*, 9–13.
52. Pankratova, G.; Pankratov, D.; Di Bari, C.; Goñi-Urtiaga, A.; Toscano, M. D.; Chi, Q.; Pita, M.; Gorton, L.; De Lacey, A. L. Three-Dimensional Graphene Matrix-Supported and Thylakoid Membrane-Based High-Performance Bioelectrochemical Solar Cell. *ACS Appl. Energy Mater.* **2018**, *1*, 319–323.
53. Saboe, P. O.; Conte, E.; Farell, M.; Bazan, G. C.; Kumar, M. Biomimetic and Bioinspired Approaches for Wiring Enzymes to Electrode Interfaces. *Energy Environ. Sci.* **2017**, *10*, 14–42.
54. Laftsoglou, T.; Jeuken, L. J. C. Supramolecular Electrode Assemblies for Electrochemistry. *Chem. Commun.* **2017**, *53*, 3801–3809.
55. Gutiérrez-Sánchez, C.; Olea, D.; Marques, M.; Fernández, V. M.; Pereira, I. A.; Vélez, M.; De Lacey, A. L. Oriented Immobilization of a Membrane-Bound Hydrogenase onto an Electrode for Direct Electron Transfer. *Langmuir* **2011**, *27*, 6449–6457.
56. Valente, F. M. A.; Oliveira, A. S. F.; Gnadt, N.; Pacheco, I.; Coelho, A. V.; Xavier, A. V.; Teixeira, M.; Soares, C. M.; Pereira, I. A. C. Hydrogenases n *Desulfovibrio vulgaris* Hildenborough: Structural and Physiologic Characterisation of the Membrane-Bound [NiFeSe] Hydrogenase. *J. Biol. Inorg. Chem.* **2005**, *10*, 667–682.
57. Gutiérrez-Sanz, Ó.; Tapia, C.; Marques, M. C.; Zacarias, S.; Vélez, M.; Pereira, I. A.; De Lacey, A. L. Induction of a Proton Gradient Across a Gold-Supported Biomimetic Membrane by Electroenzymatic H_2 Oxidation. *Angew. Chem. Int. Ed.* **2015**, *127*, 2722–2725.
58. Raj, C. R.; Kitamura, F.; Ohsaka, T. Electrochemical and *in Situ* FTIR Spectroscopic Investigation on the Electrochemical Transformation of 4-Aminothiophenol on a Gold Electrode in Neutral Solution. *Langmuir* **2001**, *17*, 7378–7386.
59. Elston, T.; Wang, H.; Oster, G. Energy Transduction in ATP Synthase. *Nature* **1998**, *391*, 510–513.

60. Gutiérrez-Sanz, Ó.; Natale, P.; Márquez, I.; Marques, M. C.; Zacarias, S.; Pita, M.; Pereira, I. A.; López-Montero, I.; De Lacey, A. L.; Vélez, M. H_2-Fueled ATP Synthesis on an Electrode: Mimicking Cellular Respiration. *Angew. Chem. Int. Ed.* **2016**, *128*, 6324–6328.
61. Milhiet, P.-E.; Gubellini, F.; Berquand, A.; Dosset, P.; Rigaud, J.-L.; Le Grimellec, C.; Lévy, D. High-Resolution AFM of Membrane Proteins Directly Incorporated at High Density in Planar Lipid Bilayer. *Biophys. J.* **2006**, *91*, 3268–3275.
62. Karamohamed, S.; Guidotti, G. Bioluminometric Method for Real-Time Detection of ATPase Activity. *Biotechniques* **2001**, *31*, 420–425.
63. Gutiérrez-Sanz, O.; Olea, D.; Pita, M.; Batista, A. P.; Alonso, A.; Pereira, M. M.; Vélez, M.; De Lacey, A. L. Reconstitution of Respiratory Complex I on a Biomimetic Membrane Supported on Gold Electrodes. *Langmuir* **2014**, *30*, 9007–9015.
64. Walker, J. E. The NADH:Ubiquinone Oxidoreductase (Complex I) of Respiratory Chains. *Quarterly Rev. Biophysics* **1992**, *25*, 253–324.
65. Pfeiffer, K.; Gohil, V.; Stuart, R. A.; Hunte, C.; Brandt, U.; Greenberg, M. L.; Schagger, H. Cardiolipin Stabilizes Respiratory Chain Supercomplexes. *J. Biol. Chem.* **2003**, *278*, 52873–52880.
66. Yano, T.; Dunham, W. R.; Ohnishi, T. Characterization of the $\Delta\mu$+-Sensitive Ubisemiquinone Species (Sqnf) and the Interaction with Cluster N2: New Insight into the Energy-Coupled Electron Transfer in Complex I. *Biochemistry* **2005**, *44*, 1744–1754.
67. Batista, A. P.; Marreiros, B. C.; Pereira, M. M. Decoupling of the Catalytic and Transport Activities of Complex I from *Rhodothermus marinus* by Sodium/Proton Antiporter Inhibitor. *ACS Chem. Biol.* **2011**, *6*, 477–483.
68. Pelster, L. N.; Minteer, S. D. Mitochondrial Inner Membrane Biomimic for the Investigation of Electron Transport Chain Supercomplex Bioelectrocatalysis. *ACS Catal.* **2016**, *6*, 4995–4999.

Chapter 11

Recent Advances on Metal Organic Framework–Derived Catalysts for Electrochemical Oxygen Reduction Reaction

Shaik Gouse Peera,[1] Hyuk Jun Kwon,[1] Tae Gwan Lee,[1] Jayaraman Balamurugan,[2] and A. Mohammed Hussain[*,3]

[1]Department of Environmental Science and Engineering, Keimyung University, Daegu 42602, Republic of South Korea

[2]Department of BIN Convergence Technology, Chonbuk National University, Jeonju 54896, Republic of Korea

[3]University of Maryland, Energy Research Center, College Park, Maryland 20740, United States

[*]E-mail: uzzain@umd.edu.

The catalysts for electrochemical oxygen reduction reaction (ORR) is earning great attention due to their significance importance in energy conversion devices like fuel cells and metal-air batteries. The traditional highly efficient platinum (Pt)-supported carbon catalysts hinder the commercialization of the fuel cells and metal-air batteries due to their high cost, scarcity, and poor durability. Hence, it is desirable to develop durable catalysts with inexpensive and more abundantly available resources. Transition metal-based catalysts represent the best alternative and cheap catalysts for ORRs. Among the various methods to synthesize the transition metal-based catalysts, metal organic framework (MOF)–derived catalysts represent the easies way to tune the metal catalyst activity and the porosity of the catalysts with large density of the active sites. MOFs are a class of materials where the metal ions are coordinated with the organic ligands. As the MOF contains carbons, metal ions, and heteroatoms, their pyrolysis can generate potential catalysts with ORR active sites. Due to their unique properties, they have been given attention recently in terms of their utilization as catalysts for various catalytic reactions. In this chapter, we review highly active ORR catalysts derived from MOFs and various currently adopted synthesis protocols, including pore size tunability, finely controlled metal nanoparticles, MOF-hybrid catalysts, and their interface properties. Of special interest, we also discuss the process parameters involved in controlling the catalytic properties and factors responsible for their electrocatalytic activity.

© 2020 American Chemical Society

Introduction

One of the key issues of our modern society is the aggressive energy demand for various growing needs (e.g., transportation and other mobile needs). Due to the increased population and growth of modern technologies, depletion of non-renewable fossil fuels, drastic negative effects of increased CO_2 levels are impacting the climate through global warming. The need for sustainable sources of energy is critical to control the disasterous emissions. Recent research efforts resulted in identifying renewable energy sources such as geothermal heat, solar power, wind power, and hydroelectric power. However, these energy technologies are not yet sufficient to meet the growing energy needs of the population. As a well-known and promising technology, fuel cells and metal-air batteries are considered to be the prefered choices due to their high power and energy densities. Among the two technologies, fuel cells, particularly low temperature operating ones such as proton exchange membrane fuel cells (PEMFCs) anion exchange membrane fuel cells (AEMFCs), or alkaline fuel cells (AFCs), are promising (*1–6*). Several efforts have also been made with other types of fuel cells like high temperature fuel cells (HT-PEMCs) (*7–10*) and solid oxide fuel cells (SOFCs) (*11–21*). However, low-temperature fuel cells (PEMFCs and AEMFCs) possess many advantages mainly due to their rapid start-up possibilities and lower emissions; a clean H_2 or reformate H_2 is utilized as a fuel, and atmospheric air and O_2 are utilized as oxidants.

PEMFCs have advantages over other types of fuel cells, such as quick their start-up, low operating temperatures, high power density, and smaller, more compact stacks. There are presently three fuel cell vehicles that are commercially available and in usage: Toyota Mirai, Hyundai Tucson fuel cell/ix35 fuel cell, and Honda Clarity (*22, 23*). However, even with the current technological advancements, the cost of the fuel cells is still high. This is mainly due to the platinum (Pt)-based catalysts utilized in PEMFCs for both hydrogen oxidation reaction (HOR) (anode electrode) and oxygen reduction reaction (ORR) (cathode electrode). Even with pure Pt, the ORR is about five orders of magnitude lower than the HOR; thus, the cathode electrode activity decides the overall performance of a fuel cell stack. Because of this, the cathode electrode in particular contains higher loading of Pt compared to the anode. Such high loading of the Pt supported on carbon (Pt/C) catalysts occupies 80%–90% of the total Pt contributing to the fuel cell cost, significantly increasing the overall system cost. U.S. Department of Energy (DOE) has put forward a target total of 0.125 mg for the Pt/C (both anode andcathode) catalyst with a mass activity of 0.44 A/mg_{PGM} @ 0.9 $V_{iR\text{-}free}$ by 2020 (*24*). However, recent statements by General Motors (GM) considered that the catalyst loading should be reduced to at least 0.0625 mg/cm^2 in order to truly compete with the current internal combustion engine–based vehicles. This is a great future challenge that has to be tackled by the scientific community (*25*). Developing highly efficient catalysts with low Pt loading is a subject of interest for research groups throughout the world. It is clear that high catalytic activity can only be achieved by Pt-based catalysts at present. Efforts are being made to enhance the ORR activity of Pt-based catalysts by: (i) controlling the dispersion of nanosized Pt nanoparticles on the high surface area carbon supports (*26–29*); (ii) alloying with transition metals (*30–32*); and (iii) inner-transition metals (*33, 34*) which modify the electronic band structure of Pt and induce the strain on the Pt alloy structure. Another way of enhancing the activity of Pt and simultaneously reducing the Pt content is by synthesizing core shell nanoparticles with non-precious metals as the core and Pt as the shell, which can greatly reduce the loading of Pt. Researchers have achieved >15 times the activity with Pt-alloys, especially with a Pt_3Ni system in half-cell configuration (RDE) (*35–38*). However, obtaining similar activity is a great challenge in an areal MEA configuration, due to the complex electrode system and mass transfer limitations of reactants and other fuel cell stack issues. Hence, it is indeed highly necessary to optimize both the catalyst and electrode design.

It is well known that the ORR is a multistep reaction process that leads to the reduction of O_2 to H_2O. The initial reaction that occurs on the surface of the electrocatalyst is the adsoption of O_2. The adsorbed oxygen further undergoes a series of reactions in the presence of protons, hydroxyl ions, and electrons transforming into H_2O, either in associative or dissociative reduction pathways. These two possible pathways differ in whether the bond between the two oxygen atoms breaks before or after the initial reduction. The oxygen coverage and nature of the catalysts influences whether the reduction step undergoes an associative or dissociative pathway to H_2O as the end product. In addition to the two previously mentioned pathways, the ORR also follows a pathway without breaking the bond between two oxygen atoms. It is said that if the reduction of O_2 undergoes an O-O bond break, it leads to the desired H_2O as the final product, and if the the reduction of O_2 occurs without an O-O bond break (4 e^- reduction process), it leads to the undesired H_2O_2 as a final product (2 e^- reduction process), which is harmful for the membrane electrode assemblies (MEAs). H_2O_2 is known to degrade the ionomer in the catalyst layer and degrade the electrolyte membrane. This contributes to the development of active catalysts, which perform high necessary and desired 4 e^- reduction processes.

Furthermore, the electronic structure of the catalyst influences the adsorption energies of the ORR intermediates (O*, OOH*, OH*). It is known that the kinetics of ORR depend on the adsorption strength of the ORR intermediates. Adsorption energies that are either too strong or too weak can slow down the ORR kinetics. If the catalyst possesses a weak binding energy, the initial proton and electron transfer to adsorbed O_2 forming OOH* becomes the rate-determining step, and if the catalyst has a strong binding energy, then the successful desorption of OOH* and OH* becomes the rate-determining step. The electronic band structure, which is the position of the d-band center relative to the fermi level, modulates the interaction of the catalysts with the intermediates, thus governing the ORR process. Currently, the Pt-based catalysts possess the best adsorption and desorption energies among ORR intermediates for faster ORR kinetics.

In spite of great achievements on Pt-based catalysts, high cost and scarcity of Pt-based noble catalysts greatly hinder the commercialization of fuel cells. Because of this, developing cheap, non-precious metal catalysts is of paramount significance to reduce the cost of fuel cells (*39–46*). Moreover, Pt-based catalysts also suffer from other issues like sensitivity to CO and CH_3OH species (*47*). esearchers around the globe have focused on the development of efficient electrocatalysts based on abundance, low cost, and facile synthesized catalysts suitable for large scale production (*48–51*). In light of this, significant achievements have been made by researchers with the non-precious metal catalyst, especially transition metal-based catalysts in combination with the doped heteroatoms (N, S, B, P, and halogens). It is well known that the carbon materials doped with heteroatoms induce the different change and spin densities around the carbon atoms, activating the carbon π electron density and inducing the metal-free ORR catalysis (*52, 53*). Accordingly, a vast amount of research has been performed to identify the efficient metal-free ORR catalysts, and many reports present comparable and even higher ORR activity of metal-free catalysts. However, most of them have reported only at RDE level studies, and very few studies have realized their application at the MEA level. Even at the MEA level, none of the metal-free catalysts showed comparable power density competitive with Pt/C based catalysts.

Researchers have identified that transition metals coordinated with N-doped carbon can generate a potentially active metal-nitrogen-carbon (M-N-C) bonds (Metals are Fe, Co, or Mn), which shows excellent ORR activity. Generally, the electrocatalytic activity of transition metal catalysts follows the order of Fe > Co > Mn > Cu > Ni (*54*, *55*). The origin of ORR activity for M-N-C bonds was revealed by the doublet and coworkers, where a direct, linear relationship between the number of coordinated Fe atoms to the N-doped carbons and ORR activity is assessed using ^{57}Fe Mossbauer spectroscopy (*56*). The high ORR activity of M-N-C bonds is thought to be due to the electronic configuration of the transition metal and the N-doped carbon structure (*57*). Adsorption and desorption of oxygen reduction intermediates on M-N-C active sites are reported by Mukerjee et al.; their study identified the direct involvement of M-N-C sites in the ORR process (*58*). Accordingly, a large variety of M-N-C based catalysts are derived and analyzed for ORR, both in acidic and alkaline electrolytes (*59*, *60*). As a general strategy, the M-N-C catalysts are synthesized from a mixture of metal sources with nitrogen or carbon sources and followed by pyrolysis. During the pyrolysis process, the carbonization of the precursor takes place with the simultaneous doping of an N species (together with the reduction of precursor ions into metallic state) to give M-N-C catalysts. Thus, a wide variety of N-containing chemical precursors are utilized to synthesize M-N-C catalysts (*61–63*). However, in a general strategy, the M-N-C catalysts can be synthesized via four different routes.

Route 1

M-N-C catalysts can be synthesized from the N-bearing precursors together with the carbon support. In this synthesis process, M, N, and C comes from three different precursors. The carbon support is essential to provide high surface area and required electronic conductivity, whereas N precursors are required to generate active N atoms to incorporate into the carbon hexagonal matrix, and metal precursors (metal salts such as sulphates, chlorides, nitrates, and acetates) are needed to form metal-centered bonds. There are a variety of carbon supports utilized, which include high surface area graphene oxide, acetylene black, Vulcan carbon, Ketjen black, carbon nanotubes (CNTs), and graphite nanofibers (*64–68*). Various N precursors are utilized as sources of N species, such as: melamine; dicyanamide; urea; triazine; 1,10-phenanthroline; pyrazine; and NH_3 gas (*69–72*). Mixtures of the carbon support, N precursor, and metal salts are further pyrolyzed in an inert atmosphere to obtain the final M-N-C catalyst. Though a large variety of pyrolysis temperatures are utilized by the researchers (i.e., 700–1100 °C), in most of the cases the optimum temperatures were found to be between 800 and 1000 °C (*73*).

Route 2

M-N-C catalysts can also be synthesized from the pyrolysis of a mixture of metal salts and a polymer substance which possess both C and N together. Sometimes, this synthesis route also contains the additional carbon support like Route 1. The pyrolysis of N-containing polymers provides the N species for doping and, in turn, form M-N-C structures in the presence of metal salts. A large variety of polymers are utilized such as electron-conducting polymers (polypyrrole, polyaniline, polythiophene, polyphenylenediamine), other polymers like polymerization of the resin urea-formaldehyde, polyamides, polyacrylonitrile, porous organic polymers, covalent organic framework, and macrocyclic-based polymers such as porphyrin and phthalocyanine (*74–81*). Among these, the conductive polymer-derived electrocatalysts are popular. In this synthesis route, a chemical polymerization of the monomer is performed either independently, or in combination with the carbon support, in presence of oxidants like hydrogen peroxide, ferric chloride, and ammonium persulfate. This is then followed by the pyrolysis of the metal-coordinated polymer complex to finally obtain M-N-C electrocatalysts.

Route 3

The M-N-C catalyst can be prepared from a hard template method usually from porous silica templates. The N-containing precursors and metal precursors are mixed with porous silica or polymerization of the monomers of the N-precursors and then pyrolyzed. After pyrolysis, the hard template silica is selectively etched without disturbing the pyrolyzed catalyst. Since the final catalyst contains the porous structure derived from the silica template, the resulting catalyst possesses a higher surface area, and usually the additional use of carbon support is prevented. Different types of silica are utilized for this purpose (i.e., SBA-15, SBA-16, KIT-6, MCM-41, MCM-48). The precursor loading on to silica can be done by using small organic molecules and N-bearing polymers along with the metal salts via wet impregnation or solvent evaporation methods (*82–85*).

Route 4

Recently, metal-organic framework(MOF-) derived catalysts have emerged as a new class of advanced materials in various fields of catalysis (*86–88*). The M-N-C catalyst can be synthesized from the new class of materials called MOFs. This new class of orderly crystalline-structured materials is made up of metal ions and organic linkers as ligands. Depending on the metal coordination and ligand type, MOFs with various topologies and geometries can be derived. As the MOF contains the ligand with heteroatoms such as N and S or metal atoms together, pyrolysis of the MOF structures generates heteroatom-doped metal catalysts simultaneously. A well-defined arrangement of metal atoms in a coordinated ligand environment allows for the possibility of various metal atoms and ligands. This, in fact, leads to the possibility of tuning porosity and better surface area; these are considered to be the unique advantages of utilizing MOF-derived catalysts. Most MOF-derived catalysts reported are with Zn^{2+} as the metal atoms and 2-methylimidazole as a ligand. In all the MOF-derived catalysts, the precursors are subjected to the pyrolysis process at the boiling point temperature of Zn (908 °C), where the Zn atoms are converted into ZnO and subsequently to metallic Zn. The metallic Zn is carried away from the catalyst by the carrier inert gas, yielding a catalyst with no Zn atoms. The main advantage of MOF-derived catalysts is that the evaporation of the metallic Zn and release of CO_2 imparts an intrinsic porous nature to the catalysts (*89*, *90*). With a suitable synthesis process, the metal atoms are coordinated with the organic ligands in 3D structures (*91*). Furthermore, pyrolysis of these coordinated structures result in transformation of metal ions into either metal nanoparticles or metal oxides together with the high surface area carbon derived from the pyrolyzed organic ligands. This process leads to the deposition of nanosized metal nanoparticles dispersed on the high surface area carbons suitable for various catalytic applications such as ORR, CO_2RR, batteries, and supercapacitors (*92–98*).

Furthermore, utilization of MOF-derived catalysts has various advantages when compared to traditional transition metal–containing porphyrins or phthalocyanines such as: (i) the choice of a wide variety of organic ligands, with one or more heteroatoms containing precursors that can give rise to a synergistic catalyst; (ii) the choice of wide range of metal ions that allow the catalysts to be derived with single metallic or bimetallic catalysts at once; (iii) the porous structure of MOFs allows for the functionalization or polymerization of the other organic ligands inside the porous structure; (iv) the relatively lower temperature synthesis; (v) the specific selection of ligands and metal atoms that can lead to a wide variety of shapes of MOFs such as nanocubes, nanoframes, nanowires, and nanorods; and (vi) unlike the three previously mentioned routes (Routes 1-3), there is no need for an additional carbon-supporting agent as organic ligands utilized in the process can produce high surface area carbons instantly (*99*). Due to these advantages, a wide variety of MOF-derived catalysts have been explored recently. In this article, we reviewed MOF-derived catalysts in three different sections: (i) MOF-derived metal-free catalysts, (ii) MOF-derived transition metal (M-N-C, M: Co, and Fe) catalysts, and (iii) MOF-hybrid catalysts for ORRs both in acidic and alkaline electrolytes.

MOF-Derived Metal-Free Catalysts

Due to chemical doping of the foreign heteroatoms, carbon atoms are replaced by the heteroatoms, which is considered to be the most effective way of band gap engineering. Such modifications enhance the electronic and surface properties of carbon nanomaterials (*100*, *101*). Various foreign heteroatoms such as nitrogen (N), sulphur (S), boron (B), phosphorus (P), and halogens such as fluorine (F), chlorine (Cl), bromine (Br), and iodine (I) are proposed in the literature as effective chemical dopants for the carbon nanomaterials (*102–109*). Due to the electronegativity difference between the carbon and heteroatoms, an induced polarization network (i.e. positive or negative charges on the adjacent carbon) occurs, thus facilitating the ORR. By doping the heteroatoms, neighboring C atoms experience different electronic charges and spin densities, which further provide surface active sites to facilitate effective electron transfer (*110*). In this regard, various carbon materials, such as CNTs, carbon nanofibers, and graphene, have been doped with heteroatoms and found to have promising catalytic activity. A large number of researchers have studied the MOF-derived heteroatoms doped with metal-free catalysts. A large number of review articles have been published on this subject (*111–113*). The metal-free catalysts derived from MOFs can be synthesized from two different routes. As previously mentioned, most MOF synthesis is performed by an organic ligand that possesses a heteroatom. The pyrolysis of such MOFs can inherit the heteroatoms from the organic ligand and dope them into the carbon matrix. Alternatively, the heteroatoms can also be doped into the carbon matrix by decomposing other heteroatom-containing compounds such as dicyanamide and NH_3 gas in a post-pyrolysis process.

For example, Wu et al. (*114*) proposed a morphological relationship among the pyrolyzed MOF precursors and their relationship with ORR activity. ZIF-7 exists in various shapes such as rod-shaped, sphere, and polyhedron, and these are synthesized by varying the contents of Zn ions and benzimidazole at different temperatures with a connection with the ORR morphology. Among the catalysts, ZIF-7-D, a derived polyhedron NC-D-800 catalyst that showed the high surface area of 538 $m^2\ g^{-1}$, compared well to spherical and rod-shaped catalysts. Moreover, the NC-D-800 catalyst showed better ORR kinetics with 0.87 V vs an RHE onset potential with 3.86 times the number of electrons transferred per O_2 molecule. Though the NC-D-800 catalyst showed good ORR activity, the overall performance was still lower than the Pt/C. To further enhance the ORR catalytic activity, the NC-D-800 catalyst was subjected to a second pyrolysis at 1000 °C in the presence of NH_3 gas (NC-D-NH_3), thus drastically enhancing the performance to 1.0 V vs an RHE onset potential that is higher than the Pt/C catalyst (0.92 V vs RHE). It is elucidated that the NC-D-NH_3 possesses a higher surface area (636 $m^2\ g^{-1}$) than the NC-D-800. However, after the second treatment, the total N content has decreased drastically from 10.30 wt.% to 2.28 wt.%. The total content of graphitic N is higher in NC-D-NH_3 and could be a reason for the increased ORR activity of the NC-D-NH_3 catalyst.

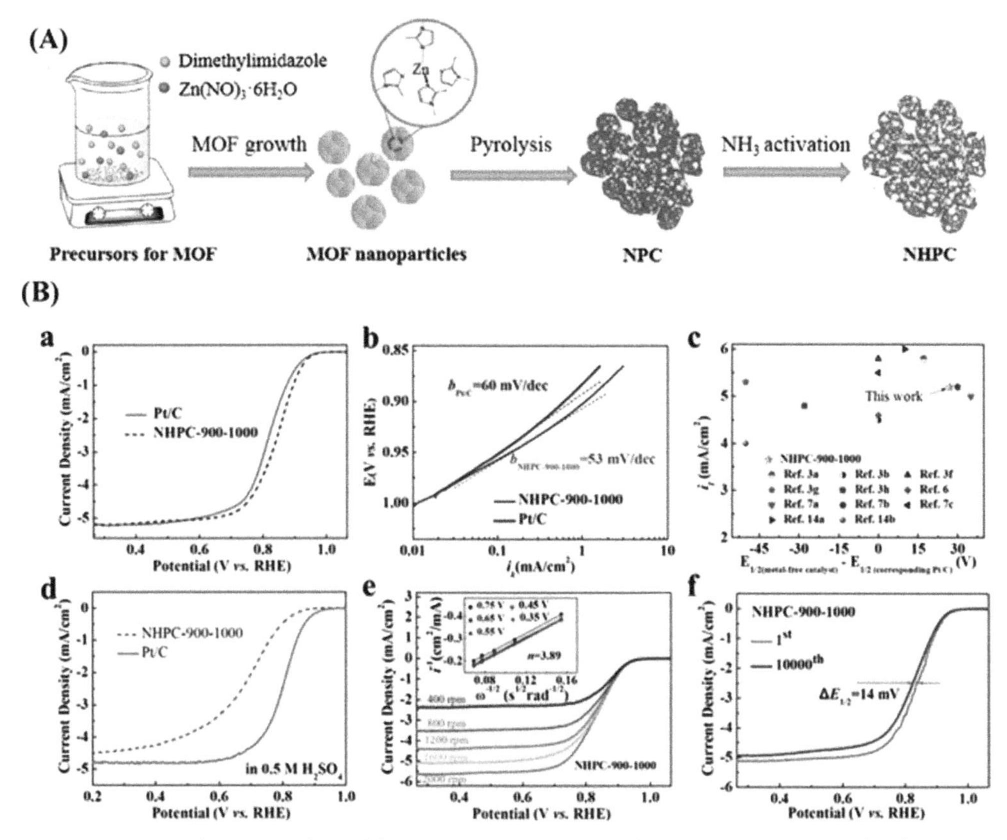

Figure 1. (A) Synthetic procedure of the NHPC. (B) (a) RDE of NHPC-900-1000 and Pt/C in O_2-saturated 0.1-M KOH at room temperature (scan rate of 10 mV/s and rotation speed of 1600 rpm); (b) mass-transfer-free i_k values of NHPC-900-1000 and commercial Pt/C catalysts derived from RDE data; (c) i_l values of several N-doped metal-free catalysts adopted from the related literature and the difference value between their $E_{1/2}$ values and that of the corresponding Pt/C; (d) RDE of NHPC-900-1000 and commercial Pt/C in O_2-saturated 0.5 M H_2SO_4; (e) RDE of the NHPC-900-1000 at various rotation speeds with the insert giving the K-L plots; and (f) RDE of NHPC-900-1000 before and after 10,000 cycles in O_2-saturated 0.1 M KOH. Reprinted (adapted) with permission from Ref. (115). Copyright 2017 American Chemical Society.

Wu (*115*) reported an NH_3 activation of the MOF-derived N-doped carbons and found the extraordinary influence of NH_3 activation on ORR activity (Figure 1A–B). Initially, the polyhedral ZIF-8 crystals were prepared and pyrolyzed at 900 °C (NPC-900). The NPC-900 catalyst was further heat treated in NH_3 to get hierarchical porous carbon NHPC-900. BET surface area analysis shows a significant increase in the mesopores in the NHPC-900 catalyst. The enhancement of the porous structure is due to the etching of carbon by the NH_3 species. During the first pyrolysis, the organic ligand decomposition generates various types of N-species like pyridinic-N, pyrrolic-N, and graphitic-N. During the second treatment, it is found that the NH_3 preferably reacts at the carbon atoms surrounding pyridinic-N and pyrrolic-N, thus converting them into more of graphitic-N types, which leads to more activity toward ORR than pyridinic-N and pyrrolic-N as per XPS analysis estimation (*116*). The ORR analysis of NHPC-900 shows a huge enhancement in the ORR activity over the NPC-900 catalyst owing to the hierarchically porous structure of the carbon and higher graphitic-N content. However, in a contradicting study on the effect of secondary NH_3 treatment on the ORR activity of ZIF-8, derived NPC catalyst is investigated by Liu et al. (*117*). However, the author claims that the NH_3 treatment of the NPC catalyst increases the pyridinic-N content and enhances the graphitization of the carbon (NPC-NH_3 600), unlike the earlier study (*115*) in which increased graphitic-N content is observed. The reason for these contrasting results comes from the secondary etching heat treatment temperatures. It is believed that at higher temperatures, the poorly stable pyridinic-N and pyrrolic-N decompose and could transform into graphitic-N based on which one can see an increased graphitic-N, whereas at lower temperatures the NH_3 reacts with edge carbon atoms enhancing particularly the pyridinic-N content. The NPC catalyst is etched at different temperatures, such as 500, 600, 700, and 800 °C. It is found that the NPC-NH_3 catalyst shows the best catalytic activity with the optimized graphitization and pyridinic-N content. The NPC-NH_3 600 catalyst shows an ORR onset potential of 0.94 V with 0.83 V of the half-wave potential in a 0.1-M KOH electrolyte.

Furthermore, the second heteroatom S is also introduced in ZIF-derived N-doped carbon (ZIF-C) by the post-treatment process. The NH_3-etched ZIF-C carbon (NH_3-C-7) is mixed with thiourea and the resulting thiourea and NH3-C-7 solid composite is pyrolyzed to obtain the N,S-NH_3-C-7 catalyst (*118*). It is found that the etched surface carbons act as active sites for the second dopant S. The S atoms increases the active site densities, and the synergistic interaction of S and N further enhances the ORR process. Similar to the other studies, the N,S-NH_3-7 catalyst possesses a higher proportion of pyridinic-N than the ZIF-C and NH_3C-7 catalysts; because of this, the ORR activity of N,S-NH_3-C-7 catalyst was found to be much higher than the other catalysts. XPS analysis showed two configurations of S bonding (-S-C-S and S-C-N). The DFT calculations also reveal a specific role of S in the N,S-doped carbons. The calculated thermodynamic overpotentials for N-doped carbons, N and S co-doped isolated carbons (N and S bonded to two different carbons), and N,S co-doped coupled carbons (N and S bonded to same carbon atoms) were found to be 0.403, 0.270, and -0.004 V, respectively. This clearly suggests that both the isolated and coupled doping configurations of N and S dopants, works better than the single N doping configurations. Furthermore, the Bader charge analysis reveals that the N,S co-doped coupled carbons possess the greater Bader charge, which can be transferred to the adsorbed O compared to other two configurations, thus indicating a specific role of S and synergistic effects between N and S toward enhanced ORR activity. The attributed synergistic effects of N and S are also evident from the ORR activity with the highest oxygen reduction kinetics observed for N,S-NH_3-C-7 catalyst compared to N, S-ZIF-C, NH_3-C-7, and ZIF-C catalysts. This study provided a solid theoretical and experimental evidence for synergistic effects of N and S. In another study by Tan et al. (*119*), a single crystal MOF, [Zn_2 $(TDC)_2$(DABCO)]·4DMF, bearing the N and S ligands was pyrolyzed to give N- and S-doped mesoporous carbon (NSMC-900). Single-crystal XRD revealed a perfect crystallographic arrangement of the ligands around the Zn atoms. In the proposed MOF structure, $(TDC)_2$ ligands coordinated with the Zn atoms create 2D structures and DAMCO ligands acts like a "pillar layer" extending the 2D structure into a 3D structure. The NSMC-900 catalyst shows the highest ORR activity with 0.98 V and is comparable with the commercial 40 wt.% of the Pt/C catalyst.

Considering the advantages of MOFs in synthesizing metal-free catalysts, a number of catalysts have been prepared and investigated for ORRs, both in acidic and alkaline electrolytes (*120–124*).

However, one should admit that although the metal-free catalysts show promising ORR activity in alkaline and comparable activity in acidic electrolytes, scientifically it is challenging to realize their practical applications due to their failure to deliver the required power density for the desired transportation applications. However, the heteroatom dopant acts as a strong coordination site for cheap, earth-abundant, transition metal atoms and helps to design potential M-N-C active sites. The following section describes various M-N-C catalysts synthesized and investigated for ORR catalysis.

MOF-Derived Transition Metal–Based Catalyst

The first MOF-derived catalyst, Co-N_4, is reported by Liu et al. with cobalt (Co) imidazole frameworks where a single Co atom coordinates with 4 N atoms of imidazole organic ligands. The solvothermal synthesized purple color Co-imidazolate crystals were further pyrolyzed at 750 °C to produce Co-N_4 catalyst. The Co-N_4 catalyst showed enhanced ORR activity with an ORR onset potential of 0.83 V in acidic electrolytes (*125*). Based on this work, a number of researchers have proposed MOF-derived catalysts. ZIF-67, a sub class of MOF with cobalt-2-methylimidazole organic ligands, has been utilized for a synthesis of highly efficient Co-N-C catalysts. It was reported that the ORR activity of the ZIF MOF-derived catalysts are size-dependent based on the precursors. Xia et al. (*126*) synthesized the Co-ZIF-67 in various sizes ranging from bulk, 1.7 µm, 800 nm, and 300 nm sizes by altering the solvent and reaction temperatures. The room temperature synthesis of initial 300 nm–sized MOFs was further synthesized by elevating temperatures, and further large macro-sized MOFs are synthesized by solvothermal procedures to obtain the MOFs of different sizes. It was found that the 300 nm–sized MOFs delivered the best ORR activities, which suggests that it is essential to start with the small–sized MOFs, which can provide more accessible ORR active sites.

Similar size-dependent activity of ZIF-derived Fe catalysts was reported by Zhang et al. (*127*). Different sizes of Fe-doped ZIFs were prepared ranging from 20 to 1000 nm. The synthesized ZIF precursors show a well-defined dodecahedron shape, which was merely maintained even after single-step thermal activation at 100 °C with a porous structure. Fe-free ZIF-derived catalysts showed poor ORR activity, as this catalyst contains no Fe-N active sites. In contrast, Fe-ZIF catalysts showed remarkable improvement due to their Fe-N_x active sites. Furthermore, the ORR activity of the Fe-ZIF catalysts is highly dependent on the size. As the particle size decreased from 1000 nm in size, the ORR activity enhanced with an optimum size of 50 nm. With a further decrease in size to 20 nm, the activity decreased possibly due to the agglomeration of small size particle. The Fe-ZIF catalyst (50 nm) shows a half wave potential of 0.85 V in 0.5 M of H_2SO_4 with less than 1% of H_2O_2 formed during the ORR reaction [Figure 2]. The electrochemical surface area derived from the double layer capacitance shows that the Fe-ZIF catalyst (50 nm) show the best ECSA for ORR to happen faster and kinetically. Furthermore, temperature activation also plays a key role in enhancing the ORR activity of the catalyst. In this study, authors varied the temperature from 800 to 1100 °C and found that a minimum temperature of 800 °C is essential to form the ORR active sites. With an increase in temperature from 800 to 1100 °C, the ORR activity of the Fe-ZIF catalyst increased. The highest ORR activity is obtained for the catalyst thermally activated at 1100 °C. A further increase in the temperature is not possible according to the phase diagram of Fe-carbon, as the eutectic point is 1147 °C where above this temperature C and Fe exit in liquid states. It is generally believed that the N content of the catalyst tends to decrease with the increase in temperature due to the decomposition of pyrrolic-N and pyridinic-N. However, an elevated temperature helps in increasing the graphitic-N content. Similar observations were also concluded with this study. However, researchers claim that the total Fe-N content remained unchanged in spite of the higher temperature of 1100 °C. Thus, the higher ORR activity of the Fe-ZIF catalyst is due to the synergistic effects of FeN_4 active sites and graphitic-N as observed in the experimental results.

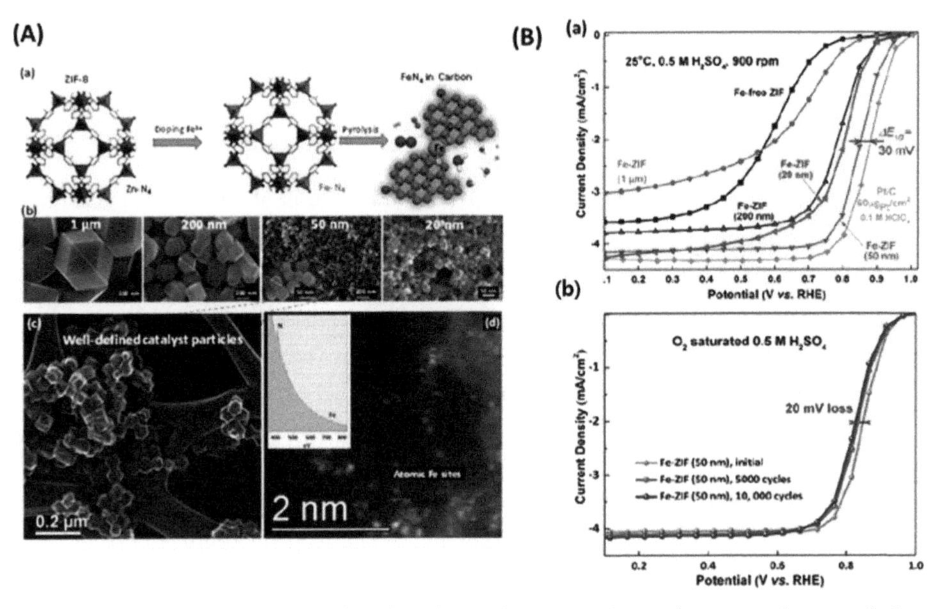

Figure 2. (A) (a) Synthesis principles of Fe-doped ZIF-derived catalysts; (b) accurately controlled sizes of the Fe-ZIF catalysts from 20 to 1000 nm; and (c, d) HAADF-STEM images of the best performing Fe-doped ZIF catalyst (50 nm) and EELS analysis (the inset of d). (B) (a) ORR polarization plots for Fe-ZIF-derived catalysts in 0.5 M H_2SO_4 and Pt/C catalysts (60 μgPt/cm²) in 0.1-M $HClO_4$ at 25 °C and 900 rpm; and (b) stability AST by cycling the potential 0.6−1.0 V in O_2-saturated 0.5-M H_2SO_4. Reprinted (adapted) with permission from Ref. (127). Copyright 2017 American Chemical Society.

Guo et al. (*128*) proposed that ZIF-derived catalysts occasionally suffer from the agglomeration of the metal nanoparticles due to their relatively short metal-metal distance in the ZIF structures. Rarely, the intact porous structure also collapses giving rise to microporous structures that are unfavorable for the reactant's mass transfer. In order to enhance the porosity and surface active FeN_x sites, Guo et al. (*128*) synthesized a gradient Zn-rich and Fe-rich cylindrical morphological Fe and Zn bimetallic ZIF (BMZIF) structures. These gradient structures provide surface-exposed FeN_x ORR active sites as shells with hierarchical porous structure as a core, generating an active porous FeN_x ORR catalyst (Figure 3A–C). The cylindrical Fe and Zn-ZIF precursors are synthesized from the Fe^{2+}, Zn^{2+}, and benzimidazole ligands under solvothermal conditions followed by the pyrolysis of the precursor to Fe,N-HPCC. Elemental mapping of the precursor by HR-TEM shows a unique gradient distribution of the Zn and Fe atoms. It was observed that the concentration of Fe increases gradually from the core to the outer shell mainly composed of Zn. This observation indicates that the Fe atoms are more enriched at the surface compared to the center. Moreover, it is also observed that Fe atoms mediate the growth of ZIF. The Fe-free Zn-ZIF shows an irregular polyhedral structure, whereas FeZn-ZIF shows cylindrical and spherical structures with increasing Fe content. It is believed that the difference in the bond lengths between Zn-BIM and Fe-MIM leads to the evolution of these structural morphologies from irregular polyhedrons to cylindrical structures. It is also found that the type of solvent and ligand also play key roles in obtaining a cylindrical structure. Changing the solvent from methanol to ethanol, DMF, or water and changing the ligand from BIM to 2-methylimidazole prevented gradient cylindrical structures. Pyrolysis of Fe and Zn-ZIF produces an Fe,N-HPCC catalyst. Gradient distribution of Zn and Fe produces a hierarchical porous core and surface-rich Fe shell–generating FeN_x active sites. An Fe,N-HPCC catalyst showed significantly enhanced ORR activity resulting in enhanced performance by 50 mV when compared to the standard Pt/C catalyst.

Recently, Wang et al. (*129*) synthesized atomically dispersed CoN_4 catalytic ORR active sites derived from ZIF nanocrystals. The authors in this study prepared catalyst with varying Co wt.% with respect to Zn ions in a methanol solution followed by the pyrolysis at 1100 °C. Among the synthesized catalysts, the 20Co-NC-1100 shows the most homogenously distributed Co nanoparticles on the surface of N-doped carbon. Various techniques such as HAADF-STEM, EEL point spectra, X-ray absorption spectroscopy (XAS), X-ray absorption fine structure (EXAFS) spectra are utilized to identify the atomically dispersed Co nanoparticles located in planar CoN_4 coordination sites. The 20Co-NC-1100 catalyst shows the onset potential of 0.93 V, showing comparable activity to that of a traditional polyaniline-derived Fe catalyst and a standard Pt/C catalyst with a H_2O_2 yield of about 5%. Furthermore, the 20Co-NC-1100 catalyst showed a 30-mV loss in half wave potential after 10,000 cycles, compared to 80 mV of traditional polyaniline-derived Fe catalyst. This study indicates that Co-based catalysts are better than Fe-based catalyst. Fe-based catalysts typically suffer from the Fenton-based reactions involving the redox reaction of $Fe^{3+}$and Fe^{2+}. This in turn reacts with H_2O_2 which can generate the hydroxyl radicals resulting in degradation of the ionomer in the catalyst layer or can even damage the Nafion membrane in the fuel cell stack. Furthermore, the 20Co-NC-1100 catalyst was also evaluated in an H_2O_2 fuel cell. The MEAs were fabricated with 20Co-NC-1100 as the cathode catalyst with catalyst loading of 4.0 mg cm^{-2} (=0.08 mg of Co cm^{-2}). Utilizing 20Co-NC-1100, a peak power density of 560, 280 mW cm^{-2} was achieved in H_2O_2 and H_2-Air. It is to be noted that the H_2-Air performance of the 20Co-NC-1100 catalyst is much higher than the PANI-derived Fe and Co Ketjen black catalysts. The 20Co-NC-1100 catalyst was also subjected to a stability test of 0.7 V for 100 h, showing a degradation of just 60 mV after the 100-h test. This study shows a remarkable performance of atomically dispersed Co catalysts and a promising future direction for Co-based catalysts in fuel cell applications.

In spite of the development of ZIF-67–based catalysts, the issues of low surface area and insufficient porosity still persist for the metal-derived catalysts (*130, 131*). MOFs also offer high surface area carbon-derived catalysts that are identified by combining the ZIF-67 with its isostructural MOFs of Zn-based ZIF-8. Chen et al. (*132*) synthesized a series of BMZIFs that were synthesized by varying the ratio of Zn and Co precursors. It was found that an optimal ratio of ZIF-8 and ZIF-67 (i.e., P-CNCo-20) can generate a catalyst that possesses synergetic properties of the MOFs with excellent surface area (1225 m^2 g^{-1}) and highly dispersed CoN_x and N species. Moreover, the high graphitization of the carbons collectively produces an extraordinary catalyst for ORR. P-CNCo-20 catalysts also present excellent durability with nearly zero degradation after 10,000 potential cycles. Wang et al. (*133*) synthesized a bimetallic Fe-N-Co@C800-AL catalyst in a sequential reaction. The Fe-N-Co@C800-AL catalyst shows slightly better ORR kinetics compared to Co@C-800-Al, with 3.98 electrons transferred per O_2 molecule. In a single-cell fuel cell, the Fe-N-Co@C800-AL (loading = 3mg cm^{-2}) catalyst showed 137 mW cm^{-2} in an AEMFC fuel cell.

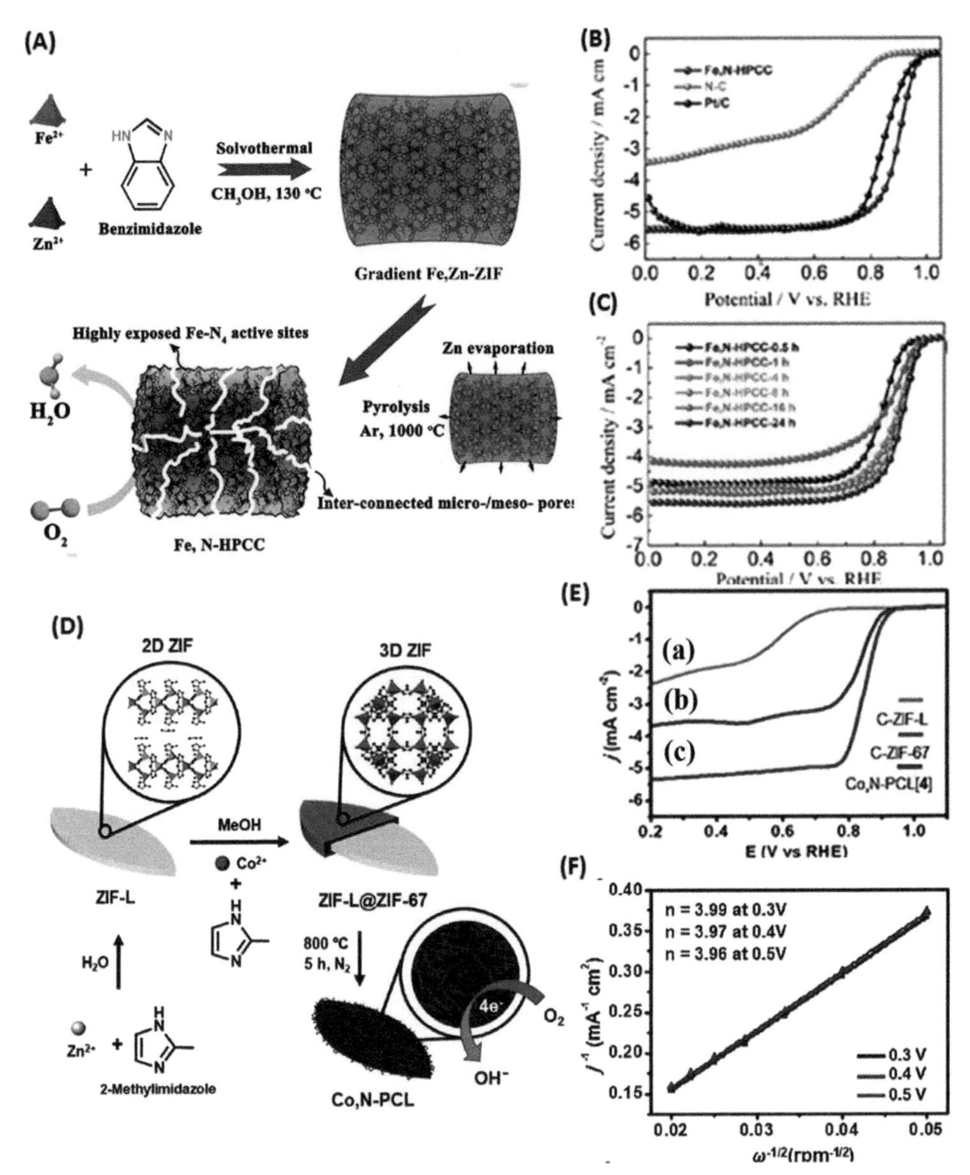

Figure 3. (A) Illustration of the synthesis of gradient Fe,Zn-ZIF and Fe,N-HPCC. (B) LSV curves of N-C, Fe,N-HPCC, and Pt/C electrocatalysts in O_2-saturated 0.1-M KOH (rotation rate of 1600 rpm). (C) LSV curves of Fe,N-HPCC electrocatalysts synthesized with Fe,Zn-ZIF precursors prepared with different reaction durations. Reprinted with permission from Ref. (128). Copyright 2019 Elsevier. (D) Schematic representation for the formation of leaf-shape ZIF-L and leaf-shape core shell type ZIF-L@ZIF-67 and its transformation to thin Co,NPCL; (E) LSV in O_2-saturated 0.1-M KOH electrolyte of (a) C-ZIF-L, (b) C-ZIF-67, and (c) Co,N-PCL[4] at the electrode rotation rate of 1600 rpm. [F] K–L plots of Co,N-PCL[4]. Reprinted with permission from ref. (139). Copyright 2019 Elsevier.

Although the traditional ZIF-based catalysts are active, their 3D structures tend to agglomerate, resulting in limited exposure of surface reactions. Li et al. (*134*) proposed a "shape-transforming method in an aqueous system" for synthesis of Co-N_x and C from the MOF nanoleaves as a sacrificial template. Most of the MOF synthesis utilizes various organic solvents which are hazardous and most importantly unsuitable for pilot scale and large scale productions. Hence, it is highly desirable to replace the MOF synthesis with either environmentally friendly and aqueous solvents or non-solvent and mechanochemical synthesis routes. Several efforts have been made to synthesize the MOFs from water-based solvents (*135–138*). Most of these synthesis routes result in 3D nanoparticles. Li et al. (*134*) reported a water-based synthesis protocol to synthesize ZIF-derived nanoleaves of Co_x-N and C for ORR. A separate solution of Zn and Co dissolved in water is mixed with a solution of 2-methylimidazole. It has been shown that hydrogen bonding occurs between the N group of 2-mIm and the H_2O molecules, generating the sodalite layers without bonding to the transition metal ions and forming a 2D network of nanoleaves. Further pyrolysis of these ZIF nanoleaves produces a Co-N and C catalyst. In this study, authors thought that Zn was atomically dispersed in the ZIF nanoleave precursors. This allows the Zn to evaporate at relatively lower temperatures (800 °C) than the bulk Zn (907 °C). A $Co_{0.6}$–N and C-800 catalyst showed that the ORR onset potential of 0.916 V is very similar to the Pt/C catalyst. For comparison, a methanol solution–derived 3D-$Co_{0.6}$-N and C-800 catalyst was synthesised and compared to the 2D-$Co_{0.6}$-N and C-800 catalyst based on their ORR activity. It was found that the 2D catalyst ORR activity is higher than that of the 3D catalyst, which indicates the significance of 2D MOFs for ORR. Rotating ring disc electrode (RRDE) measurements show a higher peroxide percentage for the 3D catalyst compared to the 2D catalyst. This indicates that a mesoporous 2D-$Co_{0.6}$-N and C-800 catalyst helps in improving the mass transfer of reactants and products.

A similar strategy was also proposed by Park et al. (*139*), who developed a Co,N-PCL catalyst from a hybrid 2D- and 3D-derived core shell (ZIF-L@ZIF-67) structure. Firstly, a 2D leave morphological template (ZIF-L) and similar such structures were synthesized from Zn^{2+} and 2-methylimidazole in a methanolic solution (Figure 3D–F). The ZIF-L structures were mixed with ZIF-67 crystals, synthesized in a separate solution, and generated a hybrid core shell ZIF-L@ZIF-67 structure. Furthermore, pyrolysis of ZIF-L@ZIF-67 structures produces a Co,N-PCL catalyst. Morphological analysis of the Co,N-PCL catalyst indicates that the template ZIF-L produces a 2D-leave structure onto which the Co nanoparticle–encapsulated graphitic-carbon are observed. Moreover, the 2D leaves also show the development of CNTs, and the density of the CNTs increases with the increase in ZIF-67 content. Based on this study, the catalysts derived from the ZIF-L@ZIF-67 core shell structures show higher ORR activity than their control samples (C-ZIF-L and C-ZIF-67). It is believed that flat and thin morphology of C-ZIF-L, together with its porosity, enhances the mass transfer of the reactants, the electron transfer to the ORR active sites, and the overall ORR kinetics of the Co-N-PCL catalyst.

Most of the MOF-derived catalysts are based on a single transition element. It is well known that the presence of two or more elements can increase the activity of the catalyst by inducing special electronic and synergistic effects. This in turn greatly reduces the work function of the catalyst thereby enhancing the catalytic activity (*140*, *141*). Due to many of these reasons, recent studies are conducted to develop MOFs with two or more transition metals (*142*). For instance, Ning et al. (*143*) synthesized the CoNi alloy encapsulated by an N-doped carbon as an efficient bifunctional catalyst for ORR and OER through a co-precursor strategy (Figure 4A–C). The bimetallic precursor (Co_xNi_y-ZIF) is synthesized by mixing a separate solution of Co^{2+} and Ni^{2+} solutions with 2-methylimidazole ligands, then hydrothermally treating it at 100 °C for 12 h, followed by performing pyrolysis to obtain the Co_xNi_y@N-C catalyst. It was also found that the morphology and catalytic activity of the Co_xNi_y@N-C catalyst is highly dependent on the contents of the individual transition metals that can be easily optimized by changing the molar ratio of the individual elements taken during the synthesis. Among the catalysts with various molar ratios, the Co_1Ni_1@N−C catalyst showed a similar ORR activity to the commercial Pt/C catalyst. It is believed that the efficient d-electron transfer from Ni atoms to Co atoms increases the electronic density around the CoNi alloy, enhancing the ORR activity in addition to the CoNi-N_x active sites and N-doped carbon (*144*). It is worth mentioning that a bimetallic Co-Ni@NC catalyst was also reported widely in the research (*145–147*).

It is well known that the greatest hindrance of non-precious metal utilization in real fuel cells is associated with high catalytic loading, increasing the thickness of the catalyst layer and imposing greater mass transfer resistance. Besides developing active non-precious metal catalysts, the porosity of the catalysts should be tuned to the desired level to overcome the mass transfer issues associated with the higher catalyst thickness (*148, 149*). There are efforts taken to balance the porosity and activity of the catalysts. However, most of the catalysts still posses a micropore structure that is inefficient to balance the mass transfer issues (*150–152*). Recently, introducing a surfactant strategy was proposed to induce the mesoporosity to the MOF-derived catalysts. For example, Dong et al. utilized block co-polymers that could be successfully introduced in micro- and two mesochannels, whereas Qiu et al. used CTAB and 1,3,5 trimethylbenzene. Recently, Meng et al. (*153*) synthesized a Co-N-C catalyst from a BMZIF that self-assembles on Pluronic P123 micelle structures. The catalysts are synthesized in a three-step process where initially the P123 micelle structures were obtained in an aqueous solution and 2-methylimidazole were self-assembled on P12 micelles by hydrogen bonding, followed by in-situ growth of Zn-Co bimetallic MOFs (Figure 4D–E). Pyrolysis of the surfactant-bimetallic MOF precursor produces the porous catalyst (HC-5Co95Zn), whereas pyrolysis assists in thermally removing the surfactant together with the sublimation of Zn occurs. Barrett-Joyner-Halenda (BJH) pore size analysis of the HC-5Co95Zn catalyst showed micro- and mesopores that are beneficial to enhancing the mass transfer of reactants during the ORR reaction. Transmission electron microscope (TEM) measurements of HC-5Co95Zn catalysts show a very well interconnected pore. The ORR analysis of the HC-5Co95Zn catalyst in acidic electrolytes show enhanced ORR activity with 0.88 V and 0.78 V of onset and halfwave potentials, but still inferior to the Pt/C catalyst. The beneficial effect of surfactant-induced porosity is also established with fuel cell performance analysis. The HC-5Co95Zn catalyst exhibits a peak power density of 412 mW cm^{-2}, much higher than the catalyst synthesized without P125 (catalyst loading of 4 mg cm^{-2}), which shows a power output of 245 mW cm^{-2}. This study suggests that tuning the catalyst activity together with pore size tuning could effectively deliver a higher power density to overcome the problems of mass transport issues associated with the higher catalyst layer thickness.

Single-Atom Catalysts Derived from MOFs

It is well known that the activity of M-N-C active sites are much higher compared to their metallic and metal-carbides sites; thus, the electrocatalytic activity of the catalyst is dependent on the density of these M-N-C sites in the catalyst (*154, 155*). As previously mentioned when discussing the synthesis of various methods to obtain M-N-C catalysts, pyrolysis of the mixture of metal salts and precursor will result in aggregated metallic nanoparticles with a low density of M-N-C active sites. Recently, single-atom catalysts (SACs) have attracted increased attention due to their highly dispersed atomic levels. The atomic level dispersion greatly enhances the metal active sites, and as a result, the maximum utilization of the catalyst can be realized (*156–158*). Well-separated and well-defined atomic dispersion of SACs allows for the identification of precise reaction kinetics and reaction pathways when compared to a homogenous nanoparticle (*159, 160*). MOF-derived

SACs can be derived from two different routes. First, the single metal atoms can be stabilized or incorporated on MOFs, where the original structure of the MOF can be well maintained to a greater level. Secondly, there is the pyrolysis of MOFs to form SACs (*161–163*). However, during the pyrolysis of MOFs there is a possibility of metal nanoparticle aggregation due to the extremely high surface energy of single atoms in SACs. It is believed that the N-atoms could effectively prevent the aggregation process by means of M-N bonds during the pyrolysis; thus, the SACs can be of great importance to a variety of catalytic applications. To realize the atomic dispersion of metal nanoparticles, the metal ion content, N-content in MOFs, pyrolysis temperature, and type of atmosphere during the pyrolysis process are significant.

SACs can also have the advantage of low metal loading and can be regarded as inexpensive catalysts. For the practical applications, the required metal loadings are much higher than the currently achieved catalysts with maximum metal loading of 4 wt%. One of the problems associated with the attempt to increase the metal loading is that SACs can form clusters and agglomerated nanoparticles. Because of this, the proper selection of synthetic procedures are nessessary to balance the high metal loading while simultaneously avoiding the formation of clusters and nanoparticles. Some attempts have been made to dissolve the small clusters and small nanoparticles in the catalyst by the process of acid leaching. In many cases, acid leaching is found to be an essential and crucial step for obtaining the SACs. The acid leaching successfully dissolves the selected clusters and nanoparticles converting the heterogeneous catalyst structure to homogeneous catalysts made purely of SACs. The following section discusses some of the synthetic protocols in which pyrolysis and acid washing is performed to generate the SACs.

Yin et al. (*164*) synthesized a single Co atom using a mixed metal approach. Due to the similar bonding configuration of Zn^{2+} and Co^{2+} ions with 2-methyl imidazole, it is expected that the addition of Zn^{2+} can replace the Co^{2+} sites, expanding the distance between the Co atoms in BMZIFs. By increasing the molar ratio of Zn and Co to above 1:1, a single-atom Co catalyst is synthesized. The single-atom Co sites are identified using high-angle annular dark-field-scanning transmission electron microscopy (HAADF-STEM) analysis. It is found that the Zn^{2+} sites play two important roles. First, the special distance between the two Co atoms are increased, and sublimation of Zn^{2+} generates the active N sites, which further stabilized the Co atoms to generate a single Co atom catalyst [Co SAs/N-C]. The ORR activity of Co SAs/N-C is analyzed in 0.1 M KOH. The Co SAs/N-C catalyst showed higher ORR activity with an onset potential of 0.982 V and a half-wave potential of 0.881 V, compared to the Pt/C catalyst.

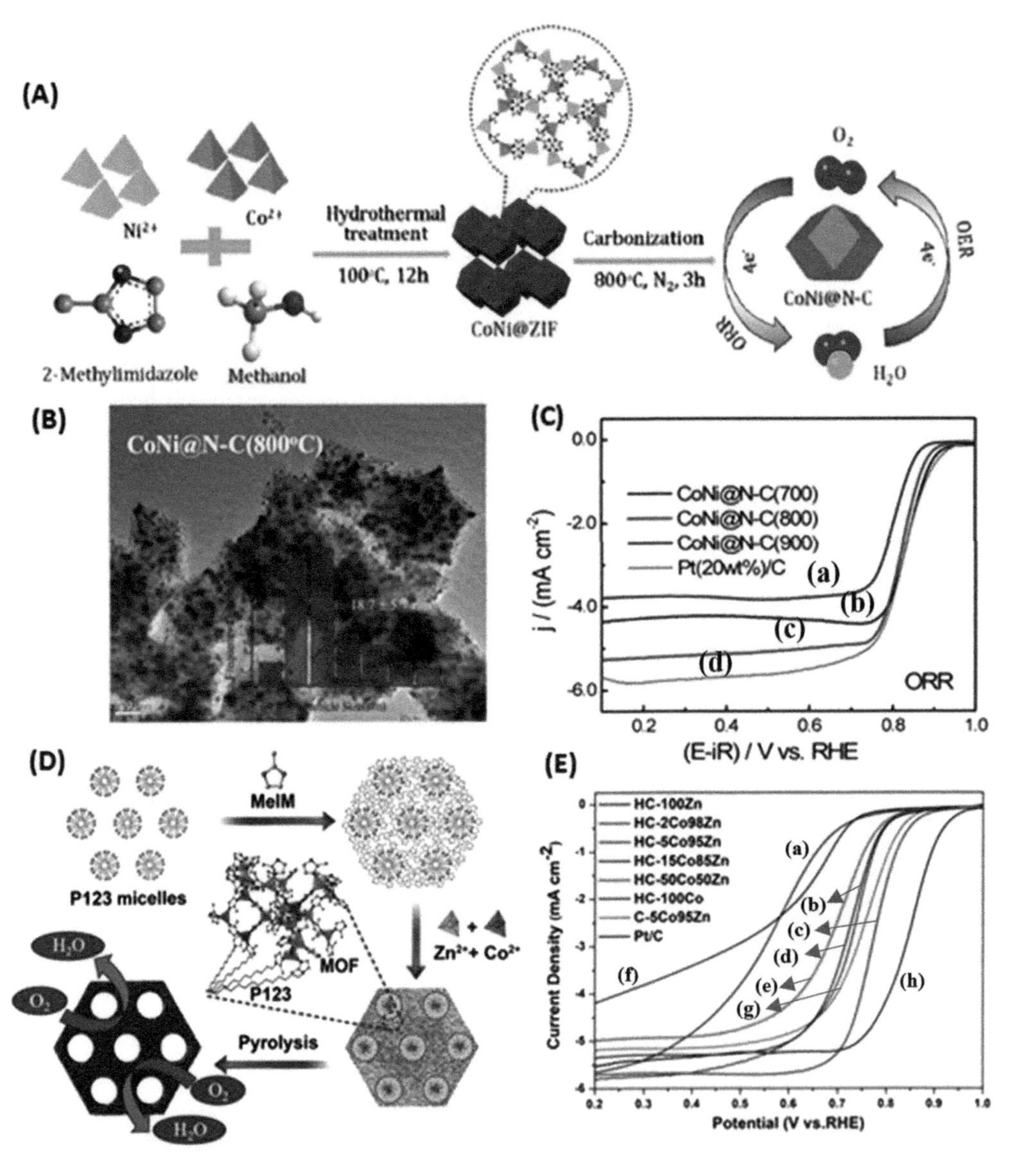

Figure 4. (A) Preparation process diagram of the parental CoxNiy−ZIF and the derived CoxNiy@N−C. (B) TEM images for the CoNi@N−C samples pyrolyzed at 800 °C. (C) Activity comparison of ORR of CoNi@N−C pyrolyzed at (a) 700, (b) 800, and (c) 900 °C, respectively, and (d) Pt/C (20 wt %, commercial). Reprinted (adapted) with permission from Ref. (143). Copyright 2019 American Chemical Society. (D) Schematic representation for the preparation process of HC-xCo(100-x)Zn. (E) LSV curves of the (a) HC-100Zn, (b) HC-2Co98Zn, (c) HC-5Co95Zn, (d) HC-15Co85Zn, (e) HC-50Co-50Zn, (f) Hc-100 Co, (g) C-5Co95Zn, and (h) Pt/C for ORR in O_2-saturated 0.1 M $HClO_4$ at 1600 rpm. Reprinted with permission from Ref. (153). Copyright 2019 Elsevier.

He et al. (*165*) developed highly active, atomically dispersed CoN_4 cathode catalysts from a surfactant-assisted MOF approach. That study involves the surface covering of ZIF-8 nanocrystals by surfactant to obtain a core shell type of structures. The initially synthesized ZIF-8 crystals are capped with the surfactant to assist in the crystallization of the ZIF-8. This also helps in controlling the size, morphology, and growth of ZIF-8 crystals. During the pyrolysis of the Co-ZIF-8@surfactant, the initially carbonized surfactant forms a carbon shell, avoiding the native pore structure and aggregation of Co nanoparticles and inducing a significant confinement effect against the neighboring Co atoms, leading to atomically dispersed Co-N-C@surfactant catalysts. Various types of surfactants are utilized such as SDS, CTAB, F127, and PVP. The F127 surfactant produces an isolated and well-dispersed ZIF of around 250 nm. BET analysis of the Co-N-C and Co-N-C@surfactant show a difference in surface area and pore characteristics. The Co-N-C@surfactant catalyst shows a higher surface area of 825 m^2g^{-1} rich in only micropores, whereas the Co-N-C catalyst showed a lower surface area of 324 m^2g^{-1} with both micro- and mesopores. It has been determined from other studies that the active M-N-C sites are generally located at the micropore of the carbon (*166*). Because of this, it is anticipated that the Co-N-C@surfactant can possess much higher ORR activity than the Co-N-C catalyst due to the abundant micropore-containing M-N-C sites. The presence of CoN_4 in the Co-N-C@surfactant is evidenced from X-ray absorption near edge structure (XANES) and EXAFS. The Co-N-C@surfactant catalyst shows the extraordinary ORR activity in an acidic electrolyte with an onset potential of 0.93 V and half-wave potential of 0.84 V, which is comparable to the best Fe-N-C catalyst reported (*167*) and surpasses the previously reported non-precious metal catalysts due to the abundant CoN_{2+2} active sites. The fuel cell analysis of the Co-N-C@surfactant catalyst shows an appreciable power density of 0.87 W cm^{-2} with a catalyst loading of 4.0 mg cm^{-2}, H_2O_2, and a 1.0 bar pressure at 100% RH. However, in the H_2-Air atmosphere, huge mass transfer losses were observed. This is due to the micropore structure of the catalysts. Further studies should be conducted to tackle the porosity and structure of the catalyst layer in order to overcome the mass transport issues.

Recently, Deng et al (*168*) synthesized an atomic Fe-doped catalyst from a ZIF-8 polyhedron crystal. The authors in this study provide a unique gaseous metal-organic doping strategy to synthesize a highly active Fe catalyst. The catalyst showed groundbreaking activity in half cell and fuel cell analysis. In this synthesis, the ferrocene and organometallic compound is used as a source of Fe and is primarily vaporized. The vaporized Fe atoms are trapped in a ZIF-8 derived porous N-doped carbon. The catalyst morphology and the ORR activity is tuned based on the amount of ferrocene precursor. The HRTEM images of the optimized C-FeZIF-1.44-950 catalyst show a highly mesoporous structure with many edge defect sites. Moreover, no visible Fe particles are observed, indicating that the Fe atoms are dispersed atomically in the catalyst and coordinated to the N species to generate Fe-N_x active sites. Furthermore, the XANES analysis suggests a clear formation of Fe-N_x active sites. The ORR activity of the C-FeZIF-1.44-950 catalyst is evaluated in 0.1 M of KOH and $HClO_4$ electrolyte, showing the 50-mV higher half-wave potential in alkaline electrolytes; whereas there are 60 mV less in acidic electrolyte in comparison to the standard Pt/C catalyst. The study is noteworthy for the C-FeZIF-1.44-950 catalyst that is also evaluated in fuel cell MEA configuration and shows the 0.775 W cm^{-2} in H_2O_2 and 0.463 W cm^{-2} in H_2-Air, while there is a fuel cell with a catalyst loading of 0.1 mg cm^{-2} at 30 psi back pressures at 80 °C with Nafion 211 membrane. This performance is considered to be the best performance results compared to many reported non-precious catalyst reports.

Xia et al. (*169*) synthesized atomic Fe-N catalysts from Fe-ZIF-8 using a controlled co-precipitation method with a controlled Fe-to-Zn ratio. This study for the first time shows the effect of heat treatment atmosphere (Ar, N_2, 7% H_2/Ar, and Ar+NH_3) on the catalytic activity of Fe-ZIF-8. Ideally, the catalysts derived from the ZIF precursors are determined by the Kirkendall-mediated metal diffusivity, and the porosity is induced by the gas diffusion into the cavity of the ZIF precursor during the pyrolysis process (*170*). It is determined that the atmosphere has no influence on the morphology of catalysts; however, they observed variations in the pore volume, BET surface area, and ORR activity. Among the investigated catalysts, the (Zn/Fe=0.95/0.5)-N-C catalyst pyrolyzed under Ar-atmosphere showed enhanced ORR activity with an onset potential of 0.95 V and a half-wave potential of 0.81 V in 0.1 M of$HClO_4$. In an alkaline electrolyte, they are much higher with 1.0 V and 0.90 V, respectively. These results indicate that the advantages of atomic dispersion of Fe, when compared to the agglomerated Fe nanoparticles or Fe carbide species, result in similar current densities and much higher onset potentials in spite of very low metal loadings. The ORR activity of the catalyst is highly dependent on the pyrolysis atmosphere and follows the order of N_2 < 7% and H_2/Ar < Ar. It is determined using Fick's law that the Ar (40 g mol^{-1}) has a much higher mass for N_2 (28 g mol^{-1}) and H_2 (2 g mol^{-1}) (*171*). The higher mass of N_2 and H_2/Ar may result in agglomeration of Fe due to the more rapid heat transfer and quicker metal diffusion rates compared to the Ar. Moreover, the diameter of Ar (3.70 Å) is slightly higher than the N_2 (3.64 Å) (*172*). The, the porous diffusivity occurs at an early stage and may lead to shrinkage of the pores in the catalyst as a result of Zn evaporation. This leads to a higher BET surface area of 1090 m^2 g^{-1} for Fe-N-C catalyst. Furthermore, it is also found that the second treatment of the catalyst Fe-N-C derived from Fe (Zn/Fe=0.95/0.05)-N-C in Ar with NH_3 did not help to further increase the ORR activity; this might be due to the effective atomic dispersion of Fe-N active sites or the unnecessary increase in overall N content. NH_3 treatment not only increases the N content, it can also etch the carbon in the catalysts and thus may harm the intrinsic porous structure of the Fe-N-C catalyst (*173*, *174*). The atomic dispersion of Fe in the Fe-N-C catalyst is identified using HAADF-STEM, and the existence of M-N-C bonds were identified by XANES and EXAFS analysis.

As previously mentioned, the development of an atomic level highly dispersed catalyst can enhance its utilization abruptly. Moreover, dispersing the active metal at the atomic scale can also decrease the metal loading to greater extent. Jiang et al. (*175*) synthesized the hierarchically porous 3D carbon–supported atomically dispersed Fe-N_4 catalyst with a metal loading as low as 0.20 wt% with ORR activity that is much higher than the commercial Pt/C catalyst. It is proposed that the encapsulated Fe phthalocyanine (FePc) on ZIF-8 crystals (FePc-x@ZIF-8) burst the ZIF-8 cage microcavity during pyrolysis, creating an edge site for mesopore active sites while the excessive Fe precursor aggregates into Fe_2O_3 were further removed in acid treatment to generate the Fe SAs-N/C catalyst. By controlling the Fe-precursor concentration, the pore diameter of the catalyst can also vary as shown in their study. The resulting Fe-SAs-N and C-20 catalyst showed extraordinary ORR activity with a half-wave potential of 0.909 V, overperforming the Pt/C catalyst by 59 mV in alkaline electrolytes, resulting in excellent durability with zero degradation under rigorous 10,000 potential cycles (Figure 5A).

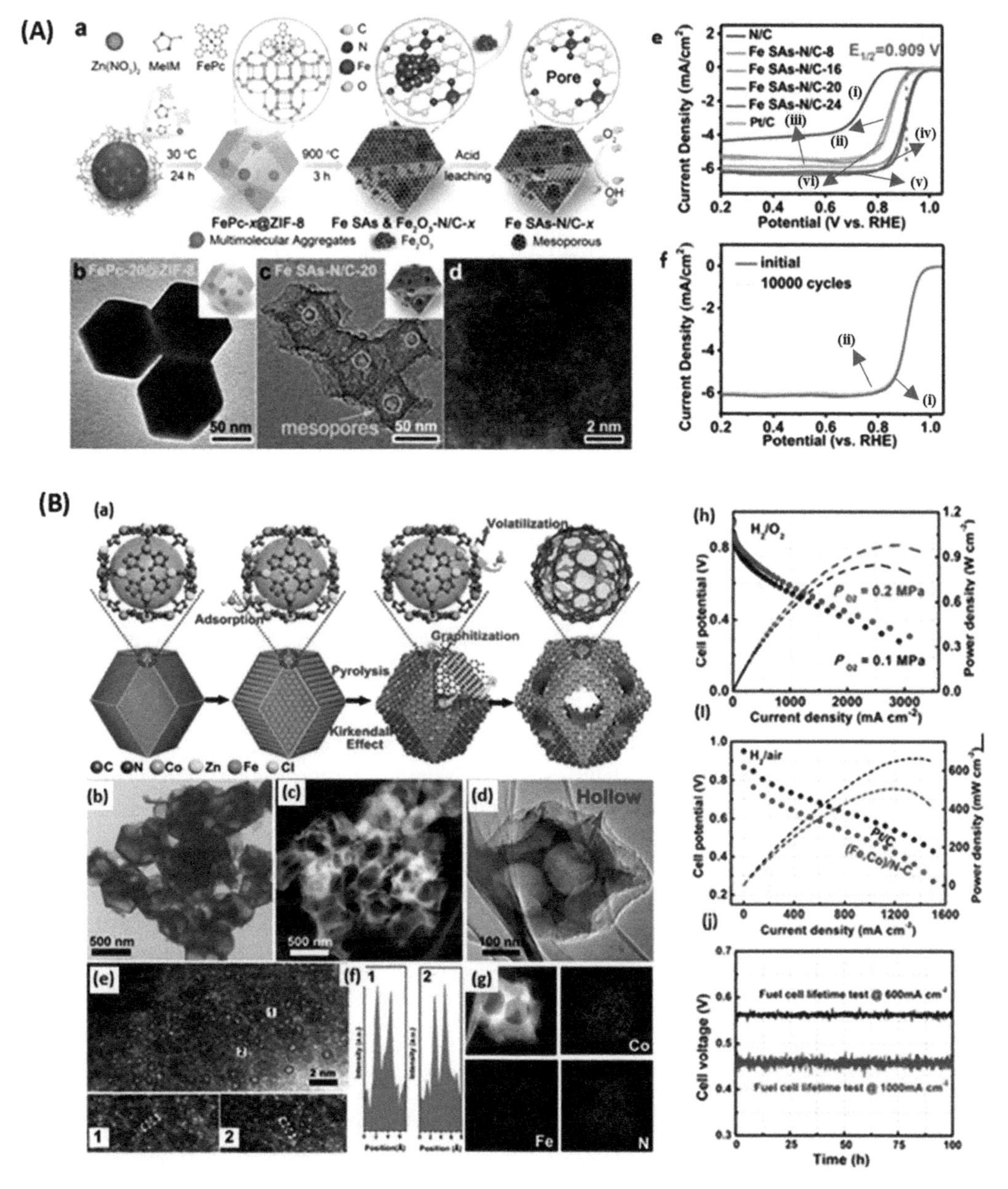

Figure 5. (A) (a) Schematic illustration. (b) TEM image of FePc-0@ZIF-8 composite. (c) HRTEM. (d) HAADF-STEM of Fe SAs-N/C-20 catalyst. (e) LSV curves of (i) N/C, (ii) Fe SAs-N/C-8, (iii) Fe SAs-N/C-16, (iv) Fe SAs-N/C-20, (v) Fe SAs-N/C-24, and (vi) Pt/C catalysts. (f) Before and after 10,000 successive cycles. Reprinted with permission from Ref. (175). Copyright 2018 American Chemical Society. (B) (a) Preparation of (Fe,Co)/N-C. (b) TEM. (c) HAADF-STEM. (d) HRTEM of (Fe,Co)/N-C. (e) Magnified HAADF-STEM of (Fe,Co)/N-C, showing Fe-Co dual sites dominant in (Fe,Co)/N-C. (f) Corresponding intensity profiles obtained on the zoomed areas in panel e. (g) Corresponding EELS mapping of Co, Fe, and N. (h) H_2/O_2 fuel cell polarization plots. Cathode is ~0.77 mg cm^{-2} of (Fe,Co)/N-C; 100% RH; and O_2, 0.1, and 0.2 MPa partial pressures. Anode is 0.1 mgPt cm^{-2} Pt/C; 100% RH; H_2 and 0.1 MPa partial pressure; cell 353 K; and 25 cm^2 electrode area. (i) H_2 and air fuel cell polarization plots. Cathode is ~0.77 mg cm^{-2} of (Fe,Co)/N-C; 100% RH; air, 0.2 MPa partial pressures. Anode is 0.1 mgPt cm^{-2} Pt/C; 100% RH; H_2, 0.1 MPa partial pressure, cell 353 K; 25 cm^2 electrode area. (j) Stability of (Fe,Co) and N-C in a H_2 and air fuel cell measured at 600 mA cm^{-2} and 1000 mA cm^{-2}. Reprinted with permission from Ref. (181). Copyright 2017 American Chemical Society.

Although most of the SACs are Fe- and Co-based catalysts, each of these catalysts have their own merits and demerits. Fe-based catalysts may present higher ORR activity, but they are criticized for their commercial scale possibilities due to their Fenton-type reactions with H_2O_2, which can catalyze to form radical species that can further degrade the electrolyte membrane and ionomer, severally affecting the catalyst's durability. Co-based catalysts were found to be logical alternatives to Fe-based catalysts, but Co-based catalysts still produce H_2O_2 during ORR. Moreover, Co-N-C sites are more poorly stabilized in the carbon matrix than in Fe-N-C sites. It is found that the Mn-based catalysts are much less prone to Fenton-based reactions due to their weaker reactivity towards H_2O_2. Because of this, exploring Mn-based catalysts would be an effective alternative mitigation strategy. Li et al. (*176*) synthesized Mn-based SACs with ZIF-8 precursors. Unlike Co and Fe, it is difficult for Mn to exchange the bonding with Zn during ZIF formation. By a two-step, doping and adsorption strategy, atomically dispersing Mn catalysts possessing Mn_4N active sites are synthesized. In the initial step, Mn ions are exchanged with Zn ions to obtain Mn-ZIFs followed by pyrolysis and acid leaching. The initially formed pyrolyzed product was used for the second adsorption of Mn and the additional N-source followed by a second thermal activation. Different sources of N are utilized such as dipicolylamine, cyanamide, phenanthroline, and melamine. Among these, cyanamide produces the best results; the XRD results show no obvious diffraction peaks of crystalline Mn in the catalyst. Furthermore, it is found that the morphology of Mn-ZIF-8 is dependent on the Mn contents utilized in the synthesis process. The Mn atomic content of 20% resulted in morphological retainment of the original Mn-ZIF-8 structures, corresponding to an overall 3.03 wt.% of the Mn in the final catalysts as per Inductively coupled plasma mass spectrometry (ICP-MS) analysis. In addition, an ORR analysis of 20Mn-NC-second catalysts showed a significant improvement in the ORR activity, especially after a second adsorption step with additional Mn and N source followed by a thermal activation step, which owe to the increased $Mn\text{-}N_4$ active sites. Acid leaching of the first host catalyst Mn-NC removes MnO phases creating micropores that are further useful for the second adsorption step. The 20Mn-NC-second catalyst has a half-wave potential of 0.80 V in an acidic electrolyte with an H_2O_2 yield of less than 4%. The 20Mn-NC-second catalyst also shows excellent durability with just a 17-mV loss in half-wave potential after 30,000 potential cycles and a loss of just 29 mV after a 100-h chronoamperometric stability test conducted at 0.7 V in a 0.5-M H_2SO_4 electrolyte. The 20Mn-NC catalyst also produces a power density of 0.46 W cm^{-2} in an H_2O_2 fuel cell. Furthermore, based on the DFT analysis, MnN_4C_{12} found adjacent-to-zigzag graphitic edges were predicted to be the active ORR site. Thus, this study encourages developing the Mn-based SACs to mitigate the issues of Fe- and Co-based catalysts for PEMFC applications. In another study, Chen et al. (*177*) synthesized Mn-based monoatomic SACs supported on 3D graphene that showed an excellent ORR catalyst in alkaline electrolytes. Pyrolysis of Mn-BTC lignad, followed by acid washing (HCl) and NH_3 activation generated atomically dispersed Mn-C-NO SACs. The half-wave potential of 0.86 V is recorded for the Mn-C-NO catalyst, which is higher than the commercial Pt/C catalyst half-wave potential of 0.82 V. The successful Mn-C-NO catalysts are utilized as ORR cathode catalysts in Zn-air batteries and show excellent power densities that are higher than in the Pt/C catalyst. Out of the three possible active sites ($Mn\text{-}N_1O_3$, $Mn\text{-}N_2O_2$, and $Mn\text{-}N_3O_1$) proposed from DFT calculations, the $Mn\text{-}N_3O_1$ configuration was found to be the most active. Unique bonding between N- and O-coordinated Mn cofactors are responsible for enhanced ORR activity, compared to the traditional N-coordinated Mn catalysts.

Very recently, Qu et al. (*178*) synthesized Cu SACs using a unique gas migration strategy. Pyrolysis of the ZIF-8 is performed in an Ar-NH_3 atmosphere together with the Cu foam and placed together. During the pyrolysis process, the sublimation of Zn atoms creates the defect sites, and the NH_3 gas picks up the Cu atoms from the Cu foam in the form of $Cu(NH_3)_x$. The generated $Cu(NH_3)_x$ is trapped in the pores of ZIF-derived N-C carbon, generating a Cu-SAs and N-C catalyst. The EXAFS analysis show a dominant Cu-N_4 coordination environment in the Cu-SAs and N-C catalyst. The resulting Cu-SAs and N-C catalysts show an enhanced ORR activity with 0.99 V, slightly higher than the Pt/C catalyst in an alkaline electrolyte.

Furthermore, dual metal catalysts are also explored as excellent catalysts compared to the individual metal atom catalysts. The synergistic interaction between the dual metal atoms creates unique active sites for efficient ORR activity (*179, 180*). Wang et al. (*181*) synthesized a Fe-Co dual site catalyst using a host-guest strategy. ZIF-8 sodalite structures formed from Zn^{2+} and Co^{2+} ions are utilized as hosts to accommodate the Fe^{3+} as guest moieties within the cavities of ZIF-8 (Figure 5B). Pyrolysis of these host-guest structures results in the (Fe,Co) and N-C catalyst. The guest Fe species acts as a catalyst to control the pore geometry of the carbon, enhancing the pyrolysis of the metal-ligand-metal bonds that induce voids and increase cavities according to the Kirkendall effect. HAADF-STEM analysis shows atomically dispersed Fe and Co atoms in the N-doped carbon support. ORR analysis of the (Fe,Co) and N-C catalyst shows an onset potential of 1.06 and a half-wave potential of 0.863 V, comparable to the Pt/C catalyst in a 0.1-M $HClO_4$ electrolyte with an H_2O_2 yield as low as 1.17%. Moreover, the (Fe,Co) and N-C catalyst shows zero degradation even after 50,000 potential cycles. Surprisingly, the (Fe,Co) and N-C catalyst shows a peak power density of 0.85 and 0.98 W cm^{-2} in the H_2O_2 fuel cell, with 0.1 MPa and 0.2 MPa back pressures, respectively, and a cathode catalyst loading of 0.77 mg cm^{-2}. In the more practical H_2-Air atmosphere, it delivers a peak power density of 0.50 W cm^{-2} with 0.2 MPa back pressure. The active area of the electrode with 25 cm^{-2} shows a promising catalyst for PEM fuel cells. Furthermore, density functional theory (DFT) analysis indicates that the Fe-Co dual active sites enhance the cleavage of O-O bonds, leading to high ORR kinetics in the (Fe,Co)/N-C catalyst. The other significantly investigated SACs can be found in the literature (*182–185*).

MOF-Hybrid Catalysts

Research was also also conducted to further improve the MOF-derived catalysts by hybridizing them with the high surface area, high graphitic carbons to further improve the electronic characteristics of MOF-derived catalysts. Some researchers believe that pyrolysis of the ZIF catalysts can suffer from morphological and structural damage in the catalyst such as formation of carbon nanoparticles with low-graphitized catalysts (*186*). To overcome these limitations, researchers improved the catalyst structure using a sandwich-type structure with high surface area carbon, especially with doped 2D graphene. It is believed that the hybrid catalyst derived in this way can possess excellent electronic conductivity owing to the close interaction of the conductive 2D graphene along with the synergistic effect between doped graphene and the MOF-derived M-N-C catalyst. Studies also propose that the presence of 2D graphene structures can effectively prevent the aggregation of ZIF crystals doping the pyrolysis process and thus offering high density ORR active sites for the faster kinetics (*187–190*).

For example, Liu et al. (*191*) synthesized a hybrid N-doped and graphene catalyst (N-PC@G-0.02) from ZIF-8 and graphene oxide (GO) followed by the pyrolysis. The morphological analysis of the catalyst indicates that the GO sheets provide excellent nucleation sites for the growth of ZIF-8 nanocrystals. It is believed that the various oxygen functional groups of GO provided nucleation sites for the coordination of Zn^{2+} ions during the ZIF crystal growth. The result of ZIF-8 hybridization with graphene resulted in a higher BET surface area of 1094 $m^2\ g^{-1}$ along with micro- and mesopores. The presence of bimodal, micro-, and mesopores can sufficiently contribute to the positive mass transfer characteristics in fuel cells (*192*). During the ORR analysis, the N-PC@G-0.02 catalyst shows higher ORR kinetics with an onset potential of 1.01 V, which is higher than the Pt/C catalyst.

In another study, Zhu et al. (*193*) synthesized the Co-N-C catalyst by combining the Co-Zn BMZIF with the high surface area 2D graphene derived from graphene oxide (GO). C-Zn bimetallic ZIFs were nucleated on the surface of graphene oxide followed by the pyrolysis process to obtain a Co-Zn-ZIF and GO-920 catalyst. It is proposed that oxygen functional groups of GO interact with the Zn^{2+} and Co^{2+}, serving as the nucleation sites for ZIF growth. The BET analysis confirms that the Co-Zn-ZIF and GO-920 catalyst has a higher surface area of 823 m^2g^{-1} due to the high surface area graphene and effective prevention of graphene aggregation by the ZIF nanoparticles during the heat treatment process, collectively resulting in a higher surface area and higher density at the ORR active sites.

Wu et al. (*194*) synthesized a hybrid MOF structure with tannic acid functionalization on the ZIF-8 to generate N and B co-doped hollow carbons (NB-HCs). The tannic acid was coated on the surface of ZIF-8, which was further used to react with the boric acid to form a phenolic-borate network. Further carbonization of these MOF-hybrid core shell structures generates NB-HC structures, which are applied as catalysts for ORR. The synergistic effect of both N and B enhanced the ORR activity of NB-HCs, compared to nitrogen doped hollow carbon (N-HC). Furthermore, first-principle calculations suggest that the energy barrier for the first electron and proton transfer, which is the rate determining step of ORR, is found to be 0.64 eV for NB-HC, much lower than the Pt-supported graphene catalyst (1.13-1.68 eV). This clearly indicates that the synergistic effect of B and N in an MOF-derived NB-HC hybrid catalyst can significantly enhance the ORR activity.

Zhang et al. (*195*) synthesized a hybrid, porous, conductive catalyst with ZIF, single-walled carbon nanohorns (Co-N-C/SWCNHs) as efficient electrocatalysts for ORR in alkaline mediums. First, the BMZIFs are synthesized with Zn^{2+} and Co^{2+}. For these BMZIFs, a mixture of SWCNHs and 2-methylimidazole solution is added to obtain the precursor, a BMZIFs/SWCNHs-n hybrid structure, after which pyrolysis of the precursor leads to the formation of a 3D conductive Co-N-C/SWCNHs catalyst. Scanning electron microscope (SEM) and TEM analysis of the Co-N-C/SWCNH catalyst confirm the attachment of SWCNHs on the surface of BMZIFs forming a 3D architecture. The resulted Co-N-C/SWCNH catalyst shows excellent ORR kinetics compared to the control Co-N-C and standard Pt/C catalyst, with an onset potential of 0.97 V and half-wave potential of 0.87 V. The enhanced ORR activity of Co-N-C/SWCNHs catalyst is due to the synergistic interaction between the SWCNHs and Co-NC structures, in which SWCNHs are expected to enhance the electronic conductivity of the catalysts.

Tang et al. (*196*) synthesized an urea-modified-Co-MOF-GO composite as an ORR catalyst. The Co-MOF/GO composite was initially synthesized from Co^{2+}, terepthalic acid, and 1,4-diazabycyclo[2,2,2]octane as organic ligands in solvothermal conditions. The synthesized Co-MOF/GO composite was mixed with a solution of urea, then dried, and followed by pyrolysis at 900 °C to obtain the Co-GO(50)-C/N(2.5). The content of GO and urea are varied to optimize the electronic conductivity of the catalysts. TEM analysis of the Co-GO(50)-C/N(2.5) presents well-organized CNT structures formed from the organic ligands and evenly distributed on the graphitic carbon matrix, derived from the GO. Moreover, additional urea etch the carbon in the catalysts to provide the porosity. It is also observed that the Co nanoparticles are surrounded by the graphitic carbon layers and are believed to enhance the electron transfer to the metal active sites and the stability of the catalysts by mitigating the metal nanoparticle leaching (*197*). The ORR analysis of the Co-GO(50)-C/N(2.5) catalyst shows excellent ORR activity with a 0.934 onset potential and 0.832-V half-wave potential, which ismuch closer to the commercial Pt/C 20 wt% catalyst. In order to tackle the microporosity associated with MOF-derived catalysts, Tang et al. (*198*) synthesized a Co/N-CNT catalyst from ZIF-67 and polydopamine (PDA) nanotubes. ZIF-67 crystals are confined to the surface of PDA nanotubes, where the ZIF-67 nanocrystals are grown in inner parts of PDA nanotube, avoiding the aggregation of ZIF-67 nanocrystals. After pysolysis and acid etching, mesoporous "niches" like carbonaceous layers are observed in the catalyst. The BET analysis further shows a mixture of micro- (1.9 nm) and mesopores (19.3 nm) in the Co/N-CNT catalyst. The Co/N-CNT catalyst shows an onset potential of 0.91 V with 3.85 electrons transferred in the ORR reaction and a peroxide yield of <8%.

Xia et al. (*199*) synthesized average-sized Co_3O_4 nanoparticles of 35 nm that are dispersed evenly on the graphene aerogel (CoO_x/NG-A). The Co_3O_4 nanoparticles are obtained from the Co MOFs grown on graphene aerogel support. It is observed that the high surface area graphene effectively mitigates the agglomeration of Co_3O_4 nanoparticles during the pyrolysis if no large-sized Co MOFs are obtained without graphene aerogel during the synthesis. Moreover, Co_3O_4 nanoparticles are observed to be highly defective and possess large edges and corners with an interior hallow structures formed from the diffusion of oxygen atoms to the core Co atoms as shown by the Kirkendall effect. Remarkably, the CoO_x and NG-A catalyst showed an ORR onset potential of 1.019 V, which almost matches the commercial Pt/C catalyst (Figure 6).

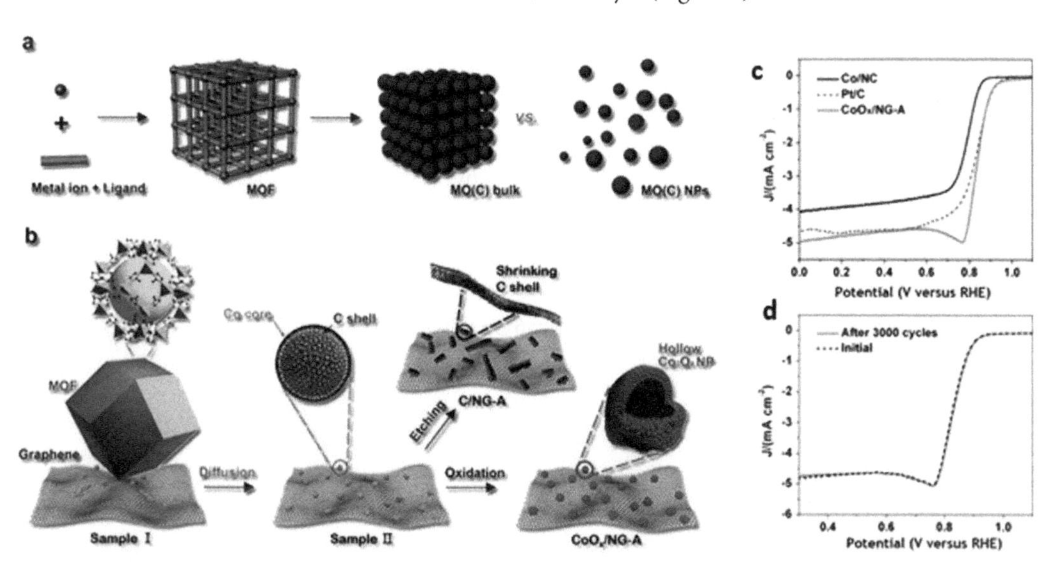

Figure 6. (a) Illustration of the concept of breaking MOF into monodisperse metal oxide NPs. (b) Schematic of the formation process of CoOx/NG-A and C/NG-A. (c) ORR polarization curves of at 10 mV s^{-1} and 1600 rpm. (d) ORR polarization curves of CoOx/NG-A in ADT measurement. Reprinted with permission from Ref. (199). Copyright 2017 American Chemical Society.

So far, the reported ZIF-based catalysts are generally pyrolyzed at temperatures well above 700–1100 °C. The pyrolysis at these temperatures often leads to a low yield (typically >70% from the precursor weight) of the final product due to the reduced hydrocarbon content. This carbon loss could also indirectly increase the total metal content of the catalyst leading to agglomoration of metal nanoparticles and necessitating the acid leaching in order to remove aggregated and unstable metal nanoparticles. Guo et al. (*200*) proposed a low temperature pyrolysis process in which the ZIF-67 is grown on the thermal-shock expanded graphene oxide (EGO) at temperatures lower than 450 °C (ZIF-67@EGO). It is found that the temperature of 500 °C would be enough to obtain the evenly distributed Co nanoparticles on the surface of the EGO. The oxygen functionalities of EGO and their slow decomposition is responsible for the mild thermolysis of the ZIF-67@EGO precursor. The ORR analysis of the Co_3O_4 and Co@N-G-450 catalyst showed an ORR onset potential of 0.962 V, similar to the commercial Pt/C catalyst. Moreover, ZIF-67 crystals are deposited on the GO/ZIF-8 to obtain a sandwich type ZIF-8/GO/ZIF-67 structures (*201*). Also, it is believed that the initially deposited ZIF-8 crystals on GO act as seed mediators, due to their similar unit and crystal structures. Generally, inhomogeneous distribution of oxygen functionalities on GO leads to weak interaction between GO and MOFs. The uncontrolled and faster growth kinetics of MOFs could lead to agglomerated nanoparticles. The process of seed growth of ZIF-67 on GO/ZIF-8 could result in homogeneous distribution of core shell Co@NC nanoparticles that are evenly distributed on the surface of GO. The catalyst GO/ZIF-8@ZIF-67-900 thus effectively ascribes the advantage of Co@NC nanoparticles and good conductivity of GO support, which leads to a higher ORR onset potential of 0.93 V, higher than the Pt/C catalyst and with similar catalyst-loading. A similar study is also conducted by Liu et al. (*202*).

A relatively new family of 2D materials called "MXenes" were discovered in 2011 by a Drexel University scientist resulting in a new field of research (*203*). MXenes are a family of layered materials composed of transition metal carbides, nitrides, and carbonitrides of $Mn_{+1}AX_n$ (MAX) where M represents the various transition metals such as Sc, Ti, Zr, Hf, V, Nb, Ta, Cr, and Mo X is carbon or nitrogen, and A represents the elements from the groups 13 and 14 on the periodic table (most commonly Aluminium [Al]). In contrast to most other 2D materials including graphene, MXenes possess hydrophilic surfaces and high metallic conductivities (~6000–8000 S cm^{-1}), showing promising performance in energy storage devices, water desalination, catalysis, electromagnetic interference shielding, transparency, conducting thin films, and other applications (*204*). Layered MXenes can be produced by selective etching the Al layer using the hydrofluoric acid (HF) treatment. Recently, composite catalysts of MOF with MXenes were explored. However, limited attention has been paid to MOF and MXene composition. For instance, Jianian et al. (*205*) synthesized Co-tipped CNTs that were integrated with Mxenes as efficient ORR catalysts in 0.1 M potassium hydroxide (KOH) electrolyte. The Co-tipped CNTs were derived from Co-based ZIF-67 and grown on the 2D Mxenes after pyrolysis at 800 °C. A CNT of 15 nm in diameter with a Co nanoparticle of around 10 nm is tipped on the end of each CNT and evenly distributed on the surface of Ti_3C_2 nanosheets. ORR activity evaluation of the Co-CNT and Ti_3C_2 catalysts showed a significant synergistic effect brtween the Co-CNTs and Ti_3C_2 support. The ORR activity of Co-CNT and Ti_3C_2 catalyst showed much lower overpotential than the Co-CNT and Pt/C catalyst. Furthermore, the ORR activity of the Co-CNT and Ti_3C_2 catalyst depends on the Ti_3C_2 content, and the best ORR performance was achieved with the Co-CNT and Ti_3C_2-60 catalyst with a half-wave potential of 0.82 V compared to the Pt/C catalyst with 3.9 electrons transferred per O_2 molecule. Strong coupling interaction and enhanced electron transfer kinetics between the Co-CNT and Ti_3C_2 is found to be the responsible factor for the enhanced ORR activity of the Co-CNT and Ti_3C_2-60 catalyst. Research on MXenes is still in its nascent state, and the research on composites of MXenes and MOFs is open for research. The other notable hybrid catalysts synthesized and evaluated using ORR can be found in the research (*206–211*).

Solvent-Free Synthesis of MOF Derived Catalysts

The great advancements and promising experimental results of MOF-derived catalysts not only facilitate the development of fuel cells but also a variety of chemical reactions, including energy storage, light-harvesting reactions, gaseous storage, CO_2 sequestration, asymmetric and heterogenous catalysis, and drug delivery. This has motivated the researchers to synthesize various MOF catalysts (*212, 213*). Yet, further research on this topic is still required to enhance the MOF structural stability and activity. Currently, many MOF-derived catalysts are synthesized at lab scale; however, efforts must be taken to develop them in large pilot scale systems in order to realize their commercial applications. Finally, industries should collectively take steps to optimize the synthetic conditions needed to obtain the MOF catalysts in large scale production. It is imperative that the large scale production and final cost of the catalysts are dependent on various factors such as precursor availability, synthetic conditions, solvents utilized in the reaction process, toxicity of the solvents in the environment, recycling of the solvents, quality and quantity of the solvents, reaction time, washing, and pre- and post-purification processes (*214*). Currently, the commercial applications of MOF-derived catalysts are limited due to the following factors: (i) utilization of non-aqueous solvents; (ii) structural maintenance of the MOFs in the final catalyst (especially the porous structure) compared to their analogue parental MOFs; (iii) toxic effects of solvents utilized in MOF synthesis; and (iv) high cost of preparation processes (*215*). Among these challenges, solvents utilized in MOFs possess special problems. Almost all the MOF synthesis utilizes a large volume of solvents, especially organic solvents such as methanol, ethanol, N,N-dimethylformamide, dimethylacetamide, and triethyl amine. These solvents are utilized in various steps such as the synthesis process and washing of the products. It is worth mentioning that no solvent-recycling protocols are currently in use, thus creating hazardous environment issues. Moreover, solution-based synthesis processes are time-consuming and tedious. Recently, Ren et al. and Barkholtz et al. described that the current hindrance to MOF-derived catalysts is not due to the availability of raw materials, but due to the solvent-based synthesis protocols (*216*). Hence, it is desirable to develop cheaper and faster synthesis protocols by completely eliminating solvents. Additionally, solid state synthesis or mechanochemical synthesis processes can be regarded as a potential alternative to the traditional solution-based synthesis protocols. Mechanochemical synthesis represents the application of mechanical force either by milling or grinding to induce chemical reactions (*217*). The advantages of mechanochemical synthesis involves faster reaction time, high yield of the products, possibility of pilot scale production, and no toxic environmental hazardous solvent by-products. The mechanochemical synthesis process is already successfully utilized in various fields; however, there are few studies available on the synthesis of MOFs (*218, 219*). This section briefly describes the catalysts derived from MOFs by utilizing mechanochemical synthesis process.

Li et al. (*220*) synthesized N and S co-doped carbons from dual ligands, namely 4,4-bypyridine and 2,5-thiophenedicarboxyllic acid. The two organic ligands and Zn acetate are ground in a mortor or pestle for 10 mins at equal molar ratios to get Zn(tdc)(bpy) crystals. Although this study utilizes no solvents during the synthesis, the resulting Zn(tdc)(bpy) crystals are washed several times with

DMF and followed by CH_3OH. The pyrolysis of the Zn(tdc)(bpy) crystals produces the N and S co-doped porous carbons (NSC-1000). The XRD analysis of the Zn(tdc)(bpy) crystals shows different diffraction patterns compared to the starting raw materials, showing the specific arrangement of ligands and Zn^{2+} in the Zn(tdc)(bpy) crystals. The BET of the NSC-1000 catalyst shows a specific surface area as high as 1999 $m^2 g^{-1}$. The ORR analysis of the NSC-1000 catalyst shows an ORR onset potential of 0.913 V, which is comparable to the commercial Pt/C catalyst in the alkaline electrolyte.

Peera et al. (*221*) synthesized Co-encapsulated nitrogen-doped carbon core shell nanostructures by a mechanochemical reaction between Zn, Co, and pyrazole as organic ligands. The precursor $Zn_xCo_{1-x}(C_3H_4N_2)$ is synthesized by a simple grinding of the reactants in a motor. A pink-colored $Zn_xCo_{1-x}(C_3H_4N_2)$ is formed by the reaction between -N groups of pyrazoles and Zn^{2+}, while other -NH groups can bind to Co^{2+} ions to finally give a MOF of $Zn_xCo_{1-x}(C_3H_4N_2)$. Moreover, the deprotonation of -NF groups are also known to offer N-M bonds. The formed $Zn_xCo_{1-x}(C_3H_4N_2)$ is pyrolyzed to get the Co@NC electrocatalyst. During the pyrolysis, the pyrazole decomposes into N-doped carbon and Co ions are reduced to Co metal with simultaneous evaporation of metallic Zn. The formation of $Zn_xCo_{1-x}(C_3H_4N_2)$ is determined based on the XRD analysis. The diffraction peaks of $Zn_xCo_{1-x}(C_3H_4N_2)$ are found to be much different from the parental compounds showing the formation of $Zn_xCo_{1-x}(C_3H_4N_2)$ MOF structures. TEM images of the Co@NC-MOF-2-900 catalyst showed core shell nanostructures in which Co core metal is surrounded by the NC shells derived from the pyrazole ligand. The Co@NC catalyst synthesized here exclusively showed mesopores of 4–6 nm, which are beneficial for mass transport of reactants for fuel cells. The ORR analysis of the Co@NC-MOF-2-900 catalyst showed a half-wave potential of 0.88 V, which is 30 mV higher than the commercial Pt/C, with 3.97 electrons transferred per O_2. The Co@NC-MOF-2-900 catalyst also explored its possible application in metal-air batteries by assessing its ORR activity in 6 M of KOH electrolyte. Even the Co@NC-MOF-2-900 catalyst with a higher electrolyte concentration shows 0.82 V of onset potential. The catalyst also showed no degradation over 5000 potential cycles with complete retention of the Co@NC nanostructures per TEM analysis.

Recently, Wang et al. (*222*) synthesized Fe and Co Zn MOFs from a molten salt–assisted solid state synthesis process. The iminodiacetic acid (IDA) is used as an organic linker and Zn^{2+} and Co^{2+} are used as coordination metal ions. The metal ions and organic linkers are mixed and then heated to melt in a muffle furnace at 200 °C to obtain a Zn (Mx) IDA MOF. The resulting MOFs are pyrolyzed at 900 °C to obtain the Co-NC and Fe-NC catalysts. The pyrolyzed catalysts show a sheet-like morphology with abundant mesoposoity and Co- or Fe-encapsulated core shell structures. ORR analysis of the catalysts show an enhanced activity with an onset potential of 0.963 V for Fe-NC catalysts followed by 0.957 V for Co-NC, which is higher than the Pt/C catalyst. Though many solid-state synthesis protocols have been explored for various catalytic reactions, their exploration of ORR catalytic reactions are rare, and the research in this direction is open for scientific society. Table 1 summarizes the ORR kinetic parameters of notable catalysts developed from MOFs taken from the literature (*114*, *125*, *127–129*, *133*, *139*, *168*, *169*, *191*, *193–195*, *223–233*).

Table 1. ORR Catalysts Derived from Metal Organic Framework Precursors

Metal Ions Ligands Solvent	*Catalyst*	*Pyrolysis Temperature (°C) and Time*	*Metal Loading on RDE Electrode ($mg\ cm^{-2}$)*	*Electrolyte*	*(i) ORR Onset (ii) Half Wave Potential (V vs RHE)*	*(i) No. of Electrons Transferred in ORR (ii) H_2O_2 Yield*	*Ref*
Zn (ac_2) benzimidazole dimethyl formamide (DMF)	NC-D-NH_3	800 – 4 h	0.49	0.1 M KOH	(i) 1.0 (ii) NR	(i) 3.8 (ii) Not reported	(*114*)
Co-$(NO_3)_2$·6H2O, 3,5-imidazolate, N,N′-dimethylacetamide	Co-N_4 non-PGM	750 – NR	0.60	0.1 M $HClO_4$	0.83 V (ii)	(i) 3.3-3.6 (ii) NR	(*125*)
Zn $(NO_3)_2.6H_2O$ $Fe(NO_3)_3$·9H2O, 2-MI Methanol	Fe-ZIF catalyst – 1100 °C	1100 – 1 h	0.80	0.5 M H_2SO_4	(i) Not mentioned (ii) 0.85 V	(i) NR (ii) <1%	(*127*)
$FeSO_4 \cdot 7H_2O$ Zn $(NO_3)_2.6H_2O$ benzimidazole (BIM) methanol	Fe-N-HPCC	1000 – NR	0.788	0.1 M KOH 0.1 M $HClO_4$	(i) 0.972 (ii) 0.898 (i) NR (ii) 0.76	(i) NR (ii) 5% (i) 3.99 (ii) 3.6%	(*128*)
Zn $(NO_3)_2.6H_2O$ $Co(NO_3)_2 \cdot 6H_2O$ methanol	20Co-NC-1100	1100 – 1 h	0.8	0.5 M H_2SO_4	(i) 0.93 (ii) 0.80	NR	(*129*)
Cobalt $(ac)_2$ 2-MI iron acetate 1,10 phenanthroline ethanol	Fe–N–Co@C-800-AL	800 – 10 min	NR	0.1 M KOH	0.91	(i) 3.98 (ii) NR	(*133*)
Zn $(NO_3)_2.6H_2O$	Co,N-PCLs	800 – 5 h	0.20	0.1 M KOH	(i) NR	(i) 3.97	(*139*)

Table 1. (Continued). ORR Catalysts Derived from Metal Organic Framework Precursors

Metal Ions Ligands Solvent	*Catalyst*	*Pyrolysis Temperature (°C) and Time*	*Metal Loading on RDE Electrode ($mg\ cm^{-2}$)*	*Electrolyte*	*(i) ORR Onset (ii) Half Wave Potential (V vs RHE)*	*(i) No. of Electrons Transferred in ORR (ii) H_2O_2 Yield*	*Ref*
$Co(NO_3)_2.6H_2O$ 2-MI water					(ii) 0.846	(ii) NR	
$Zn\ (NO_3)_2.6H_2O$ 2-MI ferrocene, methanol	C-FeZIF-1.44-950	950 – 3 h	0.50	0.5 M $HClO_4$	(i) NM (ii) 0.78	(i) 3.97 (ii) 2%	*(168)*
$Zn\ (NO_3)_2.6H_2O$ $FeSO_4{\cdot}6H_2O$ 2-MI methanol	Fe-N-C derived from Fe (Zn/Fe=0.95/0.05)-N-C in Ar	1000 – 1h	0.56	0.1 M $HClO_4$	(i) 0.95 (ii) 0.81	(i) close to 4 (ii) <2%	*(169)*
$Zn\ (NO_3)_2.6H_2O$ 2-MI GO water	N-PC@G-0.02	900 – 5 h	NR	0.1 M KOH	(i) 1.01 (ii) 0.80	(i) 3.93 (ii) NR	*(191)*
$Zn\ (NO_3)_2.6H_2O$ $Co(NO_3)_2.6H_2O$ 2-MI methanol GO	Co-Zn-ZIF/GO-920	920 – 2 h	0.26	0.1 M KOH	(i) 0.914 (ii) 0.807	(i) 3.89 (ii) 5.5 %	*(193)*
$Zn\ (NO_3)_2.6H_2O$ 2-MI, tannic acid methanol	NB-HC	900 – 1 h	NR	0.1 M KOH	-0.12 V vs Ag/AgCl	(i) 3.90-3.92 (ii) NR	*(194)*

Table 1. (Continued). ORR Catalysts Derived from Metal Organic Framework Precursors

Metal Ions Ligands Solvent	*Catalyst*	*Pyrolysis Temperature (°C) and Time*	*Metal Loading on RDE Electrode ($mg\ cm^{-2}$)*	*Electrolyte*	*(i) ORR Onset (ii) Half Wave Potential (V vs RHE)*	*(i) No. of Electrons Transferred in ORR (ii) H_2O_2 Yield*	*Ref*
$Zn(NO_3)_2.6H_2O$ $Co(NO_3)_2.6H_2O$ 2-MI	CoNC/SWCNHs-10	650 – 2 h	0.35	0.1 M KOH	(i) 0.97 (ii) 0.87	(i) 3.7 (ii) <10%	(195)
$Co(NO_3)_2·6H_2O$ 3,6-bis(1-midazolyl)-1,2,4,5-tetrazine DMF	Co/N-C-800	800 – 1 h	0.71	0.1 M KOH 0.1 M H_2SO_4	(i) 0.981 (ii) 0.872 (i) 0.852 (ii) 0.671	(i) 3.92 (ii) 3.7 % (i) 3.91 (ii) 5.9 %	(223)
$Zn(NO_3)_2.6H_2O$ $Co(NO_3)_2·6H_2O$ 2-MI water	$Co_{0.6}$–N/C-800	800 – 2 h	0.254	0.1 M KOH	(i) 0.916 (ii) 0.825	(i) 3.90 (ii) <20%	(224)
$Co(NO_3)_2.6H_2O$ 2-MI cobalt porphyrin – CoTMPP methanol	ZIF-67@CoTMPP (800)	800 – 2 h	0.60	0.1 M KOH	(i) 0.854 (ii) 0.784	(i) 3.36 (ii) 35 %	(225)
$Fe(acac)_3$ $Zn(NO_3)_2.6H_2O$ Methanol 2-MI	$Fe_{0.06}$-N/C-900	900 – 1hr	0.40	0.1 M $HClO_4$	(i) 0.84 (ii) 0.78	(i) 3.8 ~ 4.0 (ii) 0.6 %	(226)
$Co(NO_3)_2.6H_2O$ Dicyanamide Pyrazine	Co@BNCNTs-900	900 – 3 hr	0.30	0.1 M KOH	(i) 0.93 (ii) 0.82	(i) 3.85 (ii) <12%	(227)
$Co(NO_3)_2.6H_2O$ thiophene-2,	Co@SN-CNP	700 – 3 hr	0.40	0.1 M KOH	(i) 0.87 (ii) NR	(i) 3.85 (ii) NR	(228)

Table 1. (Continued). ORR Catalysts Derived from Metal Organic Framework Precursors

Metal Ions Ligands Solvent	*Catalyst*	*Pyrolysis Temperature (°C) and Time*	*Metal Loading on RDE Electrode ($mg\ cm^{-2}$)*	*Electrolyte*	*(i) ORR Onset (ii) Half Wave Potential (V vs RHE)*	*(i) No. of Electrons Transferred in ORR (ii) H_2O_2 Yield*	*Ref*
5-dicaboxylate (Tdc), 4,4′-bipyridine (Bpy) DMF-water (1:1, v/v) solution							
$CoCl_2$. $6H_2O$ 2-MI Ethanol	Co-N-C HHMTs-24	800 – 3 hr	0.10	0.1 M KOH	(i) 0.973 (ii) 0.871	(i) 3.95-4.0 (ii) NR	(229)
$Zn(NO_3)_2{\cdot}6H_2O$ tris(4-pyridyl)triazine 1,4-benzenedicarboxylate Dicyandiamide	NHMC - 900	900 – 1 hr	0.50	0.1 M KOH	(i) 1.0 (ii) 0.88	(i)3.79-3.99 (ii) NR	(230)
$Zn(Ac)_2{\cdot}2H_2O$ 2-MI 1H-1,2,3-triazole Methanol	EZIF-C	900 – 2 hr	NR	0.1 M KOH	(i) −0.030 V vs Ag/ AgCl (ii) NR	(i) 3.8-4.0 (ii) NR	(231)
Montmorillonite (MMT) $Co(NO_3)_2{\cdot}6H_2O$ 2-MI Methanol	Co/C NP	900 – 2 hr	0.10	0.1 M KOH	(i) 1.016 (ii) 0.886	(i) 3.5-3.8 (ii) NR	(232)
$Zn(NO_3)_2{\cdot}6H_2O$ 2-MI $Fe(acac)_3$, $Co(acac)_3$ Methanol	FeCo-ISAs/CN	900 – 2 hr	0.408	0.1 M KOH	(i) 0.995 (ii) 0.92	(i) 3.9-4.0 (ii) <4.7 %	(233)

Conclusions and Future Perspectives

This chapter reviews various synthesis processes of MOF-derived catalysts for ORR. It can be determined that MOF-derived catalysts show excellent catalytic properties. Versatility of MOF structures enable the synthesis of various metal-coordinated catalysts with potential M-N-C active sites, providing various choices to design materials of varying properties. The synthesis of MOFs also facilitate and can be optimized for large scale production. The perfect, pre-designed metal-nitrogen bonds offer excellent distribution of metal nanoparticles and atomic dispersion of the metal atoms in the carbon matrix. The atomically dispersed single atom catalysts have especially shown proven activity, comparable to commercial Pt/C catalyst both at RDE and MEA levels. As MOFs contain both metal and carbon sources, incorporating dual and multiple metal ions is a possibility to obtain dual or multiple metal-supported catalysts. Furthermore, the utilization of different organic ligands produces a carbon matrix with various heteroatoms and pore size distributions. The pore size tunability of MOF-derived catalysts represents great benefits as it is highly desirable to develop porous M-N-C catalysts in order to actually realize their applications at MEA level.

Nevertheless, there are still several challenges that need to be addressed and are worth investigating. Currently, there is a large variety of Co- and Fe-based catalysts derived from MOFs that have been investigated. In most cases, the catalysts outperform the commercial Pt/C catalyst in the alkaline medium, and their activity is still lower in acidic electrolytes. Even though their catalytic activity is evaluated in RDE, their actual use in a fuel cell configuration is rarely reported. Since the ORR activity evaluation by RDE is only considered as a tool to screen the catalysts, their reproducible activity is not guaranteed in an actual fuel cell system. Because of this, the first challenge would be to focus on determining their usefulness in a real fuel cell and identifying their performance and degradation rate. From the various MOF-derived catalysts, it is concluded that the preparation of precursor MOFs (before pyrolysis) determines the final catalyst. Depending on the precursors, metal ions, and organic ligands, the pore cavity of the MOF is decided, which in turn affects the porosity of the catalysts. Further studies are required to develop better strategies to obtain high density of active sites and porosity. In nearly all of the synthetic processes, the intact structures of the MOFs are unavoidably damaged during pyrolysis. This damage can include aggregation of metal nanoparticles and formation of core shell nanostructures with thick carbon layers that can hamper the density of active sites. Effective strategies are needed to control the synthetic properties. It is also worth investigating the MOFs containing other transition metals, other than Fe and Co. There is only one report available on the Mn-based MOFs. Mn-based catalysts are better options to avoid Fenton-based reactions associated with Fe-based catalysts. It is worth it to further investigate the Mn-based MOF-derived catalysts. Furthermore, dual MOFs, with Fe and Mn or Mn and Co, can effectively deal with the H_2O_2-associated issues. Among all the MOF catalysts investigated in the last decade, the SACs and atomic dispersion of metal nanocatalysts are found to be the best alternatives to Pt/C, as these catalysts have the potential to deliver the power density in a real fuel cell atmosphere. However, their atomic dispersion may lead to coalescence, and the catalytic synthetic protocols need to be carefully designed. It is also worth investigating the SACs' long-term durability in fuel cell configuration. As MOF-derived catalysts possess voids and cavities, it is easier to manipulate the porosity properties of the ORR catalyst to address the major hindrance of high catalyst layer thickness of the non-precious metal catalysts. Due to various advantages, solid-state synthesis protocols are found to be the best possible alternative for the large-scale synthesis of MOF-derived catalysts; thus, the research in this direction is critical.

Acknowledgments

This research was supported by the Keimyung University Research Grant of 2018-0742.

References

1. Banham, D.; Kishimoto, T.; Zhou, Y.; Sato, T.; Bai, K.; Ozaki, J. I.; Imashiro, Y.; Ye, S. Critical Advancements In Achieving High Power And Stable Nonprecious Metal Catalyst–Based MEAs for Real-World Proton Exchange Membrane Fuel Cell Applications. *Sci. Adv.* **2018**, *4*, eaar7180–7187.
2. Dodds, P. E.; Staffell, I.; Hawkes, A. D.; Li, F.; Grünewald, P.; McDowall, W.; Ekins, P. Hydrogen And Fuel Cell Technologies For Heating: A Review. *Int. J. Hydrog. Energy* **2015**, *40*, 2065–2083.
3. Dekel, D. R. Review Of Cell Performance in Anion Exchange Membrane Fuel Cells. *J. Power Sources* **2018**, *375*, 158–169.
4. Kumar, R.; Singh, L.; Wahid, Z.; Mahapatra, D. M.; Liu, H. Novel Mesoporous $MnCO_2O_4$ Nanorods as Oxygen Reduction Catalyst at Neutral pH in Microbial Fuel Cells? *Bioresour. Technol.* **2018**, *254*, 1–6.
5. Kumar, R.; Singh, L.; Zularisam, A. W.; Hai, F. I. Microbial Fuel Cell Is Emerging as a Versatile Technology: A Review on Its Possible Applications, Challenges and Strategies to Improve the Performances. *Int. J. Energy Res.* **2018**, *42*, 369–394.
6. Parthiban, V.; Akula, S.; Peera, S. G.; Islam, N.; Sahu, A. Proton Conducting Nafion-Sulfonated Graphene Hybrid Membranes For Direct Methanol Fuel Cells With Reduced Methanol Crossover. *Energy Fuels* **2015**, *30*, 725–734.
7. Rosli, R.; Sulong, A. B.; Daud, W. R. W.; Zulkifley, M. A.; Husaini, T.; Rosli, M. I.; Majlan, E.; Haque, M. A Review Of High-Temperature Proton Exchange Membrane Fuel Cell (HT-PEMFC) System. *Int. J. Hydrog. Energy* **2017**, *42*, 9293–9314.
8. Chandan, A.; Hattenberger, M.; El-Kharouf, A.; Du, S.; Dhir, A.; Self, V.; Pollet, B. G.; Ingram, A.; Bujalski, W. High Temperature (HT) Polymer Electrolyte Membrane Fuel Cells (PEMFC) – A Review. *J. Power Sources* **2013**, *231*, 264–278.
9. Søndergaard, T.; Cleemann, L. N.; Becker, H.; Aili, D.; Steenberg, T.; Hjuler, H. A.; Seerup, L.; Li, Q.; Jensen, J. O. Long-Term Durability Of HT-PEM Fuel Cells Based On Thermally Cross-Linked Polybenzimidazole. *J. Power Sources* **2017**, *342*, 570–578.
10. Authayanun, S.; Im-Orb, K.; Arpornwichanop, A. A Review of The Development of High Temperature Proton Exchange Membrane Fuel Cells. *Chinese J. Catal.* **2015**, *36*, 473–483.
11. Hussain, A. M.; Høgh, J. V.; Jacobsen, T.; Bonanos, N. Nickel-Ceria Infiltrated Nb-Doped SrTiO3 for Low Temperature SOFC Anodes and Analysis on Gas Diffusion Impedance. *Int. J. Hydrog. Energy* **2012**, *37*, 4309–4318.
12. Hussain, A. M.; Høgh, J. V.; Zhang, W.; Bonanos, N. Efficient Ceramic Anodes Infiltrated with Binary and Ternary Electrocatalysts for SOFCs Operating at Low Temperatures. *J. Power Sources* **2012**, *216*, 308–313.
13. Hussain, A. M.; Høgh, J. V.; Zhang, W.; Stamate, E.; Thydén, K. T.; Bonanos, N. Improved Ceramic Anodes For SOFCs With Modified Electrode/Electrolyte Interface. *J. Power Sources* **2012**, *212*, 247–253.

14. Hussain, A. M.; Høgh, J. V.; Zhang, W.; Blennow, P.; Bonanos, N.; Boukamp, B. A. Effective Improvement of Interface Modified Strontium Titanate Based Solid Oxide Fuel Cell Anodes by Infiltration With Nano-Sized Palladium and Gadolinium-Doped Cerium Oxide. *Electrochim. Acta* **2013,** *113,* 635–643.
15. Hussain, A. M.; Pan, K. J.; Robinson, I. A.; Hays, T.; Wachsman, E. D. Stannate-Based Ceramic Oxide as Anode Materials for Oxide-Ion Conducting Low-Temperature Solid Oxide Fuel Cells. *J. Electrochem. Soc.* **2016,** *163,* F1198–F1205.
16. Huang, Y. L.; Hussain, A. M.; Pellegrinelli, C.; Xiong, C.; Wachsman, E. D. Chromium Poisoning Effects on Surface Exchange Kinetics of $La_{0.6}Sr_{0.4}Co_{0.2}Fe_{0.8}O_3{}^{-}\delta$. *ACS Appl. Mater. Interfaces* **2017,** *9,* 16660–16668.
17. Huang, Y.-L.; Hussain, A. M.; Wachsman, E. D. Nanoscale Cathode Modification for High Performance and Stable Low-Temperature Solid Oxide Fuel Cells (SOFCs). *Nano Energy* **2018,** *49,* 186–192.
18. Pan, K. J.; Hussain, A. M.; Wachsman, E. D. Investigation on $Sr_{0.2}Na_{0.8}Nb_{1-x}VxO_3$ (x= 0.1, 0.2, 0.3) as New Ceramic Anode Materials for Low-Temperature Solid Oxide Fuel Cells. *J. Power Sources* **2017,** *347,* 277–282.
19. Mohammed Hussain, A.; Sudireddy, B. R.; Høgh, J. V. T.; Bonanos, N. A Preliminary Study on WO_3-Infiltrated W–Cu–ScYSZ Anodes for Low Temperature Solid Oxide Fuel Cells. *Fuel Cells* **2012,** *12,* 530–536.
20. Hays, T.; Hussain, A. M.; Huang, Y. L.; McOwen, D. W.; Wachsman, E. D. Improved Sulfur Tolerance of SOFCs Through Surface Modification of Anodes. *ACS Appl. Mater. Interfaces* **2018,** *1,* 1559–1566.
21. Huang, Y. L.; Hussain, A. M.; Robinson, I. A.; Wachsman, E. D. Nanointegrated, High-Performing Cobalt-Free Bismuth-Based Composite Cathode for Low-Temperature Solid Oxide Fuel Cells. *ACS Appl. Mater. Interfaces* **2018,** *10,* 28635–28643.
22. Kendall, M. Fuel Cell Development For New Energy Vehicles (Nevs) And Clean Air in China. *Prog Nat Sci-Mater.* **2018,** *28,* 113–120.
23. Chaudhari, N. K.; Joo, J.; Kim, B.; Ruqia, B.; Choi, S. I.; Lee, K. Recent Advances in Electrocatalysts Toward the Oxygen Reduction Reaction: The Case of PtNi Octahedra. *Nanoscale* **2018,** *10,* 20073–20088.
24. Banham, D.; Ye, S. Current Status and Future Development of Catalyst Materials and Catalyst Layers for Proton Exchange Membrane Fuel Cells: An Industrial Perspective. *ACS Energy Lett.* **2017,** *2,* 629–638.
25. Kongkanand, A.; Mathias, M. F. The Priority and Challenge of High-Power Performance of Low-Platinum Proton-Exchange Membrane Fuel Cells. *J. Phys. Chem. Lett.* **2016,** *7,* 1127–1137.
26. Peera, S. G.; Arunchander, A.; Sahu, A. Platinum Nanoparticles Supported on Nitrogen and Fluorine Co-Doped Graphite Nanofibers as an Excellent and Durable Oxygen Reduction Catalyst for Polymer Electrolyte Fuel Cells. *Carbon* **2016,** *107,* 667–679.
27. Peera, S. G.; Tintula, K.; Sahu, A.; Shanmugam, S.; Sridhar, P.; Pitchumani, S. Catalytic Activity of Pt Anchored onto Graphite Nanofiber-Poly (3, 4-Ethylenedioxythiophene) Composite Toward Oxygen Reduction Reaction in Polymer Electrolyte Fuel Cells. *Electrochim. Acta.* **2013,** *108,* 95–103.

28. Peera, S. G.; Sahu, A.; Arunchander, A.; Nath, K.; Bhat, S. Deoxyribonucleic Acid Directed Metallization of Platinum Nanoparticles on Graphite Nanofibers as a Durable Oxygen Reduction Catalyst For Polymer Electrolyte Fuel Cells. *J. Power Sources* **2015**, *297*, 379–387.
29. Arunchander, A.; Peera, S. G.; Parthiban, V.; Akula, S.; Kottakkat, T.; Bhat, S. D.; Sahu, A. K. Dendrimer Confined Pt Nanoparticles: Electro-Catalytic Activity Towards the Oxygen Reduction Reaction and Its Application in Polymer Electrolyte Membrane Fuel Cells. *RSC Adv.* **2015**, *5*, 75218–75228.
30. Jung, N.; Chung, D. Y.; Ryu, J.; Yoo, S. J.; Sung, Y. E. Pt-Based Nanoarchitecture and Catalyst Design for Fuel Cell Applications. *Nano Today* **2014**, *9*, 433–456.
31. Xiong, Y.; Xiao, L.; Yang, Y.; DiSalvo, F. J.; Abrun~a, H. C. D. High-Loading Intermetallic Pt_3Co/C Core–Shell Nanoparticles As Enhanced Activity Electrocatalysts Toward The Oxygen Reduction Reaction (ORR). *Chem. Mater.* **2018**, *30*, 1532–1539.
32. Bu, L.; Zhang, N.; Guo, S.; Zhang, X.; Li, J.; Yao, J.; Wu, T.; Lu, G.; Ma, J. Y.; Su, D. Biaxially Strained PtPb/Pt Core/Shell Nanoplate Boosts Oxygen Reduction Catalysis. *Science* **2016**, *354*, 1410–1414.
33. Peera, S. G.; Lee, T. G.; Sahu, A. K. Pt-Rare Earth Metal Alloy/Metal Oxide Catalysts for Oxygen Reduction and Alcohol Oxidation Reactions: An Overview. *Sustain Energ Fuels* **2019**, *3*, 1866–1891.
34. Stephens, I. E.; Bondarenko, A. S.; Bech, L.; Chorkendorff, I. Oxygen Electroreduction Activity and X-Ray Photoelectron Spectroscopy of Platinum and Early Transition Metal Alloys. *ChemCatChem* **2012**, *4*, 341–349.
35. Choi, S.-I.; Shao, M.; Lu, N.; Ruditskiy, A.; Peng, H. C.; Park, J.; Guerrero, S.; Wang, J.; Kim, M. J.; Xia, Y. Synthesis and Characterization of Pd@Pt–Ni Core–Shell Octahedra with High Activity Toward Oxygen Reduction. *ACS Nano* **2014**, *8*, 10363–10371.
36. Li, M.; Zhao, Z.; Cheng, T.; Fortunelli, A.; Chen, C. Y.; Yu, R.; Zhang, Q.; Gu, L.; Merinov, B. V.; Lin, Z. Ultrafine Jagged Platinum Nanowires Enable Ultrahigh Mass Activity For The Oxygen Reduction Reaction. *Science* **2016**, *354*, 1414–1419.
37. Chen, C.; Kang, Y.; Huo, Z.; Zhu, Z.; Huang, W.; Xin, H. L.; Snyder, J. D.; Li, D.; Herron, J. A.; Mavrikakis, M.; Chi, M.; More, K. L.; Li, Y.; Markovic, N. M.; Somorjai, G. A.; Yang, P.; Stamenkovic, V. R. Highly Crystalline Multimetallic Nanoframes with Three-Dimensional Electrocatalytic Surfaces. *Science* **2014**, *343*, 1339–1343.
38. Huang, X.; Zhao, Z.; Cao, L.; Chen, Y.; Zhu, E.; Lin, Z.; Li, M.; Yan, A.; Zettl, A.; Wang, Y. M.; Duan, X.; Mueller, T.; Huang, Y. High-Performance Transition Metal–Doped Pt_3Ni Octahedra for Oxygen Reduction Reaction. *Science* **2015**, *348*, 1230–1234.
39. Chen, Z.; Higgins, D.; Yu, A.; Zhang, L.; Zhang, J. A Review On Non-Precious Metal Electrocatalysts For PEM Fuel Cells. *Energy Environ. Sci.* **2011**, *4*, 3167–3192.
40. Arunchander, A.; Peera, S. G.; Giridhar, V. V.; Sahu, A. K. Synthesis of Cobalt Sulfide-Graphene as an Efficient Oxygen Reduction Catalyst in Alkaline Medium and Its Application in Anion Exchange Membrane Fuel Cells. *J. Electrochem. Soc.* **2017**, *164*, F71–F80.
41. Arunchander, A.; Vivekanantha, M.; Peera, S. G.; Sahu, A. K. MnO–Nitrogen Doped Graphene as a Durable Non-Precious Hybrid Catalyst for the Oxygen Reduction Reaction in Anion Exchange Membrane Fuel Cells. *RSC Adv.* **2016**, *6*, 95590–95600.

42. Arunchander, A.; Peera, S. G.; Sahu, A. K. Self-Assembled Manganese Sulfide Nanostructures on Graphene as an Oxygen Reduction Catalyst for Anion Exchange Membrane Fuel Cells. *ChemElectroChem* **2017**, *4*, 1544–1553.
43. Arunchander, A.; Peera, S. G.; Sahu, A. K. Synthesis of Flower-Like Molybdenum Sulfide/Graphene Hybrid as an Efficient Oxygen Reduction Electrocatalyst for Anion Exchange Membrane Fuel Cells. *J. Power Sources* **2017**, *353*, 104–114.
44. Sarkar, I. J. R.; Peera, S. G.; Chetty, R. Manganese Oxide Nanoparticles Supported Nitrogen-Doped Graphene: A Durable Alkaline Oxygen Reduction Electrocatalyst. *J. Appl. Electrochem.* **2018**, *48*, 849–865.
45. Peera, S. G.; Sahu, A. K.; Bhat, S. D.; Lee, S. C. Nitrogen Functionalized Graphite Nanofibers/Ir Nanoparticles For Enhanced Oxygen Reduction Reaction In Polymer Electrolyte Fuel Cells (PEFCs). *RSC Adv.* **2014**, *4*, 11080–11088.
46. Maiti, K.; Balamurugan, J.; Peera, S. G.; Kim, N. H.; Lee, J. H. Highly Active and Durable Core–Shell fct-PdFe@Pd Nanoparticles Encapsulated NG as an Efficient Catalyst for Oxygen Reduction Reaction. *ACS Appl. Mater. Interfaces* **2018**, *10*, 18734–18745.
47. Jung, N.; Cho, Y.-H.; Ahn, M.; Lim, J. W.; Kang, Y. S.; Chung, D. Y.; Kim, J.; Cho, Y. H.; Sung, Y. E. Methanol-Tolerant Cathode Electrode Structure Composed Of Heterogeneous Composites To Overcome Methanol Crossover Effects For Direct Methanol Fuel Cell. *Int. J. Hydrog. Energy* **2011**, *36*, 15731–15738.
48. Roger, I.; Shipman, M. A.; Symes, M. D. Earth-Abundant Catalysts for Electrochemical and Photoelectrochemical Water Splitting. *Nat. Rev. Chem.* **2017**, *1*, 1–13.
49. Seh, Z. W.; Kibsgaard, J.; Dickens, C. F.; Chorkendorff, I.; Nørskov, J. K.; Jaramillo, T. F. Combining Theory And Experiment In Electrocatalysis: Insights Into Materials Design. *Science* **2017**, *355*, eaad4998–5010.
50. Shao, M.; Chang, Q.; Dodelet, J.-P.; Chenitz, R. Recent Advances in Electrocatalysts for Oxygen Reduction Reaction. *Chem. Rev.* **2016**, *116*, 3594–3657.
51. Gong, M.; Dai, H. A Mini Review of NiFe-Based Materials as Highly Active Oxygen Evolution Reaction Electrocatalysts. *Nano Res.* **2015**, *8*, 23–39.
52. Wu, J.; Ma, L.; Yadav, R. M.; Yang, Y.; Zhang, X.; Vajtai, R.; Lou, J.; Ajayan, P. M. Nitrogen-Doped Graphene with Pyridinic Dominance as a Highly Active and Stable Electrocatalyst for Oxygen Reduction. *ACS Appl. Mater. Interfaces* **2015**, *7*, 14763–14769.
53. Srinu, A.; Shaik Gouse, P.; Akhila Kumar, S. Uncovering N, S, F Tri-Doped Heteroatoms on Porous Carbon as a Metal-Free Oxygen Reduction Reaction Catalyst for Polymer Electrolyte Fuel Cells. *J. Electrochem. Soc.* **2019**, *166*, F897–F905.
54. Rojas-Carbonell, S.; Santoro, C.; Serov, A.; Atanassov, P. Transition Metal-Nitrogen-Carbon Catalysts For Oxygen Reduction Reaction in Neutral Electrolyte. *Electrochem. Commun.* **2017**, *75*, 38–42.
55. Peng, H.; Liu, F.; Liu, X.; Liao, S.; You, C.; Tian, X.; Nan, H.; Luo, F.; Song, H.; Fu, Z.; Huang, P. Effect of Transition Metals on the Structure and Performance of the Doped Carbon Catalysts Derived From Polyaniline and Melamine for ORR Application. *ACS Catal.* **2014**, *4*, 3797–3805.
56. Kramm, U. I.; Herranz, J.; Larouche, N.; Arruda, T. M.; Lefèvre, M.; Jaouen, F.; Bogdanoff, P.; Fiechter, S.; Abs-Wurmbach, I.; Mukerjee, S.; Dodelet, J. P. Structure of The Catalytic Sites

In Fe/N/C-Catalysts For O_2-Reduction In PEM Fuel Cells. *Phys Chem Chem Phys.* **2012**, *14*, 11673–11688.
57. Choi, C. H.; Park, S. H.; Woo, S. I. N-doped Carbon Prepared by Pyrolysis of Dicyandiamide With Various $MeCl_2 \cdot xH_2O$ (Me=Co, Fe, and Ni) Composites: Effect of Type and Amount of Metal Seed on Oxygen Reduction Reactions. *Appl Catal B Environ.* **2012**, *119–120*, 123–131.
58. Jia, Q.; Ramaswamy, N.; Hafiz, H.; Tylus, U.; Strickland, K.; Wu, G.; Barbiellini, B.; Bansil, A.; Holby, E. F.; Zelenay, P.; Mukerjee, S. Experimental Observation of Redox-Induced Fe–N Switching Behavior as a Determinant Role for Oxygen Reduction Activity. *ACS Nano* **2015**, *9*, 12496–12505.
59. Zhang, H.; Osgood, H.; Xie, X.; Shao, Y.; Wu, G. Engineering Nanostructures of PGM-Free Oxygen-Reduction Catalysts Using Metal-Organic Frameworks. *Nano Energy* **2017**, *31*, 331–350.
60. Eissa, A. A.; Peera, S. G.; Kim, N. H.; Lee, J. H. g-C_3N_4 Templated Synthesis of the Fe_3C@NSC Electrocatalyst Enriched With Fe–Nx Active Sites for Efficient Oxygen Reduction Reaction. *J. Mater. Chem. A* **2019**, *7*, 16920–16936.
61. Othman, R.; Dicks, A. L.; Zhu, Z. Non-Precious Metal Catalysts for the PEM Fuel Cell Cathode. *Int. J. Hydrog. Energy* **2012**, *37*, 357–372.
62. Liu, D.; Tao, L.; Yan, D.; Zou, Y.; Wang, S. Recent Advances on Non-Precious Metal Porous Carbon-Based Electrocatalysts for Oxygen Reduction Reaction. *ChemElectroChem* **2018**, *5*, 1775–1785.
63. Osmieri, L. Transition Metal–Nitrogen–Carbon (M–N–C) Catalysts for Oxygen Reduction Reaction. Insights on Synthesis and Performance in Polymer Electrolyte Fuel Cells. *ChemEngineering* **2019**, *3*, 16–48.
64. Osmieri, L.; Escudero-Cid, R.; Armandi, M.; Monteverde Videla, A. H. A.; García Fierro, J. L.; Ocón, P.; Specchia, S. Fe-N/C Catalysts For Oxygen Reduction Reaction Supported On Different Carbonaceous Materials. Performance In Acidic And Alkaline Direct Alcohol Fuel Cells. *Appl Catal B Environ.* **2017**, *205*, 637–653.
65. Wang, X.; Wang, B.; Zhong, J.; Zhao, F.; Han, N.; Huang, W.; Zeng, M.; Fan, J.; Li, Y. Iron Polyphthalocyanine Sheathed Multiwalled Carbon Nanotubes: A High-Performance Electrocatalyst For Oxygen Reduction Reaction. *Nano Res.* **2016**, *9*, 1497–1506.
66. Park, J. C.; Choi, C. H. Graphene-Derived Fe/Co-N-C Catalyst in Direct Methanol Fuel Cells: Effects of The Methanol Concentration and Ionomer Content on Cell Performance. *J. Power Sources* **2017**, *358*, 76–84.
67. Balamurugan, J.; Peera, S. G.; Guo, M.; Nguyen, T. T.; Kim, N. H.; Lee, J. H. A Hierarchical 2D Ni–Mo–S Nanosheet@Nitrogen Doped Graphene Hybrid as a Pt-Free Cathode for High-Performance Dye Sensitized Solar Cells and Fuel Cells. *J. Mater. Chem. A* **2017**, *5*, 17896–17908.
68. Peera, S. G.; Arunchander, A.; Sahu, A. K. Cumulative Effect Of Transition Metals On Nitrogen And Fluorine Co-Doped Graphite Nanofibers: An Efficient And Highly Durable Non-Precious Metal Catalyst For The Oxygen Reduction Reaction. *Nanoscale* **2016**, *8*, 14650–14664.

69. Bayram, E.; Yilmaz, G.; Mukerjee, S. A Solution-Based Procedure For Synthesis Of Nitrogen Doped Graphene As An Efficient Electrocatalyst For Oxygen Reduction Reactions In Acidic And Alkaline Electrolytes. *Appl. Catal. B Environ.* **2016**, *192*, 26–34.
70. Choi, C. H.; Lim, H. K.; Chung, M. W.; Park, J. C.; Shin, H.; Kim, H.; Woo, S. I. Long-Range Electron Transfer Over Graphene-Based Catalyst for High-Performing Oxygen Reduction Reactions: Importance of Size, N-Doping, and Metallic Impurities. *J. Am. Chem. Soc.* **2014**, *136*, 9070–9077.
71. Tian, J.; Morozan, A.; Sougrati, M. T.; Lefèvre, M.; Chenitz, R.; Dodelet, J. P.; Jones, D.; Jaouen, F. Optimized Synthesis of Fe/N/C Cathode Catalysts for PEM Fuel Cells: A Matter of Iron–Ligand Coordination Strength. *Angew. Chem., Int. Ed.* **2013**, *52*, 6867–6870.
72. Velázquez-Palenzuela, A.; Zhang, L.; Wang, L.; Cabot, P. L.; Brillas, E.; Tsay, K.; Zhang, J. Carbon-Supported Fe–Nx Catalysts Synthesized by Pyrolysis of the Fe(II)–2,3,5,6-Tetra(2-pyridyl)pyrazine Complex: Structure, Electrochemical Properties, and Oxygen Reduction Reaction Activity. *J. Phys. Chem. C* **2011**, *115*, 12929–12940.
73. Banham, D.; Ye, S.; Pei, K.; Ozaki, J. I.; Kishimoto, T.; Imashiro, Y. A Review Of The Stability And Durability Of Non-Precious Metal Catalysts For The Oxygen Reduction Reaction In Proton Exchange Membrane Fuel Cells. *J. Power Sources* **2015**, *285*, 334–348.
74. Wu, G.; More, K. L.; Xu, P.; Wang, H. L.; Ferrandon, M.; Kropf, A. J.; Myers, D. J.; Ma, S.; Johnston, C. M.; Zelenay, P. A Carbon-Nanotube-Supported Graphene-Rich Non-Precious Metal Oxygen Reduction Catalyst With Enhanced Performance Durability. *Chem. Commun.* **2013**, *49*, 3291–3293.
75. Sulub, R.; Martinez-Millan, W.; Smit, M. A. Study of the Catalytic Activity for Oxygen Reduction of Polythiophene Modified with Cobalt or Nickel. *Int. J. Electrochem. Soc.* **2009**, *4*, 1015–1027.
76. Bashyam, R.; Zelenay, P. A Class Of Non-Precious Metal Composite Catalysts For Fuel Cells. *Nature* **2006**, *443*, 63–66.
77. Wu, G.; More, K. L.; Johnston, C. M.; Zelenay, P. High-Performance Electrocatalysts for Oxygen Reduction Derived from Polyaniline, Iron, and Cobalt. *Science* **2011**, *332*, 443–447.
78. Wu, G.; Artyushkova, K.; Ferrandon, M.; Kropf, A. J.; Myers, D.; Zelenay, P. Performance Durability of Polyaniline-Derived Non-Precious Cathode Catalysts. *ECS Transactions* **2009**, *25*, 1299–1311.
79. Sha, H. D.; Yuan, X.; Hu, X. X.; Lin, H.; Wen, W.; Ma, Z. F. Effects of Pyrrole Polymerizing Oxidant on the Properties of Pyrolysed Carbon-Supported Cobalt-Polypyrrole as Electrocatalysts for Oxygen Reduction Reaction. *J. Electrochem. Soc.* **2013**, *160*, F507–F513.
80. Fu, X.; Zamani, P.; Choi, J. Y.; Hassan, F. M.; Jiang, G.; Higgins, D. C.; Zhang, Y.; Hoque, M. A.; Chen, Z. In Situ Polymer Graphenization Ingrained with Nanoporosity in a Nitrogenous Electrocatalyst Boosting the Performance of Polymer-Electrolyte-Membrane Fuel Cells. *Adv. Mater.* **2017**, *29*, 1604456–1604464.
81. Ding, S. Y.; Wang, W. Covalent Organic Frameworks (COFs): From Design To Applications. *Chem. Soc. Rev.* **2013**, *42*, 548–568.
82. Osmieri, L.; Monteverde Videla, A. H. A.; Armandi, M.; Specchia, S. Influence Of Different Transition Metals On The Properties Of Me–N–C (Me = Fe, Co, Cu, Zn) Catalysts Synthesized Using SBA-15 As Tubular Nano-Silica Reactor For Oxygen Reduction Reaction. *Int. J. Hydrog. Energy* **2016**, *41*, 22570–22588.

83. Mun, Y.; Kim, M. J.; Park, S. A.; Lee, E.; Ye, Y.; Lee, S.; Kim, Y. T.; Kim, S.; Kim, O. H.; Cho, Y. H.; Sung, Y. E.; Lee, J. Soft-Template Synthesis Of Mesoporous Non-Precious Metal Catalyst With Fe-Nx/C Active Sites For Oxygen Reduction Reaction In Fuel Cells. *Appl. Catal. B Environ.* **2018**, *222*, 191–199.
84. Serov, A.; Artyushkova, K.; Atanassov, P. Fe-N-C Oxygen Reduction Fuel Cell Catalyst Derived from Carbendazim: Synthesis, Structure, and Reactivity. *Adv. Energy Mater.* **2014**, *4*, 1301735–1301742.
85. Serov, A.; Artyushkova, K.; Niangar, E.; Wang, C.; Dale, N.; Jaouen, F.; Sougrati, M. T.; Jia, Q.; Mukerjee, S.; Atanassov, P. Nano-Structured Non-Platinum Catalysts For Automotive Fuel Cell Application. *Nano Energy* **2015**, *16*, 293–300.
86. Yang, L.; Zeng, X.; Wang, W.; Cao, D. Recent Progress in MOF-Derived, Heteroatom-Doped Porous Carbons as Highly Efficient Electrocatalysts for Oxygen Reduction Reaction in Fuel Cells. *Adv. Funct. Mater.* **2018**, *28*, 1704537–1704558.
87. Chen, Y. Z.; Zhang, R.; Jiao, L.; Jiang, H. L. Metal–Organic Framework-Derived Porous Materials For Catalysis. *Coord. Chem. Rev.* **2018**, *362*, 1–23.
88. Fu, S.; Zhu, C.; Song, J.; Du, D.; Lin, Y. Metal-Organic Framework-Derived Non-Precious Metal Nanocatalysts for Oxygen Reduction Reaction. *Adv. Energy Mater.* **2017**, *7*, 1700363–1700382.
89. Cao, P.; Liu, Y.; Quan, X.; Zhao, J.; Chen, S.; Yu, H. Nitrogen-Doped Hierarchically Porous Carbon Nanopolyhedras Derived From Core-Shell ZIF-8@ZIF-8 Single Crystals For Enhanced Oxygen Reduction Reaction. *Catalysis Today* **2019**, *327*, 366–373.
90. Ye, L.; Chai, G.; Wen, Z. Zn-MOF-74 Derived N-Doped Mesoporous Carbon as pH-Universal Electrocatalyst for Oxygen Reduction Reaction. *Adv. Funct. Mater.* **2017**, *27*, 1606190–1606198.
91. Dhakshinamoorthy, A.; Garcia, H. Catalysis by Metal Nanoparticles Embedded on Metal–Organic Frameworks. *Chem. Soc. Rev.* **2012**, *41*, 5262–5284.
92. Jiao, L.; Wan, G.; Zhang, R.; Zhou, H.; Yu, S. H.; Jiang, H. L. From Metal–Organic Frameworks to Single-Atom Fe Implanted N-doped Porous Carbons: Efficient Oxygen Reduction in Both Alkaline and Acidic Media. *Angew. Chem., Int. Ed.* **2018**, *57*, 8525–8529.
93. Yan, C.; Li, H.; Ye, Y.; Wu, H.; Cai, F.; Si, R.; Xiao, J.; Miao, S.; Xie, S.; Yang, F.; Li, Y.; Wang, G.; Bao, X. Coordinatively Unsaturated Nickel–Nitrogen Sites Towards Selective And High-Rate CO_2 Electroreduction. *Energy Environ. Sci.* **2018**, *11*, 1204–1210.
94. Liu, X.; Zang, W.; Guan, C.; Zhang, L.; Qian, Y.; Elshahawy, A. M.; Zhao, D.; Pennycook, S. J.; Wang, J. Ni-Doped Cobalt–Cobalt Nitride Heterostructure Arrays for High-Power Supercapacitors. *ACS Energy Lett.* **2018**, *3*, 2462–2469.
95. Yan, Y.; Luo, Y.; Ma, J.; Li, B.; Xue, H.; Pang, H. Facile Synthesis of Vanadium Metal-Organic Frameworks for High-Performance Supercapacitors. *Small* **2018**, *14*, 1801815–1801823.
96. Ruan, C.; Yang, Z.; Nie, H.; Zhou, X.; Guo, Z.; Wang, L.; Ding, X.; Chen, X. A.; Huang, S. Three-Dimensional sp2 Carbon Networks Prepared by Ultrahigh Temperature Treatment for Ultrafast Lithium–Sulfur Batteries. *Nanoscale* **2018**, *10*, 10999–11005.
97. Zhu, B.; Xia, D.; Zou, R. Metal-Organic Frameworks And Their Derivatives As Bifunctional Electrocatalysts. *Coord. Chem. Rev.* **2018**, *376*, 430–448.

98. Zheng, S.; Li, X.; Yan, B.; Hu, Q.; Xu, Y.; Xiao, X.; Xue, H.; Pang, H. Transition-Metal (Fe, Co, Ni) Based Metal-Organic Frameworks For Electrochemical Energy Storage. *Adv. Energy Mater.* **2017**, 7, 1602733–1602760.
99. Yi, J. D.; Xu, R.; Chai, G. L.; Zhang, T.; Zang, K.; Nan, B.; Lin, H.; Liang, Y. L.; Lv, J.; Luo, J.; Si, R.; Huang, Y. B.; Cao, R. Cobalt Single-Atoms Anchored On Porphyrinic Triazine-Based Frameworks As Bifunctional Electrocatalysts For Oxygen Reduction And Hydrogen Evolution Reactions. *J. Mater. Chem. A* **2019**, 7, 1252–1259.
100. Wang, H.; Maiyalagan, T.; Wang, X. Review on Recent Progress in Nitrogen-Doped Graphene: Synthesis, Characterization, and Its Potential Applications. *ACS Catal.* **2012**, 2, 781–794.
101. Yang, D. S.; Song, M. Y.; Singh, K. P.; Yu, J. S. The Role Of Iron In The Preparation And Oxygen Reduction Reaction Activity Of Nitrogen-Doped Carbon. *Chem. Commun.* **2015**, *51*, 2450–2453.
102. Duan, J.; Chen, S.; Jaroniec, M.; Qiao, S. Z. Heteroatom-Doped Graphene-Based Materials for Energy-Relevant Electrocatalytic Processes. *ACS Catal.* **2015**, 5, 5207–5234.
103. Dai, L.; Xue, Y.; Qu, L.; Choi, H. J.; Baek, J. B. Metal-Free Catalysts for Oxygen Reduction Reaction. *Chem. Rev.* **2015**, *115*, 4823–4892.
104. Klingele, M.; Van Pham, C.; Fischer, A.; Thiele, S. A Review on Metal-Free Doped Carbon Materials Used as Oxygen Reduction Catalysts in Solid Electrolyte Proton Exchange Fuel Cells. *Fuel Cells* **2016**, *16*, 522–529.
105. Hu, C.; Xiao, Y.; Zou, Y.; Dai, L. Carbon-Based Metal-Free Electrocatalysis for Energy Conversion, Energy Storage, and Environmental Protection. *Electrochemical Energy Reviews* **2018**, *1*, 84–112.
106. Peera, S. G.; Sahu, A. K.; Arunchander, A.; Bhat, S. D.; Karthikeyan, J.; Murugan, P. Nitrogen and Fluorine Co-Doped Graphite Nanofibers as High Durable Oxygen Reduction Catalyst in Acidic Media for Polymer Electrolyte Fuel Cells. *Carbon* **2015**, *93*, 130–142.
107. Arunchander, A.; Peera, S. G.; Panda, S. K.; Chellammal, S.; Sahu, A. K. Simultaneous Co-Doping of N And S By A Facile In-Situ Polymerization Of 6-N,N-Dibutylamine-1,3,5-Triazine-2,4-Dithiol On Graphene Framework: An Efficient And Durable Oxygen Reduction Catalyst In Alkaline Medium. *Carbon* **2017**, *118*, 531–544.
108. Akula, S.; Parthiban, V.; Peera, S. G.; Singh, B. P.; Dhakate, S. R.; Sahu, A. K. Simultaneous Co-Doping of Nitrogen and Fluorine into MWCNTs: An In-Situ Conversion to Graphene Like Sheets and Its Electro-Catalytic Activity toward Oxygen Reduction Reaction. *J. Electrochem. Soc.* **2017**, *164*, F568–F576.
109. Srinu, A.; Peera, S. G.; Parthiban, V.; Bhuvaneshwari, B.; Sahu, A. K. Heteroatom Engineering and Co–Doping of N and P to Porous Carbon Derived from Spent Coffee Grounds as an Efficient Electrocatalyst for Oxygen Reduction Reactions in Alkaline Medium. *ChemistrySelect* **2018**, *3*, 690–702.
110. Zhang, L.; Xia, Z. Mechanisms of Oxygen Reduction Reaction on Nitrogen-Doped Graphene for Fuel Cells. *J. Phys. Chem. C* **2011**, *115*, 11170–11176.
111. Stock, N.; Biswas, S. Synthesis of Metal-Organic Frameworks (MOFs): Routes to Various MOF Topologies, Morphologies, and Composites. *Chem. Rev.* **2012**, *112*, 933–969.

112. Barkholtz, H. M.; Liu, D. J. Advancements In Rationally Designed PGM-Free Fuel Cell Catalysts Derived From Metal–Organic Frameworks. *Materials Horizons* **2017**, *4*, 20–37.
113. Shen, K.; Chen, X.; Chen, J.; Li, Y. Development of MOF-Derived Carbon-Based Nanomaterials for Efficient Catalysis. *ACS Catal.* **2016**, *6*, 5887–5903.
114. Wu, Q.; Liang, J.; Yi, J. D.; Meng, D. L.; Shi, P. C.; Huang, Y. B.; Cao, R. Unraveling The Relationship Between The Morphologies Of Metal–Organic Frameworks And The Properties Of Their Derived Carbon Materials. *Dalton Trans.* **2019**, *48*, 7211–7217.
115. Wu, M.; Wang, K.; Yi, M.; Tong, Y.; Wang, Y.; Song, S. A Facile Activation Strategy for an MOF-Derived Metal-Free Oxygen Reduction Reaction Catalyst: Direct Access to Optimized Pore Structure and Nitrogen Species. *ACS Catal.* **2017**, *7*, 6082–6088.
116. He, W.; Jiang, C.; Wang, J.; Lu, L. High-Rate Oxygen Electroreduction over Graphitic-N Species Exposed on 3D Hierarchically Porous Nitrogen-Doped Carbons. *Angew. Chem., Int. Ed.* **2014**, *53*, 9503–9507.
117. Liu, D.; Li, L.; Xu, H.; Dai, P.; Wang, Y.; Gu, X.; Yan, L.; Zhao, G.; Zhao, X. Boosting ORR Catalytic Activity by Integrating Pyridine-N Dopants, a High Degree of Graphitization, and Hierarchical Pores into a MOF-Derived N-Doped Carbon in a Tandem Synthesis. *Chem. Asian J.* **2018**, *13*, 1318–1326.
118. Song, Z.; Liu, W.; Cheng, N.; Norouzi Banis, M.; Li, X.; Sun, Q.; Xiao, B.; Liu, Y.; Lushington, A.; Li, R.; Liu, L.; Sun, X. Origin of The High Oxygen Reduction Reaction of Nitrogen and Sulfur co-Doped MOF-Derived Nanocarbon Electrocatalysts. *Mater. Horiz.* **2017**, *4*, 900–907.
119. Tan, A. D.; Wan, K.; Wang, Y. F.; Fu, Z. Y.; Liang, Z. X. N, S-containing MOF-Derived Dual-Doped Mesoporous Carbon as a Highly Effective Oxygen Reduction Reaction Electrocatalyst. *Catal. Sci. Technol.* **2018**, *8*, 335–343.
120. Li, L.; He, J.; Wang, Y.; Lv, X.; Gu, X.; Dai, P.; Liu, D.; Zhao, X. Metal–Organic Frameworks: A Promising Platform For Constructing Non-Noble Electrocatalysts For The Oxygen-Reduction Reaction. *J. Mater. Chem. A* **2019**, *7*, 1964–1988.
121. Liu, J.; Zhu, D.; Guo, C.; Vasileff, A.; Qiao, S. Z. Design Strategies toward Advanced MOF-Derived Electrocatalysts for Energy-Conversion Reactions. *Adv. Energy Mater.* **2017**, *7*, 1700518–1700544.
122. Qian, Y.; Khan, I. A.; Zhao, D. Electrocatalysts Derived from Metal–Organic Frameworks for Oxygen Reduction and Evolution Reactions in Aqueous Media. *Small* **2017**, *13*, 1701143–1701165.
123. Ren, Q.; Wang, H.; Lu, X. F.; Tong, Y. X.; Li, G. R. Recent Progress on MOF-Derived Heteroatom-Doped Carbon-Based Electrocatalysts for Oxygen Reduction Reaction. *Adv. Sci.* **2018**, *5*, 1700515–1700536.
124. Shi, P. C.; Yi, J. D.; Liu, T. T.; Li, L.; Zhang, L. J.; Sun, C. F.; Wang, Y. B.; Huang, Y. B.; Cao, R. Hierarchically Porous Nitrogen-Doped Carbon Nanotubes Derived From Core–Shell ZnO@Zeolitic Imidazolate Framework Nanorods For Highly Efficient Oxygen Reduction Reactions. *J. Mater. Chem. A* **2017**, *5*, 12322–12329.
125. Ma, S.; Goenaga, G. A.; Call, A. V.; Liu, D. J. Cobalt Imidazolate Framework as Precursor for Oxygen Reduction Reaction Electrocatalysts. *Chem. Eur. J.* **2011**, *17*, 2063–2067.

126. Xia, W.; Zhu, J.; Guo, W.; An, L.; Xia, D.; Zou, R. Well-Defined Carbon Polyhedrons Prepared From Nano Metal–Organic Frameworks For Oxygen Reduction. *J. Mater. Chem. A* **2014**, *2*, 11606–11613.
127. Zhang, H.; Hwang, S.; Wang, M.; Feng, Z.; Karakalos, S.; Luo, L.; Qiao, Z.; Xie, X.; Wang, C.; Su, D.; Shao, Y.; Wu, G. Single Atomic Iron Catalysts for Oxygen Reduction in Acidic Media: Particle Size Control and Thermal Activation. *J. Am. Chem. Soc.* **2017**, *139*, 14143–14149.
128. Guo, Z.; Zhang, Z.; Li, Z.; Dou, M.; Wang, F. Well-Defined Gradient Fe/Zn Bimetal Organic Framework Cylinders Derived Highly Efficient Iron- And Nitrogen- Codoped Hierarchically Porous Carbon Electrocatalysts Towards Oxygen Reduction. *Nano Energy* **2019**, *57*, 108–117.
129. Wang, X. X.; Cullen, D. A.; Pan, Y. T.; Hwang, S.; Wang, M.; Feng, Z.; Wang, J.; Engelhard, M. H.; Zhang, H.; He, Y.; Shao, Y.; Su, D.; More, K. L.; Spendelow, J. S.; Wu, G. Nitrogen-Coordinated Single Cobalt Atom Catalysts for Oxygen Reduction in Proton Exchange Membrane Fuel Cells. *Adv. Mater.* **2018**, *30*, 1706758–1706769.
130. You, S.; Gong, X.; Wang, W.; Qi, D.; Wang, X.; Chen, X.; Ren, N. Enhanced Cathodic Oxygen Reduction and Power Production of Microbial Fuel Cell Based on Noble-Metal-Free Electrocatalyst Derived from Metal-Organic Frameworks. *Adv. Energy Mater.* **2016**, *6*, 1501497–1501506.
131. Wang, X.; Zhou, J.; Fu, H.; Li, W.; Fan, X.; Xin, G.; Zheng, J.; Li, X. MOF Derived Catalysts For Electrochemical Oxygen Reduction. *J. Mater. Chem. A* **2014**, *2*, 14064–14070.
132. Chen, Y. Z.; Wang, C.; Wu, Z. Y.; Xiong, Y.; Xu, Q.; Yu, S. H.; Jiang, H. L. From Bimetallic Metal-Organic Framework To Porous Carbon: High Surface Area And Multicomponent Active Dopants For Excellent Electrocatalysis. *Adv. Mater.* **2015**, *27*, 5010–5016.
133. Wang, K. C.; Huang, H. C.; Chang, S. T.; Wu, C. H.; Yamanaka, I.; Lee, J. F.; Wang, C. H. Hybrid Porous Catalysts Derived from Metal-Organic-Framework for Oxygen Reduction Reaction in Anion Exchange Membrane Fuel Cell. *ACS Sustainable Chem. Eng.* **2019**, *7*, 9143–9152.
134. Li, J.; Xia, W.; Tang, J.; Tan, H.; Wang, J.; Kaneti, Y.; Bando, Y.; Wang, T.; He, J.; Yamauchi, Y. MOF Nanoleaves as New Sacrificial Templates for the Fabrication of Nanoporous Co-Nx/C Electrocatalysts for Oxygen Reduction. *Nanoscale Horiz.* **2019**, *4*, 1006–1013.
135. Shi, Z.; Yu, Y.; Fu, C.; Wang, L.; Li, X. Water-Based Synthesis Of Zeolitic Imidazolate Framework-8 For CO_2 Capture. *RSC Adv.* **2017**, *7*, 29227–29232.
136. Chen, Z.; Wang, X.; Noh, H.; Ayoub, G.; Peterson, G. W.; Buru, C. T.; Islamoglu, T.; Farha, O. K. Scalable, Room Temperature, And Water-Based Synthesis Of Functionalized Zirconium-Based Metal–Organic Frameworks For Toxic Chemical Removal. *CrystEngComm* **2019**, *21*, 2409–2415.
137. Ramos-Fernández, E. V.; Grau-Atienza, A.; Farrusseng, D.; Aguado, S. A Water-Based Room Temperature Synthesis of ZIF-93 For CO_2 Adsorption. *J. Mater. Chem. A* **2018**, *6*, 5598–5602.
138. Shieh, F. K.; Wang, S. C.; Leo, S. Y.; Wu, K. C. W. Water-Based Synthesis of Zeolitic Imidazolate Framework-90 (ZIF-90) with a Controllable Particle Size. *Chem. Eur. J.* **2013**, *19*, 11139–11142.
139. Park, H.; Oh, S.; Lee, S.; Choi, S.; Oh, M. Cobalt- and Nitrogen-Codoped Porous Carbon Catalyst Made From Core–Shell Type Hybrid Metal–Organic Framework (ZIF-L@ ZIF-67)

And Its Efficient Oxygen Reduction Reaction (ORR) Activity. *Appl. Catal. B Environ.* **2019**, *246*, 322–329.
140. Deng, J.; Ren, P.; Deng, D.; Bao, X. Enhanced Electron Penetration Through An Ultrathin Graphene Layer For Highly Efficient Catalysis Of The Hydrogen Evolution Reaction. *Angew. Chem., Int. Ed.* **2015**, *54*, 2100–2104.
141. Singh, A. K.; Xu, Q. Synergistic Catalysis Over Bimetallic Alloy Nanoparticles. *ChemCatChem* **2013**, *5*, 652–676.
142. Asokan, A.; Lee, H.; Gwon, O.; Kim, J.; Kwon, O.; Kim, G. Insights Into the Effect of Nickel Doping on ZIF-Derived Oxygen Reduction Catalysts for Zinc—Air Batteries. *ChemElectroChem* **2019**, *6*, 1213–1224.
143. Ning, H.; Li, G.; Chen, Y.; Zhang, K.; Gong, Z.; Nie, R.; Hu, W.; Xia, Q. Porous N-Doped Carbon-Encapsulated CoNi Alloy Nanoparticles Derived from MOFs as Efficient Bifunctional Oxygen Electrocatalysts. *ACS Appl. Mater. Interfaces* **2018**, *11*, 1957–1968.
144. Zhu, Q.; Lin, L.; Jiang, Y. F.; Xie, X.; Yuan, C. Z.; Xu, A. W. Carbon Nanotube/S–N–C Nanohybrids As High Performance Bifunctional Electrocatalysts For Both Oxygen Reduction And Evolution Reactions. *New J. Chem.* **2015**, *39*, 6289–6296.
145. Kim, K.; Lopez, K. J.; Sun, H. J.; An, J. C.; Shim, J.; Park, G. Carbon-Wrapped Bimetallic Co/Ni Catalysts ($C@Co_xNi_{1-x}$) Derived from Co/Ni-MOF for Bifunctional Catalysts in Rechargeable Zn-Air Batteries. *Bull. Korean Chem. Soc.* **2018**, *39*, 1357–1361.
146. Wang, N.; Li, Y.; Guo, Z.; Li, H.; Hayase, S.; Ma, T. Synthesis of Fe, Co Incorporated in P-Doped Porous Carbon Using a Metal-Organic Framework (MOF) Precursor as Stable Catalysts for Oxygen Reduction Reaction. *J. Electrochem. Soc.* **2018**, *165*, G3080–G3086.
147. He, X.; Yin, F.; Wang, H.; Chen, B.; Li, G. Metal-Organic Frameworks For Highly Efficient Oxygen Electrocatalysis. *Chinese J. Catal.* **2018**, *39*, 207–227.
148. Sun, T.; Xu, L.; Li, S.; Chai, W.; Huang, Y.; Yan, Y.; Chen, J. Cobalt-Nitrogen-Doped Ordered Macro-/Mesoporous Carbon For Highly Efficient Oxygen Reduction Reaction. *Appl. Catal. B Environ.* **2016**, *193*, 1–8.
149. Cai, S.; Meng, Z.; Tang, H.; Wang, Y.; Tsiakaras, P. 3D Co-N-Doped Hollow Carbon Spheres As Excellent Bifunctional Electrocatalysts For Oxygen Reduction Reaction And Oxygen Evolution Reaction. *Appl. Catal. B Environ.* **2017**, *217*, 477–484.
150. Zhang, Z.; Sun, J.; Wang, F.; Dai, L. Efficient Oxygen Reduction Reaction (ORR) Catalysts Based on Single Iron Atoms Dispersed on a Hierarchically Structured Porous Carbon Framework. *Angew. Chem., Int. Ed.* **2018**, *57*, 9038–9043.
151. Gong, S.; Wang, C.; Jiang, P.; Hu, L.; Lei, H.; Chen, Q. Designing Highly Efficient Dual-Metal Single-Atom Electrocatalysts For The Oxygen Reduction Reaction Inspired By Biological Enzyme Systems. *J. Mater. Chem. A* **2018**, *6*, 13254–13262.
152. You, B.; Jiang, N.; Sheng, M.; Drisdell, W. S.; Yano, J.; Sun, Y. Bimetal–Organic Framework Self-Adjusted Synthesis Of Support-Free Nonprecious Electrocatalysts For Efficient Oxygen Reduction. *ACS Catal.* **2015**, *5*, 7068–7076.
153. Meng, Z.; Cai, S.; Wang, R.; Tang, H.; Song, S.; Tsiakaras, P. Bimetallic-Organic Framework-Derived Hierarchically Porous Co-Zn-NC As Efficient Catalyst For Acidic Oxygen Reduction Reaction. *Appl. Catal. B Environ.* **2019**, *244*, 120–127.

154. Yan, X.; Liu, K.; Wang, T.; You, Y.; Liu, J.; Wang, P.; Pan, X.; Wang, G.; Luo, J.; Zhu, J. Atomic Interpretation Of High Activity On Transition Metal And Nitrogen-Doped Carbon Nanofibers For Catalyzing Oxygen Reduction. *J. Mater. Chem. A* **2017**, 5, 3336–3345.
155. Sa, Y. J.; Seo, D. J.; Woo, J.; Lim, J. T.; Cheon, J. Y.; Yang, S. Y.; Lee, J. M.; Kang, D.; Shin, T. J.; Shin, H. S. A General Approach To Preferential Formation Of Active Fe–Nx Sites In Fe–N/C Electrocatalysts For Efficient Oxygen Reduction Reaction. *J. Am. Chem. Soc.* **2016**, *138*, 15046–15056.
156. Liu, J. Catalysis By Supported Single Metal Atoms. *ACS Catal.* **2016**, 7, 34–59.
157. Liu, L.; Corma, A. Metal Catalysts For Heterogeneous Catalysis: From Single Atoms To Nanoclusters And Nanoparticles. *Chem. Rev.* **2018**, *118*, 4981–5079.
158. Fang, X.; Shang, Q.; Wang, Y.; Jiao, L.; Yao, T.; Li, Y.; Zhang, Q.; Luo, Y.; Jiang, H. L. Single Pt Atoms Confined Into A Metal–Organic Framework For Efficient Photocatalysis. *Adv. Mater.* **2018**, *30*, 1705112–1705119.
159. Han, A.; Wang, B.; Kumar, A.; Qin, Y.; Jin, J.; Wang, X.; Yang, C.; Dong, B.; Jia, Y.; Liu, J.; Xiaoming, S. Recent Advances for MOF-Derived Carbon-Supported Single-Atom Catalysts. *Small Methods* **2019**, 1800471–1800492.
160. Jiao, L.; Jiang, H. L. Metal-Organic-Framework-Based Single-Atom Catalysts For Energy Applications. *Chem.* **2019**, 5, 786–804.
161. Chung, H. T.; Cullen, D. A.; Higgins, D.; Sneed, B. T.; Holby, E. F.; More, K. L.; Zelenay, P. Direct Atomic-Level Insight Into The Active Sites of a High-Performance PGM-Free ORR Catalyst. *Science* **2017**, *357*, 479–484.
162. Zhang, H.; Ding, S.; Hwang, S.; Zhao, X.; Su, D.; Xu, H.; Yang, H.; Wu, G. Atomically Dispersed Iron Cathode Catalysts Derived from Binary Ligand-Based Zeolitic Imidazolate Frameworks with Enhanced Stability for PEM Fuel Cells. *J. Electrochem. Soc.* **2019**, *166*, F3116–F3122.
163. Yao, Z.; Shi-Zhang, Q. Metal-Organic Framework Assisted Synthesis Of Single-Atom Catalysts For Energy Applications. *Natl. Sci. Rev.* **2018**, 5, 626–627.
164. Yin, P.; Yao, T.; Wu, Y.; Zheng, L.; Lin, Y.; Liu, W.; Ju, H.; Zhu, J.; Hong, X.; Deng, Z. Single Cobalt Atoms With Precise N-Coordination As Superior Oxygen Reduction Reaction Catalysts. *Angew. Chem., Int. Ed.* **2016**, 55, 10800–10805.
165. He, Y.; Hwang, S.; Cullen, D. A.; Uddin, M. A.; Langhorst, L.; Li, B.; Karakalos, S.; Kropf, A. J.; Wegener, E. C.; Sokolowski, J. Highly Active Atomically Dispersed CoN_4 Fuel Cell Cathode Catalysts Derived From Surfactant-Assisted MOFs: Carbon-Shell Confinement Strategy. *Energy Environ. Sci.* **2019**, *12*, 250–260.
166. Proietti, E.; Jaouen, F.; Lefèvre, M.; Larouche, N.; Tian, J.; Herranz, J.; Dodelet, J. P. Iron-Based Cathode Catalyst With Enhanced Power Density In Polymer Electrolyte Membrane Fuel Cells. *Nat. Commun.* **2011**, 2, 416–425.
167. Zhang, H.; Hwang, S.; Wang, M.; Feng, Z.; Karakalos, S.; Luo, L.; Qiao, Z.; Xie, X.; Wang, C.; Su, D. Single Atomic Iron Catalysts For Oxygen Reduction In Acidic Media: Particle Size Control And Thermal Activation. *J. Am. Chem. Soc.* **2017**, *139*, 14143–14149.
168. Deng, Y.; Chi, B.; Li, J.; Wang, G.; Zheng, L.; Shi, X.; Cui, Z.; Du, L.; Liao, S.; Zang, K. Atomic Fe-Doped MOF-Derived Carbon Polyhedrons with High Active-Center Density and

Ultra-High Performance toward PEM Fuel Cells. *Adv. Energy Mater.* **2019**, *9*, 1802856–1802864.

169. Xiao, F.; Xu, G. L.; Sun, C. J.; Xu, M.; Wen, W.; Wang, Q.; Gu, M.; Zhu, S.; Li, Y.; Wei, Z. Nitrogen-Coordinated Single Iron Atom Catalysts Derived From Metal Organic Frameworks For Oxygen Reduction Reaction. *Nano Energy* **2019**, *61*, 60–68.
170. Knez, M.; Scholz, R.; Nielsch, K.; Pippel, E.; Hesse, D.; Zacharias, M.; Gösele, U. Monocrystalline Spinel Nanotube Fabrication Based on the Kirkendall Effect. *Nat. Mater.* **2006**, *5*, 627–631.
171. Webb, S. W.; Pruess, K. The use of Fick's Law For Modeling Trace Gas Diffusion In Porous Media. *Transp. Porous Media* **2003**, *51*, 327–341.
172. Pillai, R. S.; Peter, S. A.; Jasra, R. V. Adsorption Of Carbon Dioxide,Methane, Nitrogen, Oxygen And Argon in NaETS-4. *Micropor. Mesopor. Mat.* **2008**, *113*, 268–276.
173. Lai, Q.; Zheng, L.; Liang, Y.; He, J.; Zhao, J.; Chen, J. Metal–Organic-Framework-Derived Fe-N/C Electrocatalyst With Five-Coordinated Fe-N X Sites For Advanced Oxygen Reduction In Acid Media. *ACS Catal.* **2017**, *7*, 1655–1663.
174. Charreteur, F.; Jaouen, F.; Dodelet, J. P. Iron Porphyrin-Based Cathode Catalysts For PEM Fuel Cells: Influence Of Pyrolysis Gas On Activity And Stability. *Electrochim. Acta* **2009**, *54*, 6622–6630.
175. Jiang, R.; Li, L.; Sheng, T.; Hu, G.; Chen, Y.; Wang, L. Edge-Site Engineering Of Atomically Dispersed Fe–N4 by Selective C–N Bond Cleavage For Enhanced Oxygen Reduction Reaction Activities. *J. Am. Chem. Soc.* **2018**, *140*, 11594–11598.
176. Li, J.; Chen, M.; Cullen, D. A.; Hwang, S.; Wang, M.; Li, B.; Liu, K.; Karakalos, S.; Lucero, M.; Zhang, H. Atomically Dispersed Manganese Catalysts For Oxygen Reduction In Proton-Exchange Membrane Fuel Cells. *Nat. Catal.* **2018**, *1*, 935–945.
177. Yang, Y.; Kaitian, M.; Shiqi, G.; Hao, H.; Guoliang, X.; Zhiyu, L.; Peng, J.; Changlai, W.; Hui, W.; Chen, Q. O-, N-Atoms-Coordinated Mn Cofactors Within a Graphene Framework as Bioinspired Oxygen Reduction Reaction Electrocatalysts. *Adv. Mater.* **2018**, *30*, 1801732–1801742.
178. Qu, Y.; Li, Z.; Chen, W.; Lin, Y.; Yuan, T.; Yang, Z.; Zhao, C.; Wang, J.; Zhao, C.; Wang, X. Direct Transformation Of Bulk Copper Into Copper Single Sites Via Emitting And Trapping Of Atoms. *Nat. Catal.* **2018**, *1*, 781–786.
179. Wei, L.; Karahan, H. E.; Zhai, S.; Liu, H.; Chen, X.; Zhou, Z.; Lei, Y.; Liu, Z.; Chen, Y. Amorphous Bimetallic Oxide–Graphene Hybrids As Bifunctional Oxygen Electrocatalysts For Rechargeable Zn–Air Batteries. *Adv. Mater.* **2017**, *29*, 1701410–1701420.
180. Chao, T.; Luo, X.; Chen, W.; Jiang, B.; Ge, J.; Lin, Y.; Wu, G.; Wang, X.; Hu, Y.; Zhuang, Z. Atomically Dispersed Copper–Platinum Dual Sites Alloyed With Palladium Nanorings Catalyze The Hydrogen Evolution Reaction. *Angew. Chem., Int. Ed.* **2017**, *56*, 16047–16051.
181. Wang, J.; Huang, Z.; Liu, W.; Chang, C.; Tang, H.; Li, Z.; Chen, W.; Jia, C.; Yao, T.; Wei, S. Design of N-Coordinated Dual-Metal Sites: A Stable And Active Pt-Free Catalyst For Acidic Oxygen Reduction Reaction. *J. Am. Chem. Soc.* **2017**, *139*, 17281–17284.
182. Sun, T.; Xu, L.; Wang, D.; Li, Y. Metal Organic Frameworks Derived Single Atom Catalysts For Electrocatalytic Energy Conversion. *Nano Res.* **2019**, *12*, 1187–1192.

183. Zhu, Q. L.; Xia, W.; Zheng, L. R.; Zou, R.; Liu, Z.; Xu, Q. Atomically Dispersed Fe/N-Doped Hierarchical Carbon Architectures Derived From A Metal–Organic Framework Composite For Extremely Efficient Electrocatalysis. *ACS Energy Lett.* **2017**, *2*, 504–511.
184. Wan, X.; Chen, W.; Yang, J.; Liu, M.; Liu, X.; Shui, J. Synthesis and Active Site Identification of Fe−N−C Single-Atom Catalysts for the Oxygen Reduction Reaction. *ChemElectroChem* **2019**, *6*, 304–315.
185. Sun, T.; Zhao, S.; Chen, W.; Zhai, D.; Dong, J.; Wang, Y.; Zhang, S.; Han, A.; Gu, L.; Yu, R. Single-Atomic Cobalt Sites Embedded In Hierarchically Ordered Porous Nitrogen-Doped Carbon As A Superior Bifunctional Electrocatalyst. *Proc. Natl. Acad. Sci. USA* **2018**, *115*, 12692–12697.
186. Wang, J.; Han, G.; Wang, L.; Du, L.; Chen, G.; Gao, Y.; Ma, Y.; Du, C.; Cheng, X.; Zuo, P. ZIF-8 with Ferrocene Encapsulated: A Promising Precursor to Single-Atom Fe Embedded Nitrogen-Doped Carbon as Highly Efficient Catalyst for Oxygen Electroreduction. *Small* **2018**, *14*, 1704282.
187. Zhong, H. X.; Wang, J.; Zhang, Y. W.; Xu, W. l.; Xing, W.; Xu, D.; Zhang, Y. F.; Zhang, X. B. ZIF-8 Derived Graphene-Based Nitrogen-Doped Porous Carbon Sheets As Highly Efficient And Durable Oxygen Reduction Electrocatalysts. *Angew. Chem., Int. Ed.* **2014**, *53*, 14235–14239.
188. Wei, J.; Hu, Y.; Liang, Y.; Kong, B.; Zhang, J.; Song, J.; Bao, Q.; Simon, G. P.; Jiang, S. P.; Wang, H. Nitrogen-Doped Nanoporous Carbon/Graphene Nano-Sandwiches: Synthesis And Application For Efficient Oxygen Reduction. *Adv. Funct. Mater.* **2015**, *25*, 5768–5777.
189. Ge, L.; Yang, Y.; Wang, L.; Zhou, W.; De Marco, R.; Chen, Z.; Zou, J.; Zhu, Z. High Activity Electrocatalysts From Metal–Organic Framework-Carbon Nanotube Templates For The Oxygen Reduction Reaction. *Carbon* **2015**, *82*, 417–424.
190. Zhang, H.; Liu, X.; Wu, Y.; Guan, C.; Cheetham, A. K.; Wang, J. MOF-Derived Nanohybrids For Electrocatalysis And Energy Storage: Current Status And Perspectives. *Chem. Commun.* **2018**, *54*, 5268–5288.
191. Liu, S.; Zhang, H.; Zhao, Q.; Zhang, X.; Liu, R.; Ge, X.; Wang, G.; Zhao, H.; Cai, W. Metal-Organic Framework Derived Nitrogen-Doped Porous Carbon@Graphene Sandwich-Like Structured Composites As Bifunctional Electrocatalysts For Oxygen Reduction And Evolution Reactions. *Carbon* **2016**, *106*, 74–83.
192. Li, Y.; Zhang, H.; Wang, Y.; Liu, P.; Yang, H.; Yao, X.; Wang, D.; Tang, Z.; Zhao, H. A Self-Sponsored Doping Approach For Controllable Synthesis of S and N Co-Doped Trimodal-Porous Structured Graphitic Carbon Electrocatalysts. *Energy Environ. Sci.* **2014**, *7*, 3720–3726.
193. Zhu, Z.; Chen, C.; Cai, M.; Cai, Y.; Ju, H.; Hu, S.; Zhang, M. Porous Co-NC ORR Catalysts Of High Performance Synthesized With ZIF-67 Templates. *Mater. Res. Bull.* **2019**, *114*, 161–169.
194. Wu, M.; Li, C.; Zhao, J.; Ling, Y.; Liu, R. Tannic Acid-Mediated Synthesis Of Dual-Heteroatom-Doped Hollow Carbon From A Metal–Organic Framework For Efficient Oxygen Reduction Reaction. *Dalton Trans.* **2018**, *47*, 7812–7818.
195. Zhang, J.; Wu, C.; Huang, M.; Zhao, Y.; Li, J.; Guan, L. Conductive Porous Network of Metal–Organic Frameworks Derived Cobalt-Nitrogen-doped Carbon with the Assistance of

Carbon Nanohorns as Electrocatalysts for Zinc–Air Batteries. *ChemCatChem* **2018**, *10*, 1336–1343.

196. Tang, B.; Wang, S.; Li, R.; Gou, X.; Long, J. Urea Treated Metal Organic Frameworks-Graphene Oxide Composites Derived N-Doped Co-Based Materials As Efficient Catalyst For Enhanced Oxygen Reduction. *J. Power Sources* **2019**, *425*, 76–86.
197. Long, J.; Li, R.; Gou, X. Well-Organized Co-Ni@NC Material Derived From Hetero-Dinuclear MOFs As Efficient Electrocatalysts For Oxygen Reduction. *Catal. Commun.* **2017**, *95*, 31–35.
198. Tang, F.; Liu, L.; Wang, H.; Gao, X.; Jin, Z. The Combination Of Metal-Organic Frameworks And Polydopamine Nanotubes Aiming For Efficient One-Dimensional Oxygen Reduction Electrocatalyst. *J Colloid Interf. Sci.* **2019**, *552*, 351–358.
199. Xia, W.; Qu, C.; Liang, Z.; Zhao, B.; Dai, S.; Qiu, B.; Jiao, Y.; Zhang, Q.; Huang, X.; Guo, W. High-Performance Energy Storage And Conversion Materials Derived From A Single Metal–Organic Framework/Graphene Aerogel Composite. *Nano Lett.* **2017**, *17*, 2788–2795.
200. Guo, J.; Gadipelli, S.; Yang, Y.; Li, Z.; Lu, Y.; Brett, D. J.; Guo, Z. An Efficient Carbon-Based ORR Catalyst From Low-Temperature Etching Of ZIF-67 With Ultra-Small Cobalt Nanoparticles And High Yield. *J. Mater. Chem. A* **2019**, *7*, 3544–3551.
201. Wei, J.; Hu, Y.; Liang, Y.; Kong, B.; Zheng, Z.; Zhang, J.; Jiang, S. P.; Zhao, Y.; Wang, H. Graphene Oxide/Core–Shell Structured Metal–Organic Framework Nano-Sandwiches And Their Derived Cobalt/N-Doped Carbon Nanosheets For Oxygen Reduction Reactions. *J. Mater. Chem. A* **2017**, *5*, 10182–10189.
202. Liu, S.; Wang, Z.; Zhou, S.; Yu, F.; Yu, M.; Chiang, C. Y.; Zhou, W.; Zhao, J.; Qiu, J. Metal–Organic-Framework-Derived Hybrid Carbon Nanocages as a Bifunctional Electrocatalyst for Oxygen Reduction and Evolution. *Adv. Mater.* **2017**, *29*, 1700874.
203. Naguib, M.; Mochalin, V. N.; Barsoum, M. W.; Gogotsi, Y. 25th Anniversary Article: MXenes: A New Family of Two-Dimensional Materials. *Adv. Mater.* **2014**, *26*, 992–1005.
204. Anasori, B.; Lukatskaya, M. R.; Gogotsi, Y. 2D Metal Carbides and Nitrides (MXenes) for Energy Storage. *Nat. Rev. Mater.* **2017**, *2*, 16098–16115.
205. Jianian, C.; Xiaolei, Y.; Fenglei, L.; Qixuan, Z.; Huicheng, H.; Qi, P.; Qiao, Z. Integrating Mxene Nanosheets With Cobalt-Tipped Carbon Nanotubes For An Efficient Oxygen Reduction Reaction. *J. Mater. Chem. A* **2019**, *7*, 1281–1286.
206. Niu, Q.; Guo, J.; Chen, B.; Nie, J.; Guo, X.; Ma, G. Bimetal-Organic Frameworks/Polymer Core-Shell Nanofibers Derived Heteroatom-Doped Carbon Materials As Electrocatalysts For Oxygen Reduction Reaction. *Carbon* **2017**, *114*, 250–260.
207. Deng, Y.; Chi, B.; Tian, X.; Cui, Z.; Liu, E.; Jia, Q.; Fan, W.; Wang, G.; Dang, D.; Li, M. g-C_3N_4 Promoted MOF Derived Hollow Carbon Nanopolyhedra Doped With High Density/Fraction Of Single Fe Atoms As An Ultra-High Performance Non-Precious Catalyst Towards Acidic ORR And PEM Fuel Cells. *J. Mater. Chem. A* **2019**, *7*, 5020–5030.
208. Li, X.; Fang, Y.; Lin, X.; Tian, M.; An, X.; Fu, Y.; Li, R.; Jin, J.; Ma, J. MOF Derived Co_3O_4 Nanoparticles Embedded In N-Doped Mesoporous Carbon Layer/MWCNT Hybrids: Extraordinary Bi-Functional Electrocatalysts For OER and ORR. *J. Mater. Chem. A* **2015**, *3*, 17392–17402.
209. Lv, L.; Kang, S.; Li, X.; Shen, J.; Liu, S. ZIF-Derived Carbons As Highly Efficient And Stable ORR Catalyst. *Nanotechnology* **2018**, *29*, 485402.

210. Fang, H.; Huang, T.; Sun, Y.; Kang, B.; Liang, D.; Yao, S.; Yu, J.; Dinesh, M. M.; Wu, S.; Lee, J. Y. Metal-Organic Framework-Derived Core-Shell-Structured Nitrogen-Doped Cocx/Feco@C Hybrid Supported By Reduced Graphene Oxide Sheets As High Performance Bifunctional Electrocatalysts For ORR and OER. *J. Catal.* **2019**, *371*, 185–195.
211. Pang, J.; Jiang, F.; Wu, M.; Liu, C.; Su, K.; Lu, W.; Yuan, D.; Hong, M. A Porous Metal-Organic Framework With Ultrahigh Acetylene Uptake Capacity Under Ambient Conditions. *Nat. Commun.* **2015**, *6*, 7575–7582.
212. Bingjun, Z.; Ruqiang, Z.; Qiang, X. Metal–Organic Framework Based Catalysts for Hydrogen Evolution. *Adv. Energy Mater.* **2018**, *8*, 1801193–1801226.
213. Lee, J.; Farha, O.; Roberts, J.; Scheidt, K. A.; Nguyen, S. T.; Hupp, J. T. Metal-Organic Framework Materials As Catalysts. *Chem. Soc. Rev.* **2009**, *38*, 1450–1459.
214. Julien, P. A.; Mottillo, C.; Friščić, T. Metal–Organic Frameworks Meet Scalable And Sustainable Synthesis. *Green Chem.* **2017**, *19*, 2729–2747.
215. Kumar, P.; Vellingiri, K.; Kim, K. H.; Brown, R. J.; Manos, M. J. Modern Progress In Metal-Organic Frameworks And Their Composites For Diverse Applications. *Micropor. Mesopor. Mat.* **2017**, *253*, 251–265.
216. Ren, J.; Dyosiba, X.; Musyoka, N. M.; Langmi, H. W.; Mathe, M.; Liao, S. Review On The Current Practices And Efforts Towards Pilot-Scale Production Of Metal-Organic Frameworks (MOFs). *Coord. Chem. Rev.* **2017**, *352*, 187–219.
217. Do, J.-L.; Friščić, T. Mechanochemistry: A Force Of Synthesis. *ACS Cent. Sci.* **2016**, *3*, 13–19.
218. Luz, I.; Soukri, M.; Lail, M. Confining Metal–Organic Framework Nanocrystals within Mesoporous Materials: A General Approach via "Solid-State" Synthesis. *Chem. Mater.* **2017**, *29*, 9628–9638.
219. Chen, J.; Shen, K.; Li, Y. Greening The Processes Of Metal–Organic Framework Synthesis And Their Use In Sustainable Catalysis. *ChemSusChem* **2017**, *10*, 3165–3187.
220. Li, Y.; Xu, H.; Huang, H.; Gao, L.; Zhao, Y.; Ma, T. Facile Synthesis Of N, S Co-Doped Porous Carbons From A Dual-Ligand Metal Organic Framework For High Performance Oxygen Reduction Reaction Catalysts. *Electrochim. Acta.* **2017**, *254*, 148–154.
221. Peera, S. G.; Balamurugan, J.; Kim, N. H.; Lee, J. H. Sustainable Synthesis of Co@NC Core Shell Nanostructures from Metal Organic Frameworks via Mechanochemical Coordination Self-Assembly: An Efficient Electrocatalyst for Oxygen Reduction Reaction. *Small* **2018**, *14*, 1800441–1800456.
222. Wang, Y.; Pan, Y.; Zhu, L.; Yu, H.; Duan, B.; Wang, R.; Zhang, Z.; Qiu, S. Solvent-Free Assembly Of Co/Fe-Containing MOFs Derived N-Doped Mesoporous Carbon Nanosheet For ORR and HER. *Carbon* **2019**, *146*, 671–679.
223. Bai, Y.; Yang, D.; Yang, M.; Chen, H.; Liu, Y.; Li, H. Nitrogen/Cobalt Co-Doped Mesoporous Carbon Microspheres Derived from Amorphous Metal-Organic Frameworks as a Catalyst for the Oxygen Reduction Reaction in Both Alkaline and Acidic Electrolytes. *ChemElectroChem* **2019**, *6*, 2546–2552.
224. Wang, Y.; Pan, Y.; Zhu, L.; Yu, H.; Duan, B.; Wang, R.; Zhang, Z.; Qiu, S. Solvent-Free Assembly of Co/Fe-Containing MOFs Derived N-Doped Mesoporous Carbon Nanosheets for ORR and HER. *Carbon* **2019**, *146*, 671–679.

225. Wang, L.; Jin, X.; Fu, J.; Jiang, Q.; Xie, Y.; Huang, J.; Xu, L. Mesoporous Non-Noble Metal Electrocatalyst Derived From ZIF-67 And Cobalt Porphyrin For The Oxygen Reduction In Alkaline Solution. *J. Electroanal. Chem.* **2018**, *825*, 65–72.
226. Tran, T. N.; Shin, C. H.; Lee, B. J.; Samdani, J. S.; Park, J. D.; Kang, T. H.; Yu, J. S. Fe–N-Functionalized Carbon Electrocatalyst Derived From A Zeolitic Imidazolate Framework For Oxygen Reduction: Fe And NH_3 Treatment Effects. *Catal. Sci. Technol.* **2018**, *8*, 5368–5381.
227. Ma, L.; Wang, R.; Li, Y.-H.; Liu, X. F.; Zhang, Q. Q.; Dong, X. Y.; Zang, S. Q. Apically Co-Nanoparticles-Wrapped Nitrogen-Doped Carbon Nanotubes From A Single-Source MOF For Efficient Oxygen Reduction. *J. Mater. Chem. A* **2018**, *6*, 24071–24077.
228. Han, H.; Bai, Z.; Wang, X.; Chao, S.; Liu, J.; Kong, Q.; Yang, X.; Yang, L. Highly Dispersed Co Nanoparticles Inlayed In S, N-Doped Hierarchical Carbon Nanoprisms Derived From Co-MOFs As Efficient Electrocatalysts For Oxygen Reduction Reaction. *Catal. Today* **2018**, *318*, 126–131.
229. Ahn, S. H.; Manthiram, A. Self-Templated Synthesis of Co- and N-Doped Carbon Microtubes Composed of Hollow Nanospheres and Nanotubes for Efficient Oxygen Reduction Reaction. *Small* **2017**, *13*, 1603437–1603445.
230. Tan, A. D.; Wang, Y. F.; Fu, Z. Y.; Tsiakaras, P.; Liang, Z. X. Highly Effective Oxygen Reduction Reaction Electrocatalysis: Nitrogen-Doped Hierarchically Mesoporous Carbon Derived From Interpenetrated Nonporous Metal-Organic Frameworks. *Appl. Catal. B Environ.* **2017**, *218*, 260–266.
231. Zhao, R.; Xia, W.; Lin, C.; Sun, J.; Mahmood, A.; Wang, Q.; Qiu, B.; Tabassum, H.; Zou, R. A Pore-Expansion Strategy To Synthesize Hierarchically Porous Carbon Derived From Metal-Organic Framework For Enhanced Oxygen Reduction. *Carbon* **2017**, *114*, 284–290.
232. Dong, Y.; Yu, M.; Wang, Z.; Zhou, T.; Liu, Y.; Wang, X.; Zhao, Z.; Qiu, J. General Synthesis Of Zeolitic Imidazolate Framework-Derived Planar-N-Doped Porous Carbon Nanosheets For Efficient Oxygen Reduction. *Energy Storage Materials* **2017**, *7*, 181–188.
233. Zhang, D.; Chen, W.; Li, Z.; Chen, Y.; Zheng, L.; Gong, Y.; Li, Q.; Shen, R.; Han, Y.; Cheong, W. C. Isolated Fe and Co Dual Active Sites On Nitrogen-Doped Carbon For A Highly Efficient Oxygen Reduction Reaction. *Chem. Commun.* **2018**, *54*, 4274–4277.

Editors' Biographies

Dr. Lakhveer Singh

Dr. Lakhveer Singh is presently working as associate professor at the Faculty of Civil Engineering and Earth Resources at the University Malaysia Pahang, Malaysia. Prior to his joining UMP, he worked as a research associate at the Department of Biological and Ecological Engineering, Oregon State University, Corvallis, USA. His main areas of interest are energy production, bioelectrochemical systems, wastewater treatment, and nano material synthesis for energy and water treatment applications. Dr. Singh is an author or coauthor of 4 edited books, 7 book chapters, and over 55 peer-reviewed journal publications with over 1140 citations and an H-index of 20 (as of October 1, 2019).

Dr. Durga Madhab Mahapatra

Dr. Durga Madhab Mahapatra is presently working as a research associate at the Department of Biological and Ecological Engineering at Oregon State University, Corvallis, USA. His research in the United States is on the development of sustainable bioprocess and bioproduct recovery. He received his PhD in environmental bioprocess engineering from the Indian Institute of Science (IISc), Bangalore in 2015. He has extensive experience in research- and industry-oriented consulting projects. He has more than 50 peer-reviewed publications such as peer-reviewed journal articles, scientific reports, book chapters, and invited articles. His work has been cited in Google Scholar more than 990 times with an H index of 13 and i10 index of 16 (as of October 1, 2019).

Dr. Hong Liu

Dr. Hong Liu is a professor of Biological and Ecological Engineering (BEE) at Oregon State University (OSU). Her main research efforts are in bioenergy and bioproduct generation and the development of energy sustainable water and wastewater treatment systems. Dr. Liu is the author or coauthor of 7 book chapters and over 85 peer-reviewed journal publications with over 17,450 SCI citations and an H-index of 42 (as of October 1, 2019). She was named as a Highly Cited Researcher and listed among the 3000 worldwide researchers in The World's Most Influential Scientific Minds in 2014, 2015, and 2016 by Thomson Reuters. She received her PhD in 2003 from the University of Hong Kong. Prior to joining the faculty at OSU in 2005, she was a postdoctoral researcher at Pennsylvania State University in the Department of Environmental Engineering.

© 2020 American Chemical Society

Indexes

Author Index

Subject Index

I

L

M

P

Printed in the USA/Agawam, MA
February 22, 2021